Nucleic Acids and Molecular Biology

Volume 36

Series Editor

Allen W. Nicholson, Department of Biology, Temple University, Philadelphia, PA, USA

Since its inception in 1987, Nucleic Acids and Molecular Biology has emerged as a high-visibility series that brings to the scientific community in-depth analyses of cutting-edge topics in the field. Each volume offers a focused set of reviews, written by leading experts, that provides a comprehensive, multi-perspective analysis of the chosen topic. As such, Nucleic Acids and Molecular Biology serves as an enduring research resource for scientists, academicians, and students alike.

Spiridoula Matsika • Andrew H. Marcus
Editors

Nucleic Acid Photophysics and Photochemistry

Springer

Editors
Spiridoula Matsika
Department of Chemistry
Temple University
Philadelphia, PA, USA

Andrew H. Marcus
Department of Chemistry and Biochemistry
University of Oregon
Eugene, OR, USA

ISSN 0933-1891 ISSN 1869-2486 (electronic)
Nucleic Acids and Molecular Biology
ISBN 978-3-031-68806-5 ISBN 978-3-031-68807-2 (eBook)
https://doi.org/10.1007/978-3-031-68807-2

This Springer imprint is published by the registered company Springer Nature Switzerland AG
The registered company address is: Gewerbestrasse 11, 6330 Cham, Switzerland

Preface

Background and Introduction

Understanding the photophysical properties of nucleic acids is an area of long-standing and active research. Over the years, the field has greatly benefitted from steady advances in spectroscopic techniques and computational methods to study molecular excited states, which have facilitated detailed studies of the behavior of nucleic acids and their components after they absorb light. Experiments performed on ultrafast time scales (femtoseconds–picoseconds) have permitted the accurate determination of excited state lifetimes, while computational studies have provided detailed microscopic information about the mechanisms involved. As our understanding of the fundamental photophysics of nucleic acids has advanced, current studies now focus on systems of higher complexity and introduce novel optical techniques to investigate the interactions between nucleic acids and proteins. Spectroscopic studies of nucleic acids, particularly nucleic acid constructs labeled with optical probes, can yield richly detailed information important to molecular biology, biochemistry, and biophysics. This volume focuses on recent advancements in our current understanding of nucleic acid photochemistry and its relationship to biologically relevant phenomena.

Areas Covered

The chapters presented in this volume describe significant advancements in both computational and experimental approaches to better understand nucleic acid photophysics and photochemistry. These studies include experiments conducted in both the gas phase and condensed phases and employ a broad range of experimental methodologies to provide complementary information about nucleic acid photophysics. Moreover, extensive theoretical studies performed over the last two decades provide a detailed understanding of the dynamics and relaxation pathways of the optically excited electronic states of nucleic acids and their building blocks, including the canonical nucleobases and modified nucleobases. Following are brief summaries of these chapters.

Computational Methods

- Chapter 1 reviews how computational chemistry has revealed the mechanisms of photophysical and photochemical processes in canonical nucleobases, nucleosides, nucleotides, and base pairs.
- Chapter 2 extends the discussion of computational treatments of nucleotides to oligonucleotides of different size and complexities, including single-stranded oligonucleotides, complementary duplexes, quadruple helices, and i-motifs.

Experimental Studies of Nucleobases in the Gas Phase

- Chapter 3 describes time-resolved photo-electron spectroscopy to examine the excited state dynamics of nucleobases and thionated nucleic acid base analogs. Thio-substitution of nucleobases leads to enhanced intersystem crossing and increased excited state lifetimes.
- Chapter 4 describes the potential critical role played by photochemistry in the selection of heterocyclic compounds as the fundamental building blocks of nucleic acids in prebiotic conditions in which there is no ozone layer to provide protection from solar radiation.

Ultrafast Laser Experiments of Nucleic Acids in Solution

- Chapter 5 describes ultrafast laser spectroscopic studies of the photophysics and photochemistry of DNA bases and oligonucleotides in complexation with silver(I) ions. The last two chapters focus on other spectroscopic tools for understanding protein–nucleic acid interactions, structure, and function.

Spectroscopic Approaches to Study Nucleic Acid Structure and Protein–Nucleic Acid Interactions

- Chapter 6 reviews fluorescence-detected circular dichroism (FDCD) spectroscopy and how this method has been applied to study nucleic acid structure and function.
- Chapter 7 describes single-molecule and ensemble spectroscopic techniques to study the mechanisms of protein–DNA interactions at site-specific positions within model DNA scaffolds. These methods utilize DNA constructs that contain excitonically coupled fluorescent probes, which can act as "sensors" of the local environment of bases and sugar-phosphate backbones immediately adjacent to the probes

Philadelphia, PA, USA Spiridoula Matsika
Eugene, OR, USA Andrew H. Marcus

Contents

Chapter 1
Computational Studies of the Photophysics and Photochemistry of Nucleic Acid Constituents

Tolga N. V. Karsili and Barbara Marchetti

Abstract Nucleic acids consist of nucleobases characterized by unique structures crucial for maintaining their genomic functions and overall structure. This includes a conjugated π-system, a key element supporting the helical structure of double-helix DNA through π–π interactions. Despite this advantageous structural feature, the presence of π-electrons in organic molecules inevitably results in a bathochromic shift in their electronic absorption spectra compared to their saturated counterparts. Consequently, nucleobases may absorb wavelengths that penetrate Earth's atmosphere, leading to electronic excitations that promote nucleic acid damage and potential cancer. Computational chemistry has proven indispensable in complementing experimental observations and elucidating the active mechanisms through which nucleic acid constituents manage the energy afforded from electronic excitation. A variety of computational methods, spanning static electronic structure calculations to multi- and full-dimensional molecular dynamics simulations, have been employed to offer accurate insights into the photophysics and photochemistry of active excited states in nucleic acid constituents. This chapter provides an overview of some widely used computational approaches for unraveling the fate of excited states in nucleic acid constituents, along with a review of how computational chemistry has revealed the active chemical mechanisms and dynamics involved in photoexcited nucleic acid constituents.

Keywords Computational chemistry · Molecular dynamics · DNA/RNA photochemistry · Conical intersections · Electronic excitation · DNA damage · Single-reference time-dependent DFT · Multi-reference quantum chemistry

T. N. V. Karsili (✉) · B. Marchetti
University of Louisiana at Lafayette, Lafayette, LA, USA
e-mail: tolga.karsili@louisiana.edu

S. Matsika, A. H. Marcus (eds.), *Nucleic Acid Photophysics and Photochemistry*, Nucleic Acids and Molecular Biology 36,
https://doi.org/10.1007/978-3-031-68807-2_1

1.1 Introduction

Nucleic acids play an essential role in maintaining the lifecycle of cells and viruses. They consist of polymers of nucleotides which comprise a 5-carbon sugar, a phosphate group, and a nucleobase. These nucleobases are heteroatom containing organic molecules with unicyclic or bicyclic structures, commonly referred to as pyrimidine and purine nucleobases, respectively (Soukup 2003). The pyrimidine bases include thymine, cytosine, and uracil, while the purine nucleobases include adenine and guanine. These are depicted in Fig. 1.1.

One of the many distinctive structural features of nucleobases is the presence of a π-system created by the presence of p-electrons that are orthogonal to the molecular plane. The resulting π-orbitals promote π–π interactions between neighboring nucleobases in a nucleic acid sequence (Soukup 2003). This feature is essential for maintaining the helical structure of (deoxy)ribose nucleic acids (DNA/RNA) and is one of the key attributes that enable structural reorganization following DNA/RNA transcription/translation.

Despite this key structure property, the presence of orthogonal p-electrons introduces additional molecular orbitals with π and π* character, thus increasing the density of molecular orbitals and reducing the gap between the occupied and unoccupied molecular orbitals (Marchetti et al. 2016). As such, the promotion of an electron from the highest occupied molecular orbital (HOMO) to the lowest unoccupied molecular orbital (LUMO) requires much less energy than a fully saturated organic molecule analog (Marchetti et al. 2016). This lower energy gap leads to a bathochromic shift in the electronic absorption spectrum of conjugated organic molecules. Additionally, the promotion of electrons from a bonding π orbital to an anti-bonding π* is characteristically strong due to the high spatial overlap between these orbitals. As such, the resulting ππ* states typically have large absorption cross sections (Voet et al. 1963; Callis 1983).

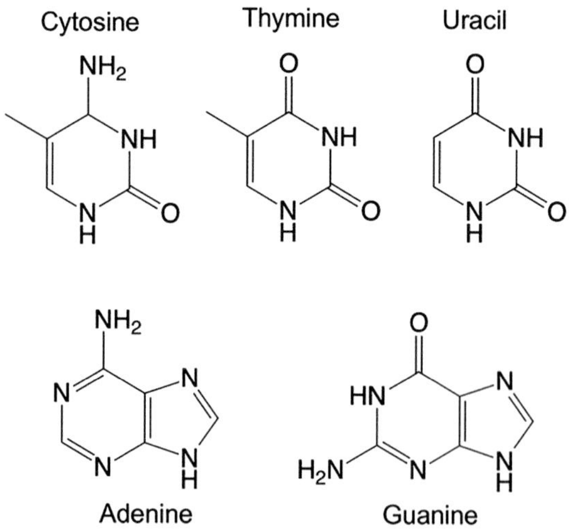

Fig. 1.1 Canonical DNA/RNA nucleobases

Before we discuss the implications of these properties on nucleic acids, we will briefly review fundamental concepts in photophysics/photochemistry. Electronic excitation, the promotion of electron(s) from a stable ground state configuration to a higher energy unoccupied orbital(s), may occur in a multitude of different ways. Absorption of electromagnetic radiation is by far the most common. When a given molecule is irradiated by electromagnetic radiation, the electric field vector of the high energy incident light may interact with the transition dipole moment vector associated with one or more of its electronic transitions. The subsequent electron promotion leads to the formation of an electronically excited-state molecule with a total energy that is higher than the ground state. The nascent excited state is metastable and typically has a finite lifetime that is limited by multiple possible decay paths that are classically described in textbooks using a Jablonski diagram (Jablonski 1933). Radiative decay is a form of electronic transition in which the metastable molecule ultimately transitions back to the ground state by the emission of a photon. This type of transition occurs by either fluorescence or phosphorescence, distinguishable by whether the radiative decay is, respectively, spin-allowed, or spin-forbidden. Alternatively, the excited electronic state may decay non-radiatively (i.e., without the release of light). Examples include internal conversion, intersystem crossing, or reaction/fragmentation. The first two are generally promoted by non-adiabatic coupling and spin–orbit coupling, respectively. It is now widely recognized that conical intersections (i.e., crossings between electronic states with common spin multiplicities) mediate much of the ultrafast phenomena that is observed in molecular photochemistry (Domcke et al. 2004, 2011). Fragmentation occurs when the incident photon energy surpasses the bond dissociation energy of one of more of the chemical bonds of the molecule. Any excess energy is then partitioned in product translational or internal (electronic, vibrational, or rotational) energy.

This simple Jablonski picture also applies to DNA/RNA nucleobases, which also form metastable electronically excited states upon photoexcitation. As outlined above, electronic excitation of DNA/RNA nucleobases occurs at bathochromic wavelengths. These wavelengths overlap with the tropospherically relevant solar irradiance and, as such, these biomolecules may become electronically excited by solar radiation. It is therefore paramount to understand how DNA/RNA nucleobases have evolved to cope with electronic excitation. Computational chemistry has been a huge asset in outlining the mechanisms and dynamics of photoexcited nucleobases (Marchetti et al. 2016; Middleton et al. 2009; Barbatti et al. 2010; Chakraborty et al. 2018; Crespo-Hernández et al. 2004; Sobolewski and Domcke 2006; Molecular mechanisms of the photostability of life 2010; Gustavsson et al. 2010; Kleinermanns et al. 2013). In this chapter, we outline the commonly used computational methods for resolving the mechanisms and dynamics of photoexcited molecules—paying particular attention to electronically excited DNA/RNA nucleobases that made up nucleic acids. We will outline the myriad of computational studies that have provided insights into the excited state fates of canonical nucleobases under isolated conditions. We will then outline studies that provide insight into how these "gas phase" mechanisms and dynamics change in the complex environment, specifically in the nucleic acid medium. We finish by outlining specific open questions that remain in the field of computational photophysics/photochemistry of nucleic acids as well as how the lessons learned from such studies could be used to understand the photochemistry/photophysics of industrially/environmentally relevant processes.

1.2 Overview of Computational Methods

Before reviewing the computational studies of nucleic acid constituents, it will be useful to briefly review the relevant computational methods that have been used for studies of the photochemistry and photophysics of nucleic acids. The purpose of this section is to give a brief overview of the methods, their strengths, and weaknesses, without overwhelming the reader with the underlying mathematical background. A more detailed mathematical description of these computational methods can be found in Szabo and Ostlund, Modern Quantum Chemistry (Szabo and Ostlund 1996).

Calculations of the electronically excited states of molecules are often difficult. For example, the orbitals involved in creating electronically excited states are typically far more diffuse than those of the ground state (Serrano-Andrés and Serrano-Pérez 2012). Basis functions that describe orbitals are in general optimized for ground state electronic configurations and thus lack the necessary full description of the spatial arrangement of electrons. The description of electron correlation for excited-state configurations is also difficult and requires sophisticated electronic structure methods that are computational costly. Often smaller basis functions (which are typically non-optimal for excited states anyway) need to be used with these higher-level methods thus reducing the inherent accuracy. An appropriate trade-off between accuracy and computational cost is required. The next two sections will describe typically used computational methods for computing excited-state energies and properties. The last section will focus on simulation methods that model the dynamics of electronically excited molecules.

1.2.1 Single-Reference Methods

Excited-state wavefunctions often require several Slater determinants in order to correctly describe their electronic properties and to ensure that the excited-state wavefunction transforms as an irreducible representation of the symmetry of a given molecule. One of the most trivial ways to describe an electronically excited state is to approximate the excited-state wavefunctions as linear combinations of the simplest excited determinants (the singles). Methods of this type include configuration-interaction singles (CIS), time-dependent Hartree–Fock (TD-HF), and time-dependent Density Functional Theory (TD-DFT) (Szabo and Ostlund 1996; Serrano-Andrés and Serrano-Pérez 2012; Merchán and Serrano-Andrés 2005;

Serrano-Andrés and Merchán 2005; Dreuw et al. 2003; Del Bene et al. 1971; Foresman et al. 1992). These methods are popular as they may be readily extended to larger molecular systems at a reasonable computational cost.

In CIS, the Hartree–Fock orbitals are used to create all singly excited determinants of the CI expansion. As such, CIS may be thought of as HF theory but for excited states. The large subset of excited-state energies and oscillator strengths may be determined both with singlet and triplet spin multiplicities. It is a relatively easy method to setup in mainstream computational chemistry programs and yields vertical excitation energies and oscillator strength far quicker than other excited-state electronic structure methods. CIS is also efficient at describing local states such as $\pi^* \leftarrow \pi$ and $\sigma^* \leftarrow \pi$ transitions and, with the availability of analytical first and second derivatives, may be used to optimize excited-state structures and determine normal mode wavenumbers. Given the advantages, however, CIS has some inherent limitations that center on its restricted ability to only describe single excitations. Electronic states that involve multiple excitations or contain components from multiple excitations usually lead to large over-estimations of the vertical excitation energies when calculated with CIS.

TDDFT is the time-dependent analog of Hohenberg–Kohn theorem (Hohenberg and Kohn 1964) that was developed in 1964 and extends ground state DFT to allow for the treatment of excited-state properties. It also offers a means by which the errors in CIS may be overcome. For non-charge transfer states with energies that are well separated from the ground state, TDDFT with various pure and hybrid functionals are known to generate vertical excitations energies and oscillator strengths in reasonable agreement with experimentally measured absorption spectra. Despite this, standard pure or hybrid functionals are usually unable to describe charge transfer, which are important in describing the photochemistry/photophysics of nucleic acid constituents (see later in this chapter) (Lange and Herbert 2007).

Another single-reference method that has been shown to produce highly accurate energies is equation of motion coupled-cluster singles and doubles (EOM-CCSD) (Stanton and Bartlett 1993). Although it is substantially more computationally expensive than TDDFT, EOM-CCSD is typically far more accurate in describing electronically excited states. It is also capable of describing valence and Rydberg state, as well as charge transfer states. Another coupled-cluster based method (CC2—second-order approximate coupled-cluster singles and doubles) has also been successfully applied to generating accurate vertical excitation energies and transition moments (Christiansen et al. 1995). CC2 is based on linear response theory and provides ground state energies at the accuracy of second-order Møller–Plesset theory (MP2) level of theory (Christiansen et al. 1995; Møller and Plesset 1934). Additionally, the Møller–Plesset partition of the Hamiltonian may be used in a diagrammatic perturbation expansion of the polarization propagator. This is the premise behind second-order algebraic diagrammatic construction (ADC(2)) theory which has also been used to determine accurate vertical excitation energies. ADC(2) is also more stable than CC2 (Schirmer 1982; Trofimov and Schirmer 1995).

All single-reference methods suffer when the energy difference between the lowest excited state and ground state becomes small and the single-determinant approximation breaks down. As such these methods are often spurious in describing how the excited-state population decays back to the ground state and the fate of the ground state molecule on the ground electronic state. Multi-reference methods are required to characterize such processes.

1.2.2 *Multi-reference Methods*

Multi-reference methods are a class of quantum chemical wavefunction based methods that accurately treat strongly electron correlated systems. They can be used to accurately compute relevant ground and excited-state properties such as minimum energy molecular structures and their energies and frequencies. They are particularly reliable in regions of the energy surface that are multi-reference (i.e., where more than a single electronic configuration dominates the electronic state character) such as those involved in highly correlated systems and when the energy difference between the ground and excited state becomes small. In modelling the photochemistry and photophysics of nucleic acid constituents, a myriad of multi-reference methods have been used to determine the fates of their excited states. One of the most popular multi-reference methods is multi-reference configuration interaction (MRCI) (Liu 1973; Knowles and Werner 1988; Knowles and Werner 1992; Szalay et al. 2012; Recent Adv 1999), wherein the eigenstates of the electronic Hamiltonian are expanded in a set of Slater determinants which correspond to excitation of ground and excited-state electronic configurations. Higher order excitations are then determined via a variational method to generate single, double, etc., excitations (MRCIS, MRCISD, etc.) from a given reference determinant. An multi-reference configuration interaction (MRCI) treatment ensures better and more balanced electron correlation for the ground and excited states. Unfortunately, MRCI based methods are however computationally expensive and at the cost of high accuracy, the method is typically computationally expensive for large molecules.

Another multi-reference method used to model the excited-state chemistry of nucleic acid constituents is complete active space self-consistent field (CASSCF) theory (Roos et al. 1980; Roos 1980). CASSCF combines self-consistent field theory with full configuration interaction of a subset of orbitals known as the active space orbitals. The method captures static electron correlation which is important for systems where more than one electronic configuration describes a given electronic state. Dynamical correlation, which considers the motion of electrons is not captured in CASSCF, but the zeroth order CASSCF reference state can be expanded in second-order perturbation theory to capture dynamical electron correlation (CASPT2) (Roos et al. 1982; Andersson et al. 1990, 1992). CASPT2 is known to yield extremely accurate ground and excited-state energies but it is inadequate at avoided crossings between electronic states. Multi-state and extended multistate CASPT2 (MS-CASPT2 and XMS-CASPT2, respectively) can be used to successfully overcome this latter issue.

1.2.3 Dynamics Methods

The methods described in the previous two sections solve molecular electronic energies using the time-independent Schrödinger equation, under which the nuclei are fixed according to the separation of nuclear and electronic timescales via the Born–Oppenheimer approximation. This eliminates the nuclear kinetic energy from the Hamiltonian operator and reduces the eigenvalue problem to one that solves for the electronic coordinates. As such, these methods do not consider how the nuclei move on the excited-state potential energy surfaces. An exact quantum treatment of the dynamics of molecules on excited-state energy surfaces requires one to solve the time-dependent Schrodinger equation, but these methods require a predetermined potential energy surface upon which a quantum wavepacket can propagate. The dynamics are then limited to the dimensionality of the potential energy surface, where a full-dimensional potential energy surface calculated with an accurate quantum chemical method provides the most accurate representation. In practice however, generating a highly accurate full-dimensional potential energy surface for medium to large molecules (i.e., where the sizes and number of degrees of freedom are high) is extremely computationally costly. To overcome this problem, one can assume that since the nuclei are so much larger than electrons, that they move in a more classical (rather than quantum) manner. This is the premise behind ab initio molecular dynamics (AIMD) wherein the nuclear motions are treated classically by solving Newton's equations. AIMD is a deterministic approach and therefore requires initial momenta and positions of nuclei. These can be generated by various sampling methods to form trajectories with initial conditions. Each of these trajectories are temporally propagated and at time step, energies and gradients of the molecular geometry at that time are calculated using a given quantum chemical method. Since the electrons are treated quantum mechanically, while the nuclei are treated classically, AIMD is often referred to as a semi-classical dynamics method. With AIMD it is possible to propagate trajectories on excited states and monitor the way in which a single (adiabatic) excited state evolves. The shortcoming of mainstream AIMD is that it does not take into the account nuclear quantum phenomena such as quantum tunnelling, nor can it account for non-radiative transitions that inevitably occur between electronic states. The latter is addressed by extending mainstream AIMD to fewest-switches surface hopping (FSSH) (Tully 1990). The basic premise behind FSSH is a typical AIMD simulation but at each time step, interstate couplings between electronic states are calculated in addition to energies and gradients. These interstate couplings are then used to calculate hopping probabilities between electronic states. When using multi-reference quantum chemical methods, interstate couplings are simply the non-adiabatic coupling matrix elements that may be explicitly calculated. When using single-reference methods such as TDDFT, interstate couplings may be calculated using the local diabatization method or time-derivative couplings. To align closer to an exact quantum dynamics treatment, FSSH can be augmented to allow trajectories to "spawn" at or near energy surface crossings. This treatment is known as the ab initio multiple spawning

(AIMS) method and more closely resembles how a quantum wavepacket bifurcates as it traverses regions of conical intersection.

1.3 Computational Studies of Photophysics and Photochemistry

1.3.1 Canonical Nucleobases

We start by discussing the photophysics of canonical nucleobases, i.e. the lowest energy configuration of the N-terminated nucleobases. The canonical nucleobases, which include 9H-adenine, 9H-guanine, 1H-thymine/uracil, and 1H-cytosine, were displayed in Fig. 1.1. In all cases, electronic photoexcitation is dominated by a $\pi^* \leftarrow \pi$ electron promotion, forming a $^1\pi\pi^*$ state (Marchetti et al. 2016; Callis 1983; Crespo-Hernández et al. 2004; Kim et al. 2000; Lobsiger et al. 2013; Nir et al. 2001, 2002a, b; Nachtigallová et al. 2010; Nix et al. 2007; Ludwig et al. 2010). This transition contains a strong absorption cross section. Following photoexcitation, various computational studies have optimized conical intersections between calculated potential energy (PE) profiles and have revealed that the dominant decay path for photoexcited nucleobases is via out-of-plane ring deformation that drives the $^1\pi\pi^*$ state toward a conical intersection with the ground state (Marchetti et al. 2016; Matsika 2004; Yamazaki et al. 2008; Delchev et al. 2010; Barbatti et al. 2011, 2012; Perun et al. 2005; Conti et al. 2009; Marian 2007; Serrano-Andrés et al. 2008; Perun et al. 2006a; Zgierski et al. 2005; Ismail et al. 2002; Blancafort et al. 2005; Kistler and Matsika 2007; Blancafort 2007; Zechmann and Barbatti 2008; Asturiol et al. 2009). Such geometries are given in Figs. 1.2 and 1.3 (for, respectively, the pyrimidine and purine nucleobases) alongside calculated CASPT2 PE profiles along the relevant deformation coordinate. Clearly, and intuitively, the DNA/RNA nucleobases are planar in the ground state in order to preserve the π-system structure. As such, motion out-of-plane leads to a rise in PE as indicated by the profiles in Figs. 1.2 and 1.3. In contrast, the electronically excited state formed upon a $\pi^* \leftarrow \pi$ electron promotion leads to a reduction in the bond-order as a π^* orbital is populated. This reduction in bond-order weakens the π-conjugation, promotes ring bending motions, and thus leads to a decrease in PE of the electronically excited molecule. The progressive decreasing and increasing PE of, respectively, the $^1\pi\pi^*$ state and the ground state inevitably lead to a crossing between these states. Such parts of the configuration space drive internal conversion of the excited-state population back to the ground state—potentially reforming the parent ground state molecule (Marchetti et al. 2016). This property is often referred to as photostability, which is a recognized phenomenon in canonical nucleobases. Such out-of-plane ring-deformation geometries can be traced back to the photochemistry of benzene which undergoes deformation to form a prefulvene conical intersection which promotes internal conversion to form fulvene. These deformations are also reminiscent of simply photoisomerisms around C=C moieties that dominate the photochemistry of aliphatic

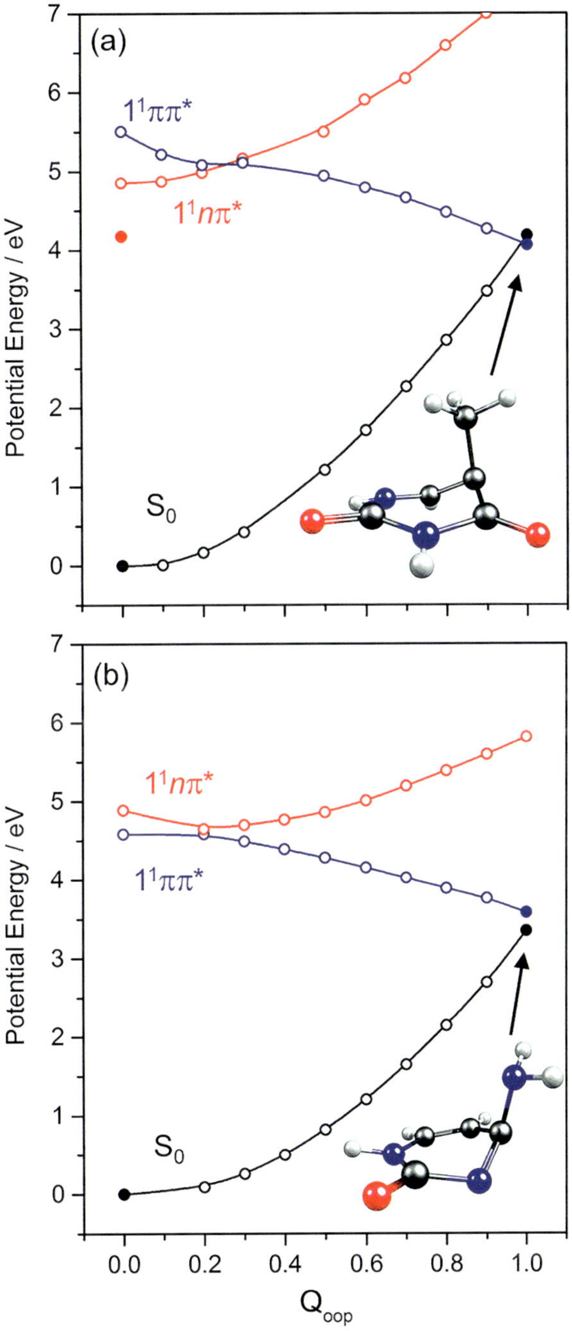

Fig. 1.2 CASPT2 potential energy profiles associated with the out-of-plane ring deformations of (**a**) thymine and (**b**) cytosine. This figure is adapted from (Marchetti et al. 2016) with permission

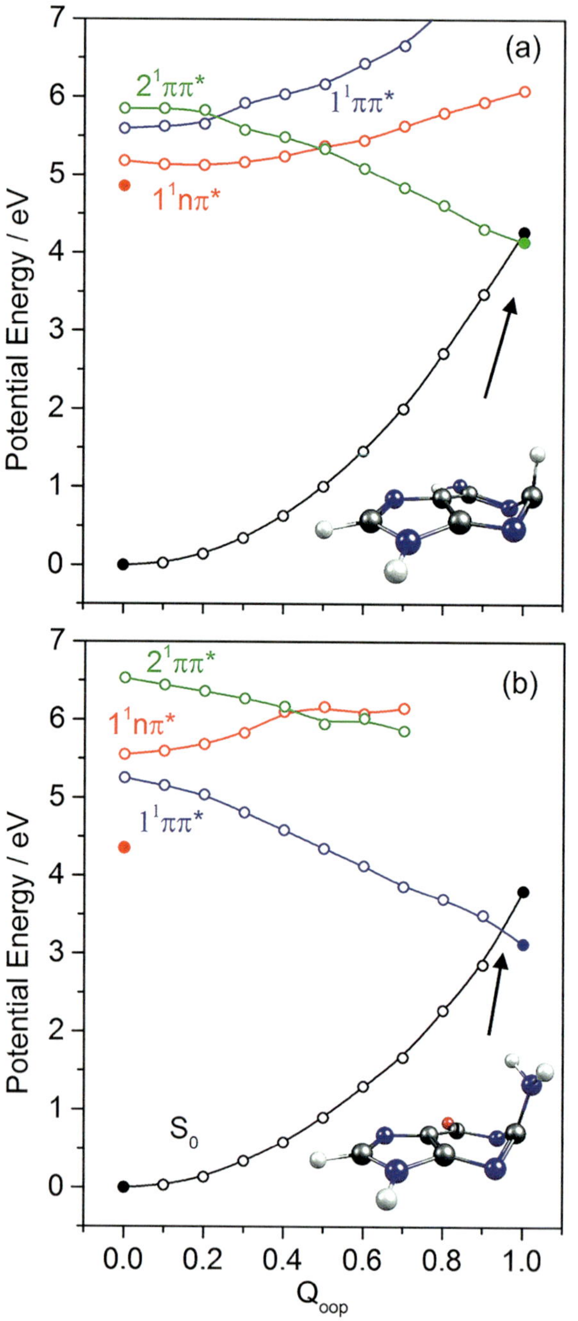

Fig. 1.3 CASPT2 potential energy profiles associated with the out-of-plane ring deformations of (**a**) adenine and (**b**) guanine. This figure is adapted from (Marchetti et al. 2016) with permission

alkenes. Such computations of the PE profiles offer detailed insight into the ways in which the pre-excited geometry is connected to the conical intersections that drive photostability, but they offer no insight into the timescales associated with such non-radiative processes.

Trajectory surface hopping (TSH) simulations offer insights into excited-state dynamics and the associated timescales (Tully 1990). On-the-fly TSH, wherein energies, gradients, and non-adiabatic couplings between electronic states are computed during the dynamics is a commonly employed method for determining the fate of electronically excited nucleobases. The inherent full dimensionality offered by on-the-fly TSH enables determination of the excited-state lifetime and, providing enough trajectories are propagated, returns yields associated with the possible excited-state outcomes.

As Table 1.1 indicates, TSH simulations reveal that the DNA/RNA nucleobases (and their derivative) undergo internal conversion back to the ground state within 1 picosecond which aligns with experimentally measured lifetimes (Barbatti et al. 2010, 2011, 2012; Asturiol et al. 2009; Lan et al. 2009; Nachtigallová et al. 2011; Barbatti and Lischka 2008; Plasser et al. 2014). The TSH simulations reveal that the nuclear motions that drive this internal conversion are the aforementioned out-of-plane ring deformations, which play a dominant role in the purine bases, adenine and guanine. The pyrimidine nucleobases uracil, thymine, and cytosine also have an analogous sub-picosecond excited-state decay, driven by analogous out-of-plane deformations. In constant to the purine nucleobases, the pyrimidine nucleobases show a second time constant with a much longer timescale (>1 picosecond) decay, which is attributable to internal conversion to the dark nπ* state—which is longer

Table 1.1 Experimental and theoretical lifetimes for excited-state decay of nucleobases in gas phase

Nucleobase	λ_{pump}/nm	τ_1/ps	τ_2/ps	τ_3/ps	τ_4/ps	References
Adenine	250 267 277	<0.05 0.10	0.75, 0.63 1.0, 0.77 ~2			Ullrich et al. (2004a, b), Canuel et al. (2005), Kang et al. (2002, 2003), Chin et al. (2007), Satzger et al. (2006)
Guanine	267	0.15	0.36, 0.8, 0.28 0.49	6.4, >2.3 5.1, 6.4 > 1.7		Canuel et al. (2005), Kang et al. (2002)
Thymine	250 267	<0.05 0.11	0.49		63	Ullrich et al. (2004a), Canuel et al. (2005), Kang et al. (2002)
Uracil	250 267	<0.05 0.13	0.53, 0.65 0.75			Ullrich et al. (2004a), Canuel et al. (2005), Kang et al. (2002)
Cytosine	250 267 314	<0.05 0.16	0.82 0.53		290	Nir et al. (2002a), Ullrich et al. (2004a), Canuel et al. (2005), Kang et al. (2002)

lived. There is still some debate about the active mechanism by which this long-lived $n\pi^*$ state decays, in part due to the restrictively long propagations timescales that are required to explore the longer timescale dynamics. Notwithstanding, static electronic structure calculations have provided some insight into the possible routes by which this $n\pi^*$ state decays—which also involves out-of-plane ring deformations as well as longer timescale intersystem crossing onto a triplet state.

In solution, there is also some debate on the deactivation mechanism of DNA/RNA nucleobases following photoexcitation (Kistler and Matsika 2010; Petersen et al. 2012; Yoshikawa and Matsika 2008; Busker et al. 2008; Blancafort and Migani 2007; Yamazaki and Kato 2007; González et al. 2012). The added complexity of the solution provides a platform by which the thermal energy is dissipated to the surrounding solvent environment, leading to vibrational energy relaxation. Modelling the solvated or bulk DNA environment is often tricky and requires methods that adequately describe the electrostatic effects of the surrounding environment. Mixed quantum/classical studies have been performed to model nucleobases in bulk solution or in the bulk DNA environment—highlighting the quenching of out-of-plane motions due to the stacking configuration and the promotion of alternative non-radiative decay paths such as intermolecular proton-transfer and photocycloaddition reactions that lead to DNA lesions (see later). While these studies represent significant strides toward understanding the photophysics of nucleobases in bulk environment, modelling nucleobases in such environments is still challenging.

1.3.2 Nucleosides and Nucleotides

In nucleic acids, the nucleobases considered in the previous section are attached to a sugar group. This sugar group may be ribose or deoxyribose, distinguishable by whether the form of the nucleic acid is RNA or DNA, respectively. If the ribose/deoxyribose moiety is terminal, the resulting structure is termed a nucleoside, while if the ribose/deoxyribose moiety is attached to a phosphate group, the resulting structure is a nucleotide. The distinction between these structures in given in Fig. 1.4. Since nucleosides contain the same chromophore as the nucleobase, the first absorption band is dominated by a $\pi^* \leftarrow \pi$ electron promotion (Stange and Temps 2018). These locally excited $\pi\pi^*$ states may then promote non-radiative decay via ring-deformation geometries which (as with the isolated nucleobases) lead to conical intersections between the S_1 and S_0 states which mediate internal conversion to reform the parent molecule and thus photostability (Marchetti et al. 2016; Tuna et al. 2014; Asami et al. 2013; Zgierski and Alavi 2006; Mansour et al. 2022, 2023; Pepino et al. 2017; Zhao et al. 2023; De Camillis et al. 2015; Zhao et al. 2020). This feature is highlighted in Figs. 1.5 and 1.6 (left panel) for the pyrimidine and purine nucleosides, respectively.

As these two Figures highlight, computational studies have shown a second plausible pathway for non-radiative decay of DNA/RNA nucleosides. The sugar moieties contain several hydroxy groups that are suitable hydrogen-bond donors. The

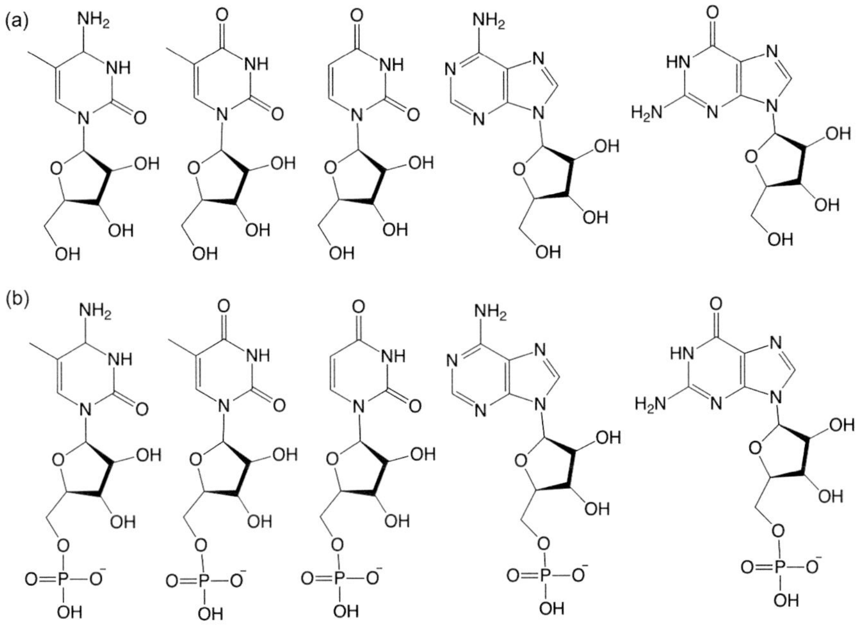

Fig. 1.4 Molecular structures of DNA/RNA (**a**) nucleosides and (**b**) nucleotides

nitrogen-heteroatoms and carbonyl oxygens contain lone-pairs that are available to participate in hydrogen bonding. The ground state minimum energy geometry of the nucleosides is with one of the hydroxyl groups of the sugar moiety hydrogen-bonded to the N-heteroatom of the purine nucleosides or the carbonyl oxygen of the pyrimidine nucleosides (as shown in Figs. 1.5 and 1.6). In this arrangement, the locally excited ππ* state may develop charge transfer character wherein the electronic configuration contains singly occupied molecular orbitals that are localized on the donor OH and accept O/N moieties. This characteristic charge transfer creates an electron–hole pair wherein the heterocyclic moiety contains a partial negative charge while the OH moiety contains a particle positive charge. This charge imbalance is countered by proton transfer from the OH donor to the O/N acceptor. As such the charge transfer state is reactive with respect to proton transfer, characterized by the decrease in energy of the red (charge transfer) states in Figs. 1.6 and 1.7. In contrast, the ground state is unreactive with respect to proton transfer and, as such, leads to a rise in PE, as indicated by the black curves in Figs. 1.6 and 1.7. As above, the respective decrease and increase in PE of the S_1 and S_0 states inevitably lead to an energy degeneracy and thus a conical intersection when orthogonal motions are considered.

The so-called excited-state proton transfer is ubiquitous across molecular photochemistry. In the context of nucleic acids, hydrogen bonding of the nucleobases with solvent water molecules has been observed to play a role in the excited-state decay. In 9H-adenine-$(H_2O)_5$ clusters (Chakraborty et al. 2018; Szabla et al. 2016),

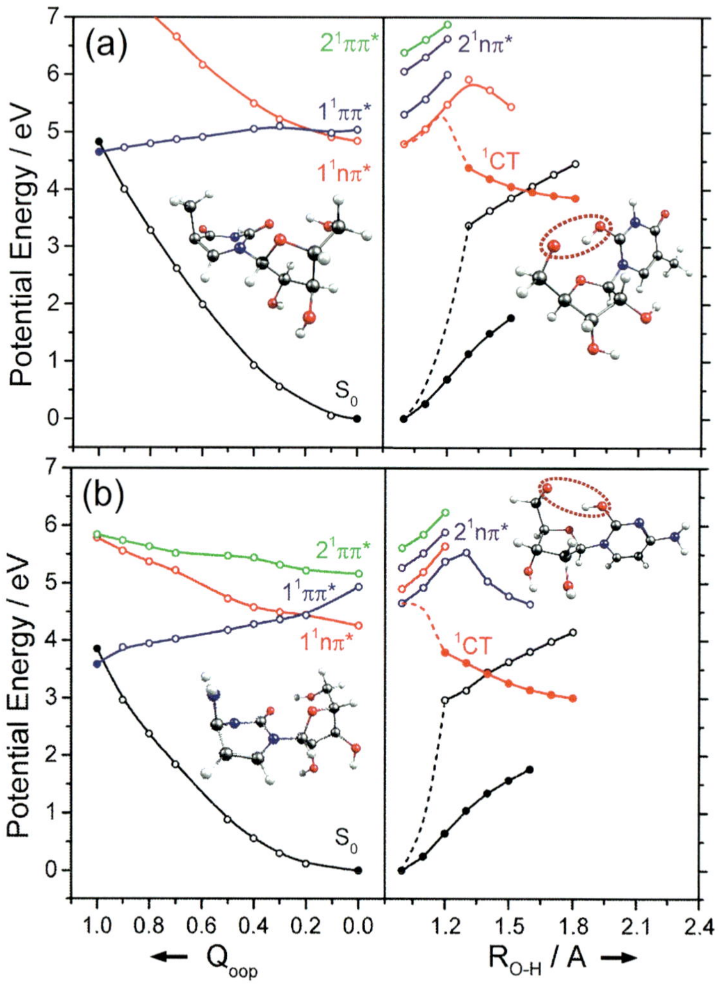

Fig. 1.5 Potential energy profiles associated with out-of-plane ring deformations (left panels) and excited-state proton transfer (right panel) for (**a**) thymidine and (**b**) Cytidine. This figure is adapted from (Marchetti et al. 2016) with permission

trajectory surface hopping simulations, where energies and gradients were calculated at the ADC(2)/aug-cc-pVDZ level of theory, have revealed that the locally excited ππ* state develops into a charge transfer state that involves an electron promotion from a water centered oxygen lone-pair, to a π* orbital of the adenine. This charge transfer state mediates an analogous out-of-plane deformation as isolated 9H-adenine which leads to a conical intersection with the ground state—motion through which leads to photostability. As with the nucleosides, such charge transfer in 9H-adenine–water clusters have also been postulated to mediate proton transfer. Due to the inherent high entropy of solvated environments, intermolecular (solvent

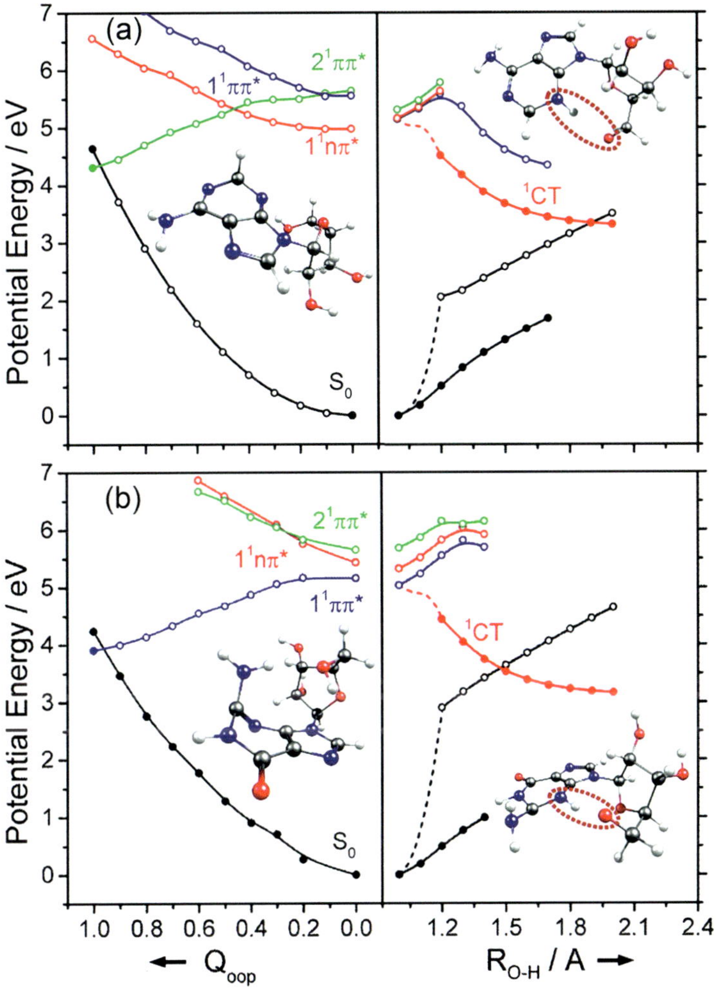

Fig. 1.6 Potential energy profiles associated with out-of-plane ring deformations (left panels) and excited-state proton transfer (right panel) for (**a**) adenosine and (**b**) guanosine. This figure is adapted from (Marchetti et al. 2016) with permission

to solute) proton transfer is unlikely to play a major role in the non-radiative decay of 9H-adenine (and other nucleobases) in bulk solution, as supported by trajectory surface hopping simulations.

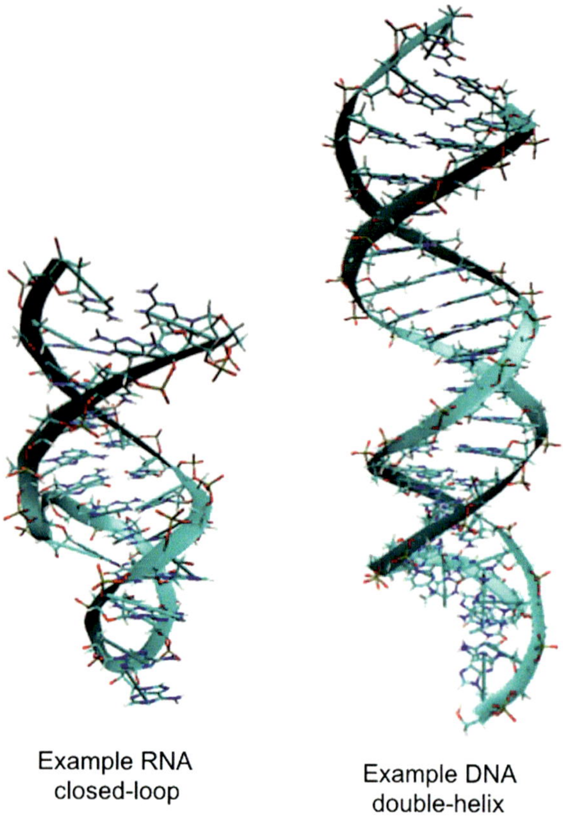

Fig. 1.7 Example closed-loop RNA (left) and double-helix DNA (right) structures

1.3.3 Base Pairs and the Bulk Biomolecular Environment

As illustrated in Fig. 1.7, complementary base pairs form in the stable conformation of double helix DNA or when RNA folds in on itself to form a complex, three-dimensional shape. The first reported computational study of the excited state dynamics was by Domcke and co-workers in 2004 (Sobolewski and Domcke 2004), where the second-order approximate singles and doubles (CC2) method was used to compute the potential energy profiles associated with the fate of the excited states of the guanine-cytosine nucleobase dimer in the gas phase. Figure 1.8b presents reproduced potential energy profiles (at the ADC(2) level of theory) along the proposed dominant non-radiative decay path—i.e., intermolecular proton transfer (Marchetti et al. 2016). As with the nucleosides and nucleobase-water clusters, such excited-state intermolecular proton transfer reactions were postulated to occur via initial excitation of a bright $^1\pi\pi^*$ state. This pre-excited state then develops into a charge separated state which involves an electron promotion between π and π^* orbitals of, respectively, the guanine and cytosine nucleobases. As with the nucleosides above, the charge transfer state creates a charge imbalance that is neutralized by the

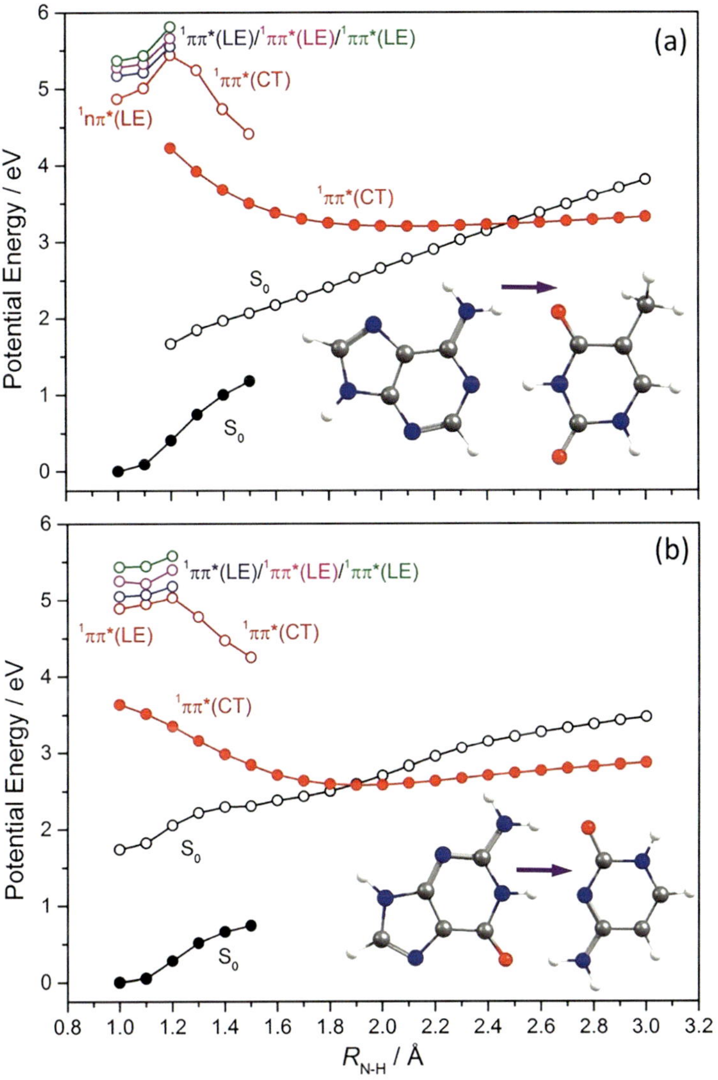

Fig. 1.8 ADC(2) potential energy profiles associated with the intramolecular excited-state proton transfer reaction for (**a**) adenine–thymine and (**b**) guanine–cytosine base pairs. This figure is adapted from (Marchetti et al. 2016) with permission

migration of a proton. In the case of the adenine-thymine base pair, this proton-transfer occurs from the NH moiety of guanine to the N heteroatom of cytosine. The favorable proton transfer in the excited charge transfer state leads to a conical intersection with the ground state, which is postulated to mediate photostability among base pair (Marchetti et al. 2016; Sobolewski and Domcke 2004; Sobolewski et al. 2005).

This same study (Sobolewski and Domcke 2004) also revealed that out-of-plane ring deformations such as those that drive photostability in the isolated nucleobases may also be active upon base pairing. We note however that in the bulk DNA helical environment, it is unlikely that out-of-plane ring deformations drive photostability, since stacking between adjacent nucleobases likely leads to steric hindrances of such ring deformations.

Later studies by the Domcke group on the adenine–thymine base pair revealed an analogous excited-state proton transfer reaction mediated by a charge transfer state (Perun et al. 2006b; Gobbo et al. 2012). The PE profiles associated with this reaction are given in Fig. 1.8a. The locally excited $\pi\pi^*$ state involves an electron promotion between π and π^* orbitals that are both localized on the adenine molecule. The charge transfer state is formed by an electron promotion from a π orbital of adenine to the π^* orbital of thymine, creating an electron–hole pair in which the adenine and thymine, respectively, carry a partial positive and negative charge. As with the aforementioned charge separated states, this charge separation is neutralized by the migration of a proton from the amine group of adenine to the carbonyl oxygen atom of the thymine. An analogous conical intersection between the charge transfer and ground states was postulated to mediated photostability of photoexcited nucleobase pairs. The same study also explored the possibilities of out-of-plane ring deformations, but without scans of the potential energy profiles. A more recent perspective article showed that non-radiative decay via out-of-plane deformations is energetically unfavorable when compared to the excited-state proton transfer path.

Later studies have also explored the photostability of nucleobase pairs with various higher levels of theory and in bulk solution; these later studies also confirm that the same excited-state proton transfer process is the active non-radiative decay path.

In the DNA double-helix environment, experimental studies have impressively identified the excited-state dynamics using femtosecond time-resolved spectroscopy (Zhang et al. 2015, 2016). One notable study is by the Kohler group (Zhang et al. 2015) who have performed UV pump, transient infrared probe studies of DNA double-helix sequences, where they compare the excited-state dynamics of common nucleobases on each strand making up the common (henceforth, config1—Fig. 1.9a) and alternating (henceforth, config2—Fig. 1.9b) nucleobases on each strand.

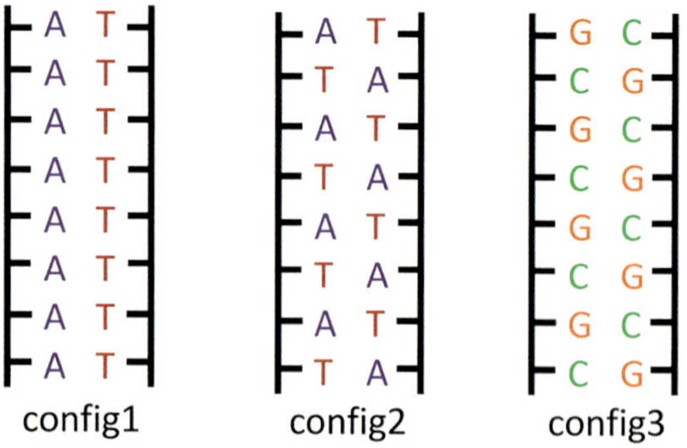

Fig. 1.9 The three combinations of double-helix DNA studied experimentally

In that study the supporting computational studies identified the major peaks in the transient infrared spectrum. The combined experimental and computational approach showed that upon photoexcitation of config1, an electron promotion occurs between a pair of π-stacked to form a ππ* charge transfer state—creating an electron–hole pair across adjacent stacked adenines. This charge separated promotes proton transfer from the thymine nucleobase on the complementary strand. This entire photoinduced process (outlined in Fig. 1.9a) happens on a timescale of ca. 150 ps. An analogous charge separation is created in config2 wherein the electron promotion occurs between pairs of π-stacked adenine and thymine nucleobase creating a charge transfer state. Specifically, the charge transfer state is formed from an electron promotion from a thymine π orbital to an adenine π* orbital. This charge transfer state then mediates proton transfer from a thymine nucleobase on the complementary strand to the adenine nucleobase. The combined photoinduced process takes ca. 300 ps timescale. The source of the difference in the lifetimes is still unclear and future computational studies are needed to discover the origin of the longer lifetime for config2 versus config1. A plausible explanation for the difference in lifetime could be attributed to a more stable charge transfer state in π-stacked configuration. In a final DNA double-helix configuration involving guanine and cytosine (config3 in Fig. 1.9), the charge separation is created by a UV-induced electron transfer from guanine to cytosine which is then neutralized by proton transfer from a neighboring guanine to the partial negatively charged cytosine. Interestingly this process occurs on ca. 30 ps—i.e., on a much shorter than config1 and config2.

While on the topic of DNA in the bulk environment, it is also worth mentioning that adjacent π-stacked thymine nucleobases can undergo reactive photochemistry upon UV-excitation (Daya-Grosjean et al. 1995; Liu et al. 2011; Lima-Bessa and Menck 2005; Mitchell et al. 1992). Upon prolonged UV-exposure, internal conversion to the long-lived nπ* state may mediate cyclization of a pair of π-stacked thymine nucleobases to form cycloadducts. Such cycloadducts are known to cause DNA damage if left unrepaired and are a source of cancers such as squamous skin carcinoma. The simplest type of photochemically induced cycloadduct is cyclobutane pyrimidine dimers (CPDs). Upon photoexcitation, CPD formation in thymine occurs via a photochemically allowed 2+2 cycloaddition of the ring centered C=C moiety to form cyclobutene as illustrated in Fig. 1.10 (Daya-Grosjean et al. 1995; Liu et al. 2011; Lima-Bessa and Menck 2005; Mitchell et al. 1992). This reaction essentially bonds two adjacent thymine molecules which disrupts the structure and function of these nucleobases in DNA. In basic terms, the promotion of CPDs can be understood by considering the orbital correlation diagram (Fig. 1.11) as outlined by the Woodward–Hoffmann rules (Woodward and Hoffmann 1965; Woodward and Hoffmann 1969), which states that the highest occupied molecular orbitals (HOMO) involved in a 2π + 2π cycloaddition creates a long-range repulsive interaction (via an anti-bonding π–π interaction). As such, the HOMO of the reactant pair correlate with the lowest unoccupied molecular orbital (LUMO) of the cycloadduct (as shown in Fig. 1.11) which leads to a bond-order of 0. In contrast, since the π* (LUMO) orbital of the reactant pair correlates with a bonding orbital of the cycloadduct, electron promotion to the LUMO results in a cycloadduct with a singly occupied π

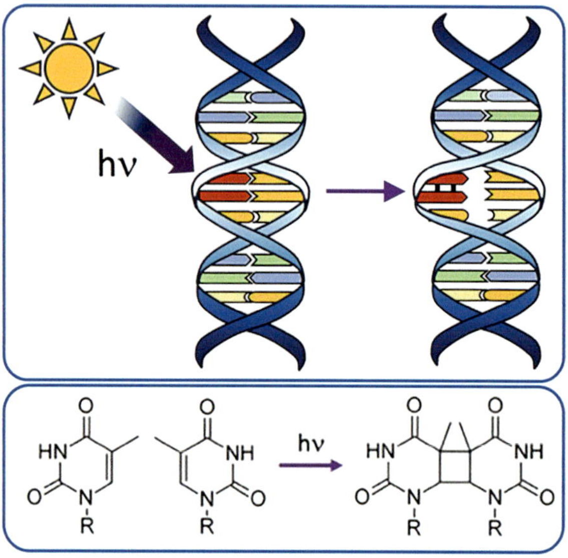

Fig. 1.10 [2+2]-photocycloaddition reaction between a pair of π-stacked thymine nucleobases

and π*—creating a total bond-order of 0.5. Quantum chemical calculations have revealed the potential energy paths associated with CPD formation following photoexcitation of a pair of π-stacked thymine nucleobases (Schreier et al. 2007; Boggio-Pasqua et al. 2007; Aida et al. 1996, 1997; Barbatti 2014; Stangl et al. 2019; Tachikawa and Kawabata 2008; Danilov et al. 2001; Lee and Matsika 2015; Giussani et al. 2018). These calculations show that CPDs are formed via a S_1/S_0 conical intersection with one of the two nascent cycloadduct C–C bonds formed. This geometry is given in Fig. 1.12 and is supported by the aforementioned bond-order of 0.5 for the S_1 state of a $2\pi + 2\pi$ configuration. Internal conversion at this conical intersection leads to a stable ground state configuration of the cycloadduct. Complementary ab initio molecular dynamics simulations and experiments reveal that the cycloaddition process to form CPDs occurs on timescales which are on a competitive timescale to the non-radiative paths that lead to photostability (Schreier et al. 2007).

As Fig. 1.12 shows, another form of DNA lesion is 6–4 photoadducts. These occur via a similar conical intersection that drive CPDs, but instead, the singly formed C–C bond is between the inequivalent carbon atoms on the neighboring thymine (i.e., between C5 of one thymine to the C6 of the neighboring thymine). In this configuration, CPD formation is sterically hindered and thus the minimum energy path is for a hydrogen atom migration from the methyl group of one thymine to the C4 atom of the neighboring thymine molecule. The resulting product is commonly known as a 6–4 lesion. Mixed Quantum Mechanics/Molecular Mechanics (QM/MM) electronic structure calculations and accompanying molecular dynamics simulations have revealed an analogous path to 6–4 lesion formation and dynamics occurring within 1 picosecond.

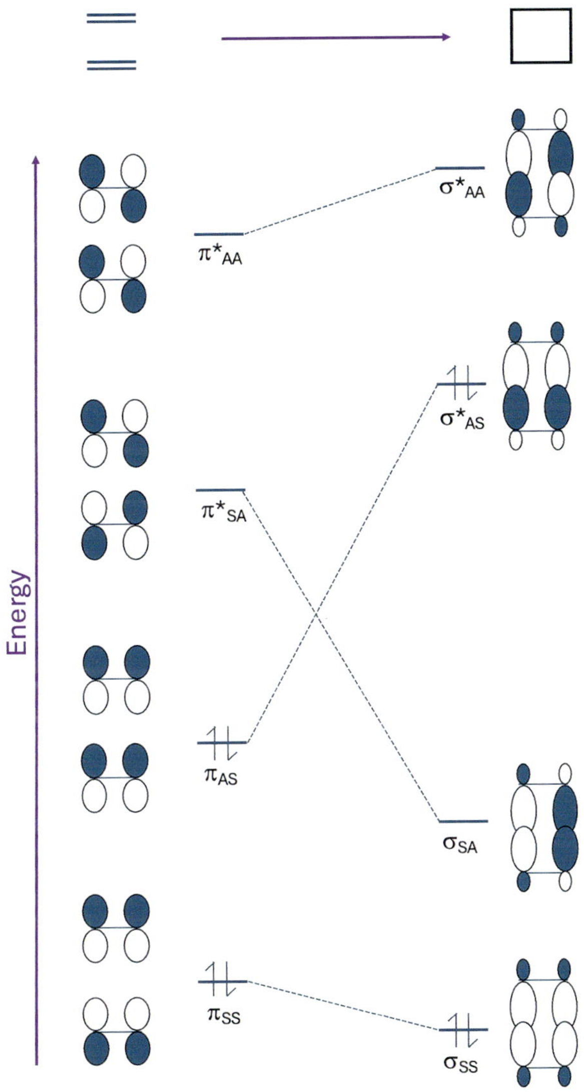

Fig. 1.11 Orbital correlation diagram showing the correlation of p-orbitals of the started alkene pair and the subsequent cycloadduct

1.4 Conclusions and Outlook

It is now apparently clear that the physical chemistry community has made significant strides in understanding the photophysics and photochemistry of nucleic acid constituents. This community has been essential in determining the detailed mechanisms by which photoexcited DNA undergoes photostability and photodamage

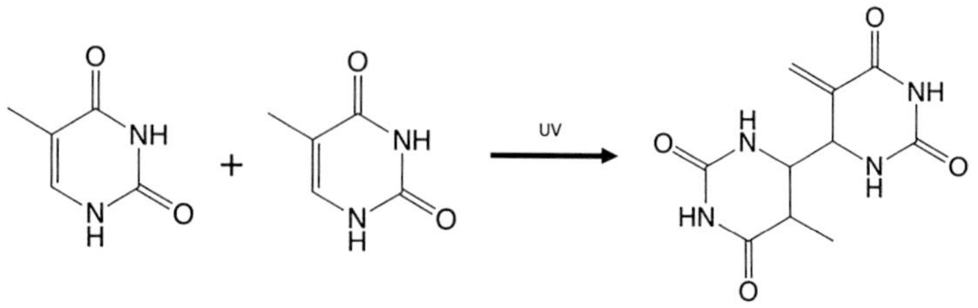

Fig. 1.12 Photoreaction path associated with the photoinduced formation of 6–4 lesions

upon prolonged UV-exposure. It has guided the biological sciences community in understanding the crucial first steps that lead to cancers, while connecting with industry to use the knowledge gained to develop more efficient sunscreen protection products. This chapter has reviewed the important contributions and highlights the overall big picture on the excited-state dynamics of nucleic acid constituents. We do however caution the reader that while this chapter provides a clear overall big picture, several aspects of the excited-state dynamics of nucleic acid constituents are still under some debate.

As the above research review highlights, computational chemistry has been essential in determining the mechanisms and timescales associated with the photophysics and photochemistry of nucleic acids. Quantum chemical calculation have been used to optimize critical points along the excited-state potential energy surface and how the excited-state population undergoes non-radiative transition.

There are however several open questions in photochemical and photophysical studies of nucleic acid constituents. While the early-time chemistry of nucleic acid constituents is fairly well understood, the longer timescale dynamics is less well understood. Experiments have identified long timescale features of nucleic acids (particularly uracil and thymine) which current computational methods attribute to population trapped in the dark $n\pi^*$ which decays on a longer timescale. Quantum chemical studies have identified some mechanisms, but the very long timescales (>100 ps) preclude the use of currently available dynamics methods in determining the ultimate fate of the $n\pi^*$ state. Additionally, there lacks computational insight into the early-time structure changes and the longer timescale implications to the bulk structure of DNA/RNA helices. Such bulk structural changes of the bulk biochemical environment would occur on several nanoseconds to microseconds. Although classical dynamics methods with empirical force fields are available, they are primarily parameterized with methods that are distinct to those that described the electronic structure of early-time dynamics. Additionally, connecting the early-time single-molecule nuclear motions to later timescale bulk structural changes with ab initio molecular dynamics is impossible given the inherent computational expense.

Both these examples relate to the inability to characterize the longer timescale dynamics. In the present day, computational chemistry is undergoing several spectacular advances that may soon be used to address these open questions. Firstly,

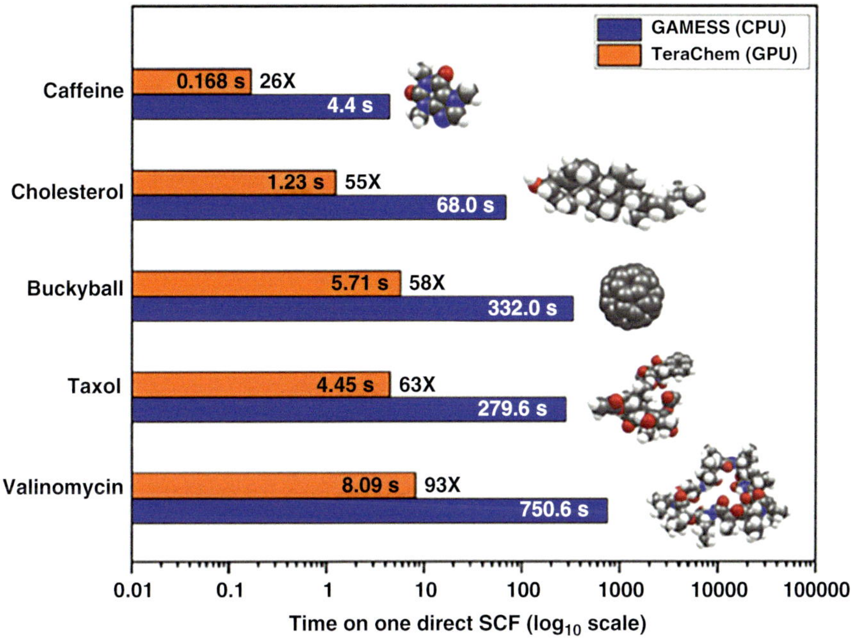

Fig. 1.13 Timings of various molecules performed on CPUs versus GPUs, adapted from http://www.petachem.com/performance.html

since the recognition that quantum chemistry can be performed on graphical processing units (GPUs) as opposed to commonly used central processing units (CPUs), a significant enhancement in computational power has been observed. As Fig. 1.13 shows a single point calculation performed on GPUs versus CPUs is between 26 and 93 times faster of GPUs compared to CPUs. Multi-reference methods, including CASPT2 have also been implemented on GPU architectures. With these advancements, it is now possible to undertake long timescale molecular simulation at a reduced computational cost (Ufimtsev and Martinez 2009a, b).

Furthermore, the integration of machine learning and computational chemistry has seen significant strides in the molecular modelling of excited states. It is now possible to train machine learned potentials that deliver ab initio molecular dynamics at the cost of force field molecular dynamics. Some strides are being made to integrate excited-state dynamics and machine learning.

References

(1999) Recent Adv. Comput Chem 4:224 (book) https://www.worldscientific.com/worldscibooks/10.1142/4030?srsltid=AfmBOopML2bCwXuIdHZpdF6eVIpyzQTfJxTNt5qckOwOaP03n3Mp7hiZ#t=toc

Aida M, Kaneko M, Dupuis M (1996) An ab initio MO study on the thymine dimer and its radical cation. Int J Quantum Chem 57:949–957

Aida M, Inoue F, Kaneko M, Dupuis M (1997) An ab initio MO study on fragmentation reaction mechanism of thymine dimer radical cation. J Am Chem Soc 119:12274–12279

Andersson K, Malmqvist PA, Roos BO, Sadlej AJ, Wolinski K (1990) Second-order perturbation theory with a CASSCF reference function. J Phys Chem 94:5483–5488

Andersson K, Malmqvist P, Roos BO (1992) Second-order perturbation theory with a complete active space self-consistent field reference function. J Chem Phys 96:1218–1226

Asami H, Yagi K, Ohba M, Urashima S, Saigusa H (2013) Stacked base-pair structures of adenine nucleosides stabilized by the formation of hydrogen-bonding network involving the two sugar groups. Chem Phys 419:84–89

Asturiol D, Lasorne B, Robb MA, Blancafort L (2009) Photophysics of the π,π* and n,π* states of thymine: MS-CASPT2 minimum-energy paths and CASSCF on-the-fly dynamics. J Phys Chem A 113:10211–10218

Barbatti M (2014) Computational reference data for the photochemistry of cyclobutane pyrimidine dimers. ChemPhysChem 15:3342–3354

Barbatti M, Lischka H (2008) Nonadiabatic deactivation of 9H-adenine: a comprehensive picture based on mixed quantum–classical dynamics. J Am Chem Soc 130:6831–6839

Barbatti M, Aquino AJA, Szymczak JJ, Nachtigallová D, Hobza P, Lischka H (2010) Relaxation mechanisms of UV-photoexcited DNA and RNA nucleobases. Proc Natl Acad Sci USA 107:21453–21458

Barbatti M, Szymczak JJ, Aquino AJA, Nachtigallová D, Lischka H (2011) The decay mechanism of photoexcited guanine – a nonadiabatic dynamics study. J Chem Phys 134:14304

Barbatti M, Lan Z, Crespo-Otero R, Szymczak JJ, Lischka H, Thiel W (2012) Critical appraisal of excited state nonadiabatic dynamics simulations of 9H-adeninea. J Chem Phys 137:22A503

Blancafort L (2007) Energetics of cytosine singlet excited-state decay paths—a difficult case for CASSCF and CASPT2†. Photochem Photobiol 83:603–610

Blancafort L, Migani A (2007) Water effect on the excited-state decay paths of singlet excited cytosine. J Photochem Photobiol A Chem 190:283–289

Blancafort L, Cohen B, Hare PM, Kohler B, Robb MA (2005) Singlet excited-state dynamics of 5-fluorocytosine and cytosine: an experimental and computational study. J Phys Chem A 109:4431–4436

Boggio-Pasqua M, Groenhof G, Schäfer LV, Grubmüller H, Robb MA (2007) Ultrafast deactivation channel for thymine dimerization. J Am Chem Soc 129:10996–10997

Busker M, Nispel M, Häber T, Kleinermanns K, Etinski M, Fleig T (2008) Electronic and vibrational spectroscopy of 1-methylthymine and its water clusters: the dark state survives hydration. ChemPhysChem 9:1570–1577

Callis PR (1983) Electronic states and luminescence of nucleic acid systems. Annu Rev Phys Chem 34:329–357

Canuel C, Mons M, Piuzzi F, Tardivel B, Dimicoli I, Elhanine M (2005) Excited states dynamics of DNA and RNA bases: characterization of a stepwise deactivation pathway in the gas phase. J Chem Phys 122:74316

Chakraborty P, Karsili TNV, Marchetti B, Matsika S (2018) Mechanistic insights into photoinduced damage of DNA and RNA nucleobases in the gas phase and in bulk solution. Faraday Discuss 207:329–350

Chin C-H, Mebel AM, Kim G-S, Baek KY, Hayashi M, Liang KK, Lin SH (2007) Theoretical investigations of spectroscopy and excited state dynamics of adenine. Chem Phys Lett 445:361–369

Christiansen O, Koch H, Jørgensen P (1995) The second-order approximate coupled cluster singles and doubles model CC2. Chem Phys Lett 243:409–418

Conti I, Garavelli M, Orlandi G (2009) Deciphering low energy deactivation channels in adenine. J Am Chem Soc 131:16108–16118

Crespo-Hernández CE, Cohen B, Hare PM, Kohler B (2004) Ultrafast excited-state dynamics in nucleic acids. Chem Rev 104:1977–2020

Danilov VI, Les A, Alderfer JL (2001) A theoretical study of the cis-syn pyrimidine dimers in the gas phase and water cluster and a tautomer-bypass mechanism for the origin of UV-induced mutations. J Biomol Struct Dyn 19:179–191

Daya-Grosjean L, Dumaz N, Sarasin A (1995) The specificity of p53 mutation spectra in sunlight induced human cancers. J Photochem Photobiol B Biol 28:115–124

De Camillis S, Miles J, Alexander G, Ghafur O, Williams ID, Townsend D, Greenwood JB (2015) Ultrafast non-radiative decay of gas-phase nucleosides. Phys Chem Chem Phys 17:23643–23650

Del Bene JE, Ditchfield R, Pople JA (1971) Self-consistent molecular orbital methods. X. Molecular orbital studies of excited states with minimal and extended basis sets. J Chem Phys 55:2236–2241

Delchev VB, Sobolewski AL, Domcke W (2010) Comparison of the non-radiative decay mechanisms of 4-pyrimidinone and uracil: an ab initio study. Phys Chem Chem Phys 12:5007–5015

Domcke W, Yarkony DR, Köppel H (2004) Conical intersections: electronic structure, dynamics and spectroscopy. Adv Ser Phys Chem 15:838

Domcke W, Yarkony DR, Köppel H (2011) Conical intersect. Theory Comput Exp 17:1–754

Dreuw A, Weisman JL, Head-Gordon M (2003) Long-range charge-transfer excited states in time-dependent density functional theory require non-local exchange. J Chem Phys 119:2943–2946

Foresman JB, Head-Gordon M, Pople JA, Frisch MJ (1992) Toward a systematic molecular orbital theory for excited states. J Phys Chem 96:135–149

Giussani A, Conti I, Nenov A, Garavelli M (2018) Photoinduced formation mechanism of the thymine–thymine (6–4) adduct in DNA; a QM(CASPT2//CASSCF):MM(AMBER) study. Faraday Discuss 207:375–387

Gobbo JP, Saurí V, Roca-Sanjuán D, Serrano-Andrés L, Merchán M, Borin AC (2012) On the deactivation mechanisms of adenine–thymine base pair. J Phys Chem B 116:4089–4097

González L, Escudero D, Serrano-Andrés L (2012) Progress and challenges in the calculation of electronic excited states. ChemPhysChem 13:28–51

Gustavsson T, Improta R, Markovitsi D (2010) DNA/RNA: building blocks of life under UV irradiation. J Phys Chem Lett 1:2025–2030

Hohenberg P, Kohn W (1964) Inhomogeneous electron gas. Phys Rev 136:B864–B871

Ismail N, Blancafort L, Olivucci M, Kohler B, Robb MA (2002) Ultrafast decay of electronically excited singlet cytosine via a π,π^* to n_O,π^* state switch. J Am Chem Soc 124:6818–6819

Jablonski A (1933) Efficiency of anti-stokes fluorescence in dyes. Nature 131:839–840

Kang H, Lee KT, Jung B, Ko YJ, Kim SK (2002) Intrinsic lifetimes of the excited state of DNA and RNA bases. J Am Chem Soc 124:12958–12959

Kang H, Jung B, Kim SK (2003) Mechanism for ultrafast internal conversion of adenine. J Chem Phys 118:6717–6719

Kim NJ, Jeong G, Kim YS, Sung J, Keun Kim S, Park YD (2000) Resonant two-photon ionization and laser induced fluorescence spectroscopy of jet-cooled adenine. J Chem Phys 113:10051–10055

Kistler KA, Matsika S (2007) Radiationless decay mechanism of cytosine: an ab initio study with comparisons to the fluorescent analogue 5-Methyl-2-pyrimidinone. J Phys Chem A 111:2650–2661

Kistler KA, Matsika S (2010) Photophysical pathways of cytosine in aqueous solution. Phys Chem Chem Phys 12:5024–5031

Kleinermanns K, Nachtigallová D, de Vries MS (2013) Excited state dynamics of DNA bases. Int Rev Phys Chem 32:308–342

Knowles PJ, Werner H-J (1988) An efficient method for the evaluation of coupling coefficients in configuration interaction calculations. Chem Phys Lett 145:514–522

Knowles PJ, Werner H-J (1992) Internally contracted multiconfiguration-reference configuration interaction calculations for excited states. Theor Chim Acta 84:95–103

Lan Z, Fabiano E, Thiel W (2009) Photoinduced nonadiabatic dynamics of 9H-guanine. ChemPhysChem 10:1225–1229

Lange A, Herbert JM (2007) Simple methods to reduce charge-transfer contamination in time-dependent density-functional calculations of clusters and liquids. J Chem Theory Comput 3:1680–1690
Lee W, Matsika S (2015) QM/MM studies reveal pathways leading to the quenching of the formation of thymine dimer photoproduct by flanking bases. Phys Chem Chem Phys 17:9927–9935
Lima-Bessa KM, Menck CFM (2005) Skin cancer: lights on genome lesions. Curr Biol 15:R58–R61
Liu B (1973) Ab initio potential energy surface for linear H3. J Chem Phys 58:1925–1937
Liu Z, Tan C, Guo X, Kao Y-T, Li J, Wang L, Sancar A, Zhong D (2011) Dynamics and mechanism of cyclobutane pyrimidine dimer repair by DNA photolyase. Proc Natl Acad Sci 108:14831–14836
Lobsiger S, Trachsel MA, Frey H-M, Leutwyler S (2013) Excited-state structure and dynamics of keto–amino cytosine: the $1\pi\pi^*$ state is nonplanar and its radiationless decay is not ultrafast. J Phys Chem B 117:6106–6115
Ludwig V, da Costa ZM, do Amaral MS, Borin AC, Canuto S, Serrano-Andrés L (2010) Photophysics and photostability of adenine in aqueous solution: a theoretical study. Chem Phys Lett 492:164–169
Mansour R, Toldo JM, Barbatti M (2022) Role of the hydrogen bond on the internal conversion of photoexcited adenosine. J Phys Chem Lett 13:6194–6199
Mansour R, Toldo JM, Mukherjee S, Pinheiro M, Barbatti M (2023) Temperature effects on the internal conversion of excited adenine and adenosine. Phys Chem Chem Phys 25:27083–27093
Marchetti B, Karsili TNV, Ashfold MNR, Domcke W (2016) A 'bottom up', ab initio computational approach to understanding fundamental photophysical processes in nitrogen containing heterocycles, DNA bases and base pairs. Phys Chem Chem Phys 18:20007–20027
Marian CM (2007) The guanine tautomer puzzle: quantum chemical investigation of ground and excited states. J Phys Chem A 111:1545–1553
Matsika S (2004) Radiationless decay of excited states of uracil through conical intersections. J Phys Chem A 108:7584–7590
Merchán M, Serrano-Andrés L (2005) II - ab initio methods for excited states. Theor Comput Chem 16:35–91
Middleton CT, De La Harpe K, Su C, Law YK, Crespo-Hernández CE, Kohler B (2009) DNA excited-state dynamics: from single bases to the double helix. Annu Rev Phys Chem 60:217–239
Mitchell DL, Jen J, Cleaver JE (1992) Sequence specificity of cyclobutane pyrimidine dimers in DNA treated with solar (ultraviolet B) radiation. Nucleic Acids Res 20:225–229
Molecular mechanisms of the photostability of life (2010) Phys Chem Chem Phys 12:4897–4898
Møller C, Plesset MS (1934) Note on an approximation treatment for many-electron systems. Phys Rev 46:618–622
Nachtigallová D, Lischka H, Szymczak JJ, Barbatti M, Hobza P, Gengeliczki Z, Pino G, Callahan MP, de Vries MS (2010) The effect of C5 substitution on the photochemistry of uracil. Phys Chem Chem Phys 12:4924–4933
Nachtigallová D, Aquino AJA, Szymczak JJ, Barbatti M, Hobza P, Lischka H (2011) Nonadiabatic dynamics of uracil: population split among different decay mechanisms. J Phys Chem A 115:5247–5255
Nir E, Janzen C, Imhof P, Kleinermanns K, de Vries MS (2001) Guanine tautomerism revealed by UV–UV and IR–UV hole burning spectroscopy. J Chem Phys 115:4604–4611
Nir E, Müller M, Grace LI, de Vries MS (2002a) REMPI spectroscopy of cytosine. Chem Phys Lett 355:59–64
Nir E, Plützer C, Kleinermanns K, de Vries M (2002b) Properties of isolated DNA bases, base pairs and nucleosides examined by laser spectroscopy. Eur Phys J D - At Mol Opt Plasma Phys 20:317–329
Nix MGD, Devine AL, Cronin B, Ashfold MNR (2007) Ultraviolet photolysis of adenine: dissociation via the $\pi 1\sigma^*$ state. J Chem Phys 126:124312
Pepino AJ, Segarra-Martí J, Nenov A, Improta R, Garavelli M (2017) Resolving ultrafast photoinduced deactivations in water-solvated pyrimidine nucleosides. J Phys Chem Lett 8:1777–1783

Perun S, Sobolewski AL, Domcke W (2005) Ab initio studies on the radiationless decay mechanisms of the lowest excited singlet states of 9H-adenine. J Am Chem Soc 127:6257–6265
Perun S, Sobolewski AL, Domcke W (2006a) Conical intersections in thymine. J Phys Chem A 110:13238–13244
Perun S, Sobolewski AL, Domcke W (2006b) Role of electron-driven proton-transfer processes in the excited-state deactivation of the adenine–thymine base pair. J Phys Chem A 110:9031–9038
Petersen J, Wohlgemuth M, Sellner B, Bonačić-Koutecký V, Lischka H, Mitrić R (2012) Laser pulse trains for controlling excited state dynamics of adenine in water. Phys Chem Chem Phys 14:4687–4694
Plasser F, Crespo-Otero R, Pederzoli M, Pittner J, Lischka H, Barbatti M (2014) Surface hopping dynamics with correlated single-reference methods: 9H-adenine as a case study. J Chem Theory Comput 10:1395–1405
Roos BO (1980) The complete active space SCF method in a fock-matrix-based super-CI formulation. Int J Quantum Chem 18:175–189
Roos BO, Taylor PR, Sigbahn PEM (1980) A complete active space SCF method (CASSCF) using a density matrix formulated super-CI approach. Chem Phys 48:157–173
Roos BO, Linse P, Siegbahn PEM, Blomberg MRA (1982) A simple method for the evaluation of the second-order-perturbation energy from external double-excitations with a CASSCF reference wavefunction. Chem Phys 66:197–207
Satzger H, Townsend D, Stolow A (2006) Reassignment of the low lying cationic states in gas phase adenine and 9-methyl adenine. Chem Phys Lett 430:144–148
Schirmer J (1982) Beyond the random-phase approximation: a new approximation scheme for the polarization propagator. Phys Rev A 26:2395–2416
Schreier WJ, Schrader TE, Koller FO, Gilch P, Crespo-Hernández CE, Swaminathan VN, Carell T, Zinth W, Kohler B (2007) Thymine dimerization in DNA is an ultrafast photoreaction. Science (80-.) 315:625–629
Serrano-Andrés L, Merchán M (2005) Quantum chemistry of the excited state: 2005 overview. J Mol Struct Theochem 729:99–108
Serrano-Andrés L, Serrano-Pérez JJ (eds) (2012) J. Leszczynski. Springer, Dordrecht, pp 483–560
Serrano-Andrés L, Merchán M, Borin AC (2008) A three-state model for the photophysics of guanine. J Am Chem Soc 130:2473–2484
Sobolewski AL, Domcke W (2004) Ab initio studies on the photophysics of the guanine–cytosine base pair. Phys Chem Chem Phys 6:2763–2771
Sobolewski AL, Domcke W (2006) The chemical physics of the photostability of life. Europhys News 37:20–23
Sobolewski AL, Domcke W, Hättig C (2005) Tautomeric selectivity of the excited-state lifetime of guanine/cytosine base pairs: the role of electron-driven proton-transfer processes. Proc Natl Acad Sci 102:17903–17906
Soukup GA (2003) In encyclopedia of life sciences
Stange UC, Temps F (2018) Ultrafast electronic deactivation of UV-excited adenine and its ribo- and deoxyribonucleosides and -nucleotides: a comparative study. Chem Phys 515:441–451
Stangl CM, Krasnomowitz JM, Apsokardu MJ, Tiszenkel L, Ouyang Q, Lee S, Johnston MV (2019) Sulfur dioxide modifies aerosol particle formation and growth by ozonolysis of monoterpenes and isoprene. J Geophys Res Atmos 124:4800–4811
Stanton JF, Bartlett RJ (1993) The equation of motion coupled-cluster method. A systematic biorthogonal approach to molecular excitation energies, transition probabilities, and excited state properties. J Chem Phys 98:7029–7039
Szabla R, Góra RW, Janicki M, Šponer J (2016) Photorelaxation of imidazole and adenine via electron-driven proton transfer along H2O wires. Faraday Discuss 195:237–251
Szabo A, Ostlund NS (1996) Modern quantum chemistry introduction to advanced electronic structure theory. Dover Publications

Szalay PG, Müller T, Gidofalvi G, Lischka H, Shepard R (2012) Multiconfiguration self-consistent field and multireference configuration interaction methods and applications. Chem Rev 112:108–181

Tachikawa H, Kawabata H (2008) A direct ab initio molecular dynamics (MD) study on the repair reactions of stacked thymine dimer. Chem Phys Lett 462:321–326

Trofimov AB, Schirmer J (1995) An efficient polarization propagator approach to valence electron excitation spectra. J Phys B Atomic Mol Phys 28:2299

Tully JC (1990) Molecular dynamics with electronic transitions. J Chem Phys 93:1061–1071

Tuna D, Sobolewski AL, Domcke W (2014) Mechanisms of ultrafast excited-state deactivation in adenosine. J Phys Chem A 118:122–127

Ufimtsev IS, Martinez TJ (2009a) Quantum chemistry on graphical processing units. 2. Direct self-consistent-field implementation. J Chem Theory Comput 5:1004–1015

Ufimtsev IS, Martinez TJ (2009b) Quantum chemistry on graphical processing units. 3. Analytical energy gradients, geometry optimization, and first principles molecular dynamics. J Chem Theory Comput 5:2619–2628

Ullrich S, Schultz T, Zgierski MZ, Stolow A (2004a) Electronic relaxation dynamics in DNA and RNA bases studied by time-resolved photoelectron spectroscopy. Phys Chem Chem Phys 6:2796–2801

Ullrich S, Schultz T, Zgierski MZ, Stolow A (2004b) Direct observation of electronic relaxation dynamics in adenine via time-resolved photoelectron spectroscopy. J Am Chem Soc 126:2262–2263

Voet D, Gratzer WB, Cox RA, Doty P (1963) Absorption spectra of nucleotides, polynucleotides, and nucleic acids in the far ultraviolet. Biopolymers 1:193–208

Woodward RB, Hoffmann R (1965) Stereochemistry of electrocyclic reactions. J Am Chem Soc 87:395–397

Woodward RB, Hoffmann R (1969) The conservation of orbital symmetry. Angew Chem Int Ed English 8:781–853

Yamazaki S, Kato S (2007) Solvent effect on conical intersections in excited-state 9H-adenine: radiationless decay mechanism in polar solvent. J Am Chem Soc 129:2901–2909

Yamazaki S, Domcke W, Sobolewski AL (2008) Nonradiative decay mechanisms of the biologically relevant tautomer of guanine. J Phys Chem A 112:11965–11968

Yoshikawa A, Matsika S (2008) Excited electronic states and photophysics of uracil–water complexes. Chem Phys 347:393–404

Zechmann G, Barbatti M (2008) Photophysics and deactivation pathways of thymine. J Phys Chem A 112:8273–8279

Zgierski MZ, Alavi S (2006) Quantum chemical study of biradical decay channels in cytidine nucleosides. Chem Phys Lett 426:398–404

Zgierski MZ, Patchkovskii S, Fujiwara T, Lim EC (2005) On the origin of the ultrafast internal conversion of electronically excited pyrimidine bases. J Phys Chem A 109:9384–9387

Zhang Y, de La Harpe K, Beckstead AA, Improta R, Kohler B (2015) UV-induced proton transfer between DNA strands. J Am Chem Soc 137:7059–7062

Zhang Y, Li X-B, Fleming AM, Dood J, Beckstead AA, Orendt AM, Burrows CJ, Kohler B (2016) UV-induced proton-coupled electron transfer in cyclic DNA miniduplexes. J Am Chem Soc 138:7395–7401

Zhao L, Zhou G, Jia B, Teng G, Zhan K, Zheng H, Luo J, Liu B (2020) New insight into the ultrafast excited state deactivation mechanism of guanosine in the gas phase. J Photochem Photobiol A Chem 401:112753

Zhao L, Geng X, Han G, Guo Y, Liu R, Chen J (2023) Revealing the excited-state dynamics of cytidine and the role of excited-state proton transfer process. Phys Chem Chem Phys 25:32002–32009

Chapter 2
Computational Studies on Photoinduced Charge Transfer Processes in Nucleic Acids: From Watson–Crick Dimers to Quadruple Helices

Lara Martínez Fernández and Roberto Improta

Abstract Absorption of ultraviolet light by nucleic acids can lead to the population of excited electronic states with significant inter-base charge transfer character, which have been recently shown to be involved in their photophysics and photochemistry. In this chapter, we shall concisely review the most recent computational papers on this topic, in oligonucleotide systems of different size and complexity, trying to highlight the most important chemical-physical effects into play. In Watson–Crick hydrogen bonded guanine–cytosine dimers, charge transfer states are populated with a large yield and can undergo inter-base proton transfer. The outcome is a proton-coupled electron transfer process which provides a very effective route for ground state recovery. This mechanism, which does not occur in adenine–thymine Watson–Crick dimer, is significantly affected by environmental effects. In single strands, inter-base charge transfer states are effectively populated for most of the possible stacked pairs. Within duplexes, both intra-strand and inter-strand charge transfer routes can be active and can lead to inter-strand proton transfer. The interplay between these two possible paths depends on the sequence. Finally, charge transfer states are populated also in non-canonical DNA structures as guanine-rich quadruple helices and cytosine-rich I-motifs.

Keywords Proton-coupled electron transfer · Hydrogen bonds · Quantum dynamics · Time-dependent DFT

L. M. Fernández
Departamento de Química Física Blas Cabrera, Consejo Superior de Investigaciones Científícas, Madrid, Spain

R. Improta (✉)
Istituto di Biostrutture e Bioimmagini-CNR (IBB-CNR), Naples, Italy
e-mail: roberto.improta@cnr.it

S. Matsika, A. H. Marcus (eds.), *Nucleic Acid Photophysics and Photochemistry*, Nucleic Acids and Molecular Biology 36,
https://doi.org/10.1007/978-3-031-68807-2_2

2.1 Introduction

The canonical nucleobases in water have vertical ionization potentials in the range 7.4–8.1 eV (references (Slavíček et al. 2009; Muñoz-Losa et al. 2015) and references therein), which decreases by 1.1~1.2 eV when the solvent is fully equilibrated with the electron density of the cationic base (B^+) (Muñoz-Losa et al. 2015). The electron affinities of the bases in water (producing the anion B^-) are weakly positive (Gu et al. 2012; Hendricks et al. 1998) and the creation of the radical ion pair (B^+B^-) between close-lying bases (stacked or hydrogen bonded) is characterized by a strong electrostatic stabilization. On this ground, we can expect that when Nucleic Acids (hereafter NA) are excited in their lowest energy absorption band (with a maximum at ~260 nm, i.e. at ~4.7 eV) the population of excited states with partial Charge Transfer (CT) character is possible. Indeed, already two decades ago, it was proposed that CT states in the gas phase are formed, with significant yield in Guanine–Cytosine (GC) hydrogen bonded pair within a Watson–Crick (WC) arrangement, and that they trigger an inter-base proton transfer (PT) process allowing sub-ps ground state recovery (Sobolewski et al. 2005; Sobolewski and Domcke 2004). The involvement of CT states in the photoactivated dynamics of NA duplexes has been more debated (Martínez Fernández et al. 2022; Improta et al. 2016; Barbatti et al. 2015; Schreier et al. 2015; Gustavsson and Markovitsi 2021; Crespo-Hernández et al. 2004; Middleton et al. 2009), but it is now generally accepted, thanks to the clear indications provided by Time Resolved (TR) Infrared Experiments (IR) (Crespo-Hernández et al. 2004; Middleton et al. 2009; Bucher et al. 2014a, b; Martinez-Fernandez et al. 2016; Zhang et al. 2014, 2015a, b, 2016a, b). This technique is indeed particularly suitable to track CT states, since the B^+ and B^- nucleobases exhibit diagnostic vibrational frequencies, quite different with respect to those of the neutral nucleobases, either in their ground or excited electronic states. From the computational side, several studies have provided fundamental contributions to assess the role of CT states in the photophysics and photochemistry of NA (Martínez Fernández et al. 2022; Improta et al. 2016; Barbatti et al. 2015), considering that these states are 'dark' (or in any case weakly absorbing and emitting species) making their identification via optical spectroscopies more difficult (Schreier et al. 2015; Gustavsson and Markovitsi 2021). In this chapter, we shall concisely review the computational studies that have dealt, at different levels, with photoactivated CT processes in NA, including Proton-Coupled Electron Transfer (PCET), and ionization (after charge ejection to the environment). We focus on studies published in the last few years, since quite comprehensive reviews exist for the papers published until 2015 (Improta et al. 2016; Barbatti et al. 2015; Crespo-Hernández et al. 2004). We shall discuss only studies of processes triggered by direct UV absorption by NA, i.e. without the participation of additional sensitizers, e.g. electron/donor acceptor, or in the presence of 'non physiological' metal ions, clusters, etc. The study of CT processes in NA has provided important indications on the accuracy of the different electronic methods (many 'standard' density functionals overestimate the stability of excited states with strong CT character) (Dreuw and Head-Gordon 2005) or on the solvation models (Improta 2008; Santoro et al. 2021). We shall, however, not extensively discuss these methodological issues. The interested readers can indeed find a concise treatment of these issues in several review papers, also considering that NA have become an important playground to test and validate new theoretical and computational approaches (Toldo et al. 2023; Wohlgemuth and Mitrić 2020; Segatta et al. 2019; Zuluaga et al. 2021; González and Lindh 2020; Mai and González 2020). We shall simply highlight some recent developments in this field.

We first discuss the main outcomes of the studies dealing with hydrogen bonded pairs and those of stacked bases in single strands. Then, we focus on larger structures, double and quadruple helices, I-motifs, where both stacking and hydrogen bond interactions are operative. Finally, we try to sketch some general perspectives for future studies in this field.

2.2 Watson–Crick Hydrogen Bonded Pairs

In some seminal papers, Sobolewski, Domcke and their collaborators (Sobolewski et al. 2005; Sobolewski and Domcke 2004) proposed that a very effective PCET process is responsible for the ultrafast, sub ps, excited state decay of the GC dimer (within a WC arrangement, Scheme 2.1) in the gas phase (Abo-Riziq et al. 2005). As sketched in Fig. 2.1, the population of the bright excited states localized on G (we shall label as G-L_a and G-L_b) or C ($C_{\pi\pi^*}$ and $C_{\pi\pi 2^*}$) can be transferred to a close-lying G → C CT state (hereafter labelled GC-CT). GC-CT can easily deactivate by intermolecular PT, involving the H1 atom of G (Scheme 2.1), leading to a crossing with the ground electronic state. Independently of the adopted method, inspection of the Potential Energy Surface (PES) suggests that this PCET path provides the most important deactivation route in the gas phase for GC and it is possible also for adenine–thymine (AT) dimers (Improta et al. 2016). Very recently, the first step of this process, i.e. the population of the inter-base CT states, has been investigated by quantum dynamical (QD) calculations, in the gas phase (Green et al. 2021a; Jouybari et al. 2021) and in chloroform solution (Green et al. 2022), exploiting a fragment (Fr) diabatization (D) procedure that provides the parameters for a linear vibronic

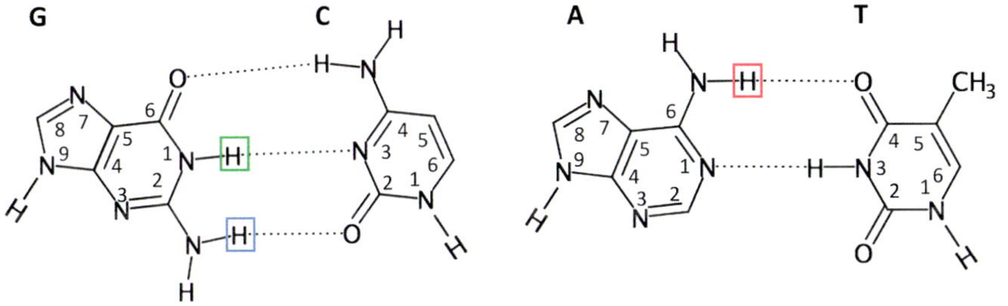

Scheme 2.1 Watson–Crick Guanine–Cytosine and Adenine–Thymine pairs. The atom labelling is also reported. H1 atom is framed in green, H21 in blue, and H61 in pink

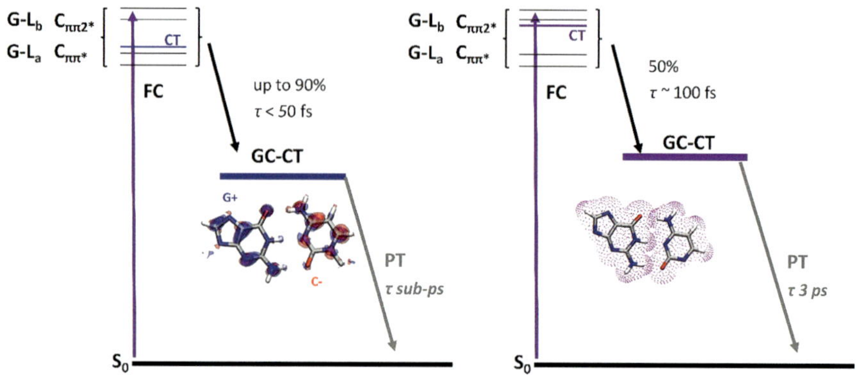

Fig. 2.1 Watson–Crick dimers. Schematic representation of the main decay pathways, percentage and characteristic times for CT population in GC WC dimer according to QD simulations in gas phase (left) (Green et al. 2021a) and chloroform (right) (Green et al. 2022). The experimental time components associated to the ground state recovery are reported in grey. See text for details

coupling (LVC) Hamiltonian (FrD-LVC). As schematically depicted in Fig. 2.1, in the gas phase, either when using CAM-B3LYP/6-31G(d) or ωB97X-D/6-31G(d) calculations to parametrize the LVC model, more than 50% and 80% of the excited population located on C and G, respectively, decays to GC-CT state in less than 50 fs (Green et al. 2021a). Interestingly, the population transfer to GC-CT is fast and effective also when this latter state is less stable by 0.2~0.3 eV than G-L$_a$ and $C_{\pi\pi*}$ in the Franck–Condon (FC) region, as obtained with ωB97X-D. Fairly large electronic couplings exist, indeed, between the bright excited states and GC-CT, which has a large reorganization energy (Green et al. 2021a). As a consequence, any transient population of GC-CT becomes quickly 'irreversible' (i.e. there is very little back transfer to the bright states). Excited states localized on G are more coupled with GC-CT than those localized on C, explaining the larger population transfer from the former states (Green et al. 2021a). A recent study of GC WC dimer, using ab initio molecular dynamics and Ehrenfest dynamics, highlights the coherences in the photoactivated CT process, with several features exhibiting oscillations with a period of ~10 fs (Zhu et al. 2023).

When a similar QD approach is applied to the study of AT dimer, the population of CT states is predicted to be negligible, at least in the fs time regime, suggesting that the ultrafast PCET decay mechanism is not operative in AT (Jouybari et al. 2021). According to CAM-B3LYP/6-31G(d) calculations, in the FC region A → T CT state is, indeed, 0.6~0.7 eV less stable than the bright excited states localized on A and T, an energy gap too large to enable a significant population transfer, despite the large electronic coupling existing between these states and the large reorganization energy of A → T CT.

The possible occurrence of CT and PCET processes in WC dimers has been investigated also in chloroform solution since in this low polarity solvent hydrogen bond dimers are the majority species (Biemann et al. 2011; Röttger et al. 2015, 2017). According to TR experiments, PCET is not a relevant process for AT dimers

(Röttger et al. 2017), in agreement with the computational indications in the gas phase discussed above. For GC dimers, instead, a significant fraction of the excited state population should undergo the CT/PCET reaction. However, the ground state recovery occurs more slowly than in the gas phase (~2.9 ps), and the excited state dynamics depends on the excitation wavelength (Röttger et al. 2015). When exciting at 290 nm (i.e. in the red-wing of the absorption band), the amount of population decaying through CT/PCET route is significantly smaller than at 260 nm (Biemann et al. 2011; Röttger et al. 2015), i.e. close to the absorption maximum, when the quantum yield for the PCET route is estimated to be ≥0.6 (Röttger et al. 2015). The PCET reaction has been mapped in chloroform by TD-DFT calculations, including solvent effects by integrating continuum and explicit models (Martinez-Fernandez et al. 2018).

More recently, the formation of the CT state for the GC dimer in chloroform has been simulated also by using QD methods (Green et al. 2022). These studies highlight that both CT and PCET reactions are significantly affected by the inclusion of solvent effects, even in a low polarity and weakly hydrogen bonding solvent as chloroform. (Green et al. 2022; Martinez-Fernandez et al. 2018) Independently of the adopted solvation model, in chloroform GC-CT is significantly destabilized with respect to the gas phase (Fig. 2.1) (Green et al. 2022; Martinez-Fernandez et al. 2018). The amount of destabilization is larger (~0.5 eV) when adopting a continuum model with the standard linear response implementation (LR-PCM), likely due to the 'differential' solvent effect on the frontier orbitals of G and C involved in the CT state (their energy gap increase by 0.15 eV with respect to the gas phase) (Green et al. 2022). However, Linear Response(LR)-PCM underestimates the stability of excited states with significant CT character, which are more accurately described either by using more accurate continuum solvation models (corrected LR, cLR-PCM or State Specific (SS)-PCM) (Improta et al. 2006, 2007; Caricato et al. 2006; Corni et al. 2005) or with explicit solvation models. These approaches, indeed, relatively stabilize GC-CT, which, nonetheless, in the FC region is still less stable than in the gas phase (Green et al. 2022). As a consequence, the amount of population transferred on the ultrafast timescale from the bright excited states to GC-CT is smaller than that predicted in the gas phase by using the same electronic method. In gas phase, at the CAM-B3LYP 6-31G(d) level, >90% of the photoexcited population decays to the G → C CT within 100 fs of the excitation, and this percentage decreases, on average, to ~50% in chloroform (Fig. 2.1) (Green et al. 2022). Also in solution, the population transfer to GC-CT is larger and faster when exciting one of the two excited states localized on G, and, therefore, will be closer to the maximum of the absorption band and smaller in the red-tail, in agreement with the experimental indications discussed above (Green et al. 2022). The calculations using explicit solvation models show that static disorder effects, i.e. the fact that solvent can be in many different configurations at the moment of the excitation, strongly modulate the energy of the CT states and, therefore, the possible population transfer from the bright states (Green et al. 2022).

No computational dynamical studies are available, to the best of our knowledge, on the PCET process in solution. SS-PCM and cLR-PCM calculations indicate that solvent significantly modulates this reaction (Martinez-Fernandez et al. 2018). Differently from the gas phase, a stable minimum is found for GC-CT state in chloroform. The solvation shell is sensitive to the presence of the G^{+} and C^{-} ionic species and rearranges according to the 'new' electronic distribution of the CT state. On the other hand, the PT reaction is exergonic also in chloroform, leading to a diradical minimum, which is not distant from a crossing region with S_0, but the energy gap is larger than in the gas phase. However, when solvent is fully equilibrated with the excited state density (a process that requires a few ps) the diradical minimum is very close to S_0 (Martinez-Fernandez et al. 2018). Dynamical solvation effects modulate also the energy barrier separating the GC-CT and the diradical minimum, which is, in any case not large (0.15~0.3 eV), suggesting that PCET is an important decay channel also in chloroform (Martinez-Fernandez et al. 2018). However, the energy barrier on the PT path, together with the role of dynamical solvation in the non-radiative decay, can explain why this process is slower (2.9 ps) (Röttger et al. 2015) in chloroform than in the gas phase (sub ps).

Finally, a recent ADC (2) study in the gas phase (Szkaradek et al. 2020) suggests that, in addition to the GC-CT state we discussed above (which involves the π HOMO orbital of G), the population of a nπ* GC-CT state (involving the Lone Pair of the G carbonyl group) is also possible. This excited state evolves towards a minimum showing a Wobble-Like hydrogen bond pattern, which can also trigger G → C PT and ground state recovery. At difference of the 'Domcke' PCET mechanism described above, in this case, however, the proton acceptor is the Carbonyl group of C (Scheme 2.1), and the PCET reaction occurs in two steps, via a stable GC-CT minimum.

The PECT involvement in DNA will be discussed in the next sections.

2.3 Stacked Oligonucleotides

Although the population of excited states with significant CT character in single-strand DNA was already proposed in 2005 based on Transient Absorption (TA) experiments (Crespo-Hernández et al. 2005), it was ~10 years later that their presence was unambiguously detected by exploiting TR-IR (Bucher et al. 2014b; Zhang et al. 2014). Moreover, in stacked bases, the CT states decay to the ground state via charge recombination (CR) in a few ps (at least), a timescale slower than that found in WC dimers, making their detection relatively easier (Bucher et al. 2014b; Zhang et al. 2014). From the computational point of view, the first studies found that, due to the large overlap between the frontier orbitals of the stacked nucleobases, their excited states are strongly coupled with exciplexes, involving two bases, with different degrees of CT character (Figs. 2.2 and 2.3) (Improta 2008; Improta and Barone 2011). This indication has been confirmed by more recent computational studies (Zuluaga et al. 2021; Saha and Quiney 2017). In an interesting benchmark study, the energy of the lowest energy excited states in 5′-TG-3′ and 5′-GT-3′ stacked bases (with the orientation as in B-DNA) in the FC region has been systematically studied by using a plethora of computational methods, both wave-function and TD-DFT based methods (Zuluaga et al. 2021). From the methodological point of view, it was found that long-range-corrected functionals, with a careful balance of the Hartree–Fock exchange included, provide a description of the relative energy of CT and bright excited states in good agreement with multi-reference methods. In 5′-GT-3′ the two bases are closer and their rings have a larger overlap, and indeed, there is an excited state with clear-cut G → T CT character falling within a few tenths of eV from the lowest energy bright excited states. A preliminary QD analysis in the gas phase, adopting the same approach used for GC WC dimer, indicates, indeed, that also for two GC stacked bases as in a B-DNA duplex, a fast and effective decay to the intra-strand G → C CT state (see Fig. 2.2). In line with these indications, in dGpT nucleosides in water, PCM/TD-DFT calculations indicate that for a stacked conformation three of the four lowest energy excited states exhibit a significant G → T CT character (Duchi et al. 2019).

The computational picture fully agrees with the ultrafast TA and TR-IR available for 2′-deoxyguanosine-3′-monophosphate-5′-thymidine, d(GpT) dinucleotide in

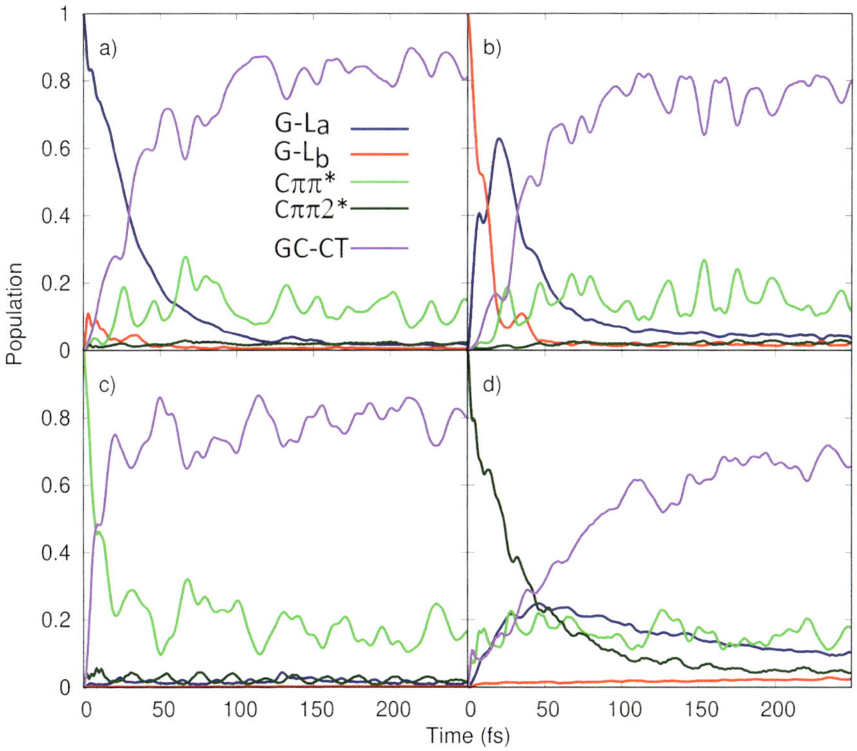

Fig. 2.2 FrD-LVC models for the GC stacked pair parameterized by ωB97X-D/6-31G(d) after initial excitation to a) G-L_a, b) G-L_b, c) $C_{\pi\pi^*}$, d) C. Five electronic states and 62 vibrational modes are considered in the LVC model

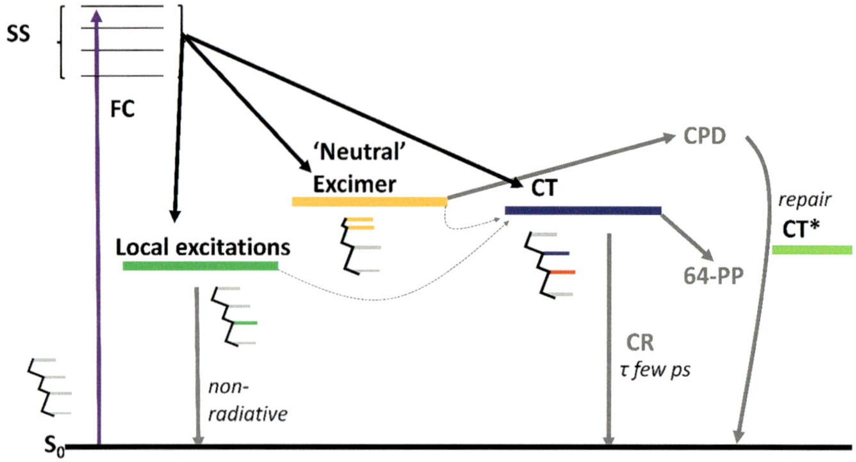

Fig. 2.3 Single strands. Schematic representation of the main decay pathways in NA single strands

water (Duchi et al. 2019). In d(G^+pT^-) exciplexes are formed on a <350 fs timescale and decay back to the ground state in ~10 ps (Duchi et al. 2019). Finally, a computational study of a GC stacked dimer, made in the framework of quantum non-adiabatic theory of electron transfer reactions, shows that solvent plays a fundamental role also in determining the CR rate of the G → C CT state (Cerezo et al. 2016).

The studies of other single-stranded NA provide indications similar to that found for GC steps. A QM/MM study of oligoadenine single strand in water, coupled with extensive sampling by Molecular Dynamics (MD) simulations, indicates that pure A → A CT states fall in the blue side of the absorption band (Nogueira et al. 2017). Nonetheless, a significant fraction of the lower in energy states has a non-negligible CT character and only about 50% of the spectral intensity is due to pure Frenkel excitons and local excitations.

The first computational study mapping the excited state processes of oligoadenine single strands also predicted that full equilibration of solvent degrees of freedom strongly stabilizes states with significant CT character, the associated minima becoming the lowest energy excited state ones, suggesting that a significant population transfer to CT states can occur (Improta and Barone 2011). A recent TA spectroscopy study (Borrego-Varillas et al. 2019) of oligo(deoxy)adenine single strands showed indeed that, after ultrafast internal conversion of the initial exciton states, the CT character of the excitation increases, and a stable minimum is reached within 3 ps. Similar indications were obtained by integrating TA and transient absorption anisotropy spectroscopy in the study of adenosine single strands of different length, which indicate that fully charge-separated CT excitons are formed in less than 0.3 ps (Bauer et al. 2022).

The photoactivated dynamics of single-strand oligoadenine $(dA)_{20}$ was studied by non-adiabatic surface hopping simulations based on CAM-B3LYP/MM calculations, where four bases are described at the QM level (Ibele et al. 2020). Within

100 fs after absorption, relaxation to monomer-like states and excimers was found. On a timescale of hundreds of femtoseconds, monomer-like states either decay non-radiatively to the ground state or (passing through additional excimer states) to CT states (Fig. 2.3), which are stable at least for 2 ps (i.e. the time window simulated in that study).

In the systems we have examined until now (i.e. GC, oxoGA, and AA steps) the CT state decays back to the ground electronic state via CR (Fig. 2.3). On the other hand, there are several examples of their involvement in the photochemistry of NA. The system studied more in detail has been TpT dinucleotide. It has been shown that on the PES of the lowest energy 5′ T → 3′ T CT state the formation of the oxetane intermediate is possible (Improta 2012), in the path leading to 6–4 pyrimidine pyrimidone adducts (64-PP), i.e. the second most-common photoproduct found in for dipyrimidine steps, after the cyclobutane pyrimidine dimer (CPD). This reaction was suggested to be favoured for certain combination of the sugar puckerings of the involved bases, which determine the stacking geometry of the bases, and being sensitive to environmental effects, which modulate the energy barrier associated to the formation of the oxetane intermediate (Improta 2012). The presence of charge-separated states in T single strands has been confirmed by TR-IR studies (Schreier et al. 2015) and their involvement in 64-PP formation was later confirmed either at the CASPT2/MM level (Giussani et al. 2018) or by a joint experimental and computational study where one of the base was substituted by a xylene moiety (Jian et al. 2017).

Excited states with significant CT character have also been connected with photodimerization reactions between thymine and other nucleobases, as C or A (Martinez-Fernandez and Improta 2017, 2018). For the former system, as for T/T steps, the formation of the 64-PP involves a 5′-ter → 3′-ter CT state, whose stability depends on both the stacking geometry (related to the combination of the puckering of the sugars) and the sequence (TpdC or dCpT) (Martínez-Fernández and Improta 2018). In detail, for dCpT, 64-PP dimerization occurs within two steps: the lowest energy excited state, which in its minimum has significant C → T CT character (S_1-CT minimum), evolves, after overpassing an energy barrier, to the oxetane intermediate that, then, leads to the final 64-PP product through, again, a sizeable energy barrier. This process was predicted to take place only for the C2endo-C1endo conformers, due to the larger stability of the CT states in the FC region. For TpdC, instead, the final 64-PP product is obtained in a single step from the S_1-CT minimum for the C2endo-C1endo conformer, and it is as stable as the CPD photoproduct (Martínez-Fernández and Improta 2018). CT states play a role also in the photophysics also of stacked T/A pairs (Martinez-Fernandez and Improta 2017; Luo et al. 2017). PCM/TD-M052X/6-31G(d) calculations located, indeed, S_1 minima with partial CT character for both dApT and TpdA (Martinez-Fernandez and Improta 2017). Subsequent relaxed scans at the same level of theory along the photodimerization coordinate connect these minima, almost barrierless, with internal conversion funnels leading to a AT/*TA* cyclobutyl adduct. However, when the path was computed considering full solvent equilibration there was a substantial increase of the energy barrier (up to 0.65 eV) (Martinez-Fernandez and Improta 2017).

Finally, TR experiments on different DNA sequences containing a CPD formed by two stacked TT bases (TT) indicated that CT states are involved also in the mechanism of the UV-induced self-repair of the DNA CPD photolesions (Bucher et al. 2016). This reaction has been studied also combining MD simulations and QM/MM computations, confirming that in a GAT = T sequence excitation of G (G*) can trigger a sequence of CT reactions, e.g. $G^{*}ATT \rightarrow G^{+}A^{-}T{=}T \rightarrow G^{+}AT{=}T^{-}$, localizing an electron in the T=T moiety. This latter species can easily undergo a ring opening reaction producing two T bases, i.e. repairing the lesion (Szabla et al. 2018). An interesting review about the effects that the flanking bases have on the formation and repair of CPD lesions has been recently published (Lee and Matsika 2023).

2.4 Double Strands

The CT processes occurring in duplex can be interpreted in the framework sketched in Sects. 2.2 and 2.3, though, obviously, considering that both inter-strand and intra-strand CT processes are simultaneously possible. Based on the available studies, the relative importance of these two processes depends on the investigated sequence, which affects the stability of the different CT states. On the other hand, due to the overlap between the molecular orbitals of stacked bases, we can expect that the coupling with the bright states is larger for intra-strand CT states (hereafter intra-CT) than for inter-strand CT states (here after inter-CT) (Ko and Hammes-Schiffer 2013). For alternated GC sequences, already several years ago, PCM/TD-M052X/6-31G(d) calculations observed that, indeed, the intra-CT state was more stable at the FC region than the inter-CT counterpart (Zhang et al. 2015a, 2016a; Ko and Hammes-Schiffer 2013). This is also true upon geometry relaxation (Fig. 2.4), with the minimum of intra-CT being ~0.5 eV more stable than the inter-CT one (Zhang et al. 2015a, 2016a; Ko and Hammes-Schiffer 2013). Their

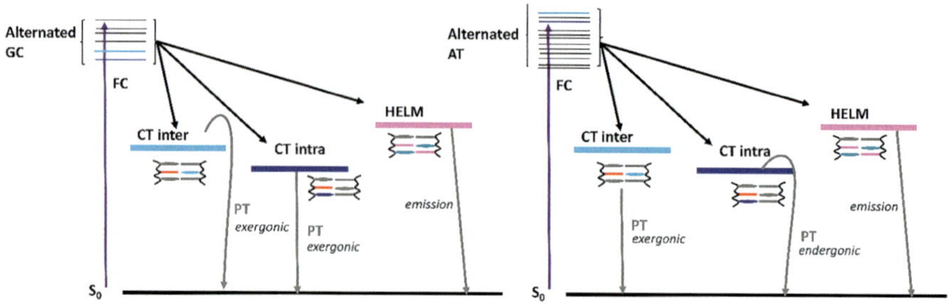

Fig. 2.4 Alternated Duplexes. Schematic representation of the main decay pathways with partial CT character in alternated GC (left) and AT (right) sequences

evolution towards PT is also different. For intra-CT a neutral base (for example, G) transfers a proton to its 'anionic' hydrogen bonded partner (C^-), and thus this reaction involves the shift of the charge between two close bases, without affecting the total formal charges of the system. For CT-inter, as discussed in Sect. 2.2, the PT reaction involves a radical ion pair (G^+ and C^-), and, thus, leads to a quenching of the charges. This can explain why the PT reaction needs to overcome a larger barrier for inter-CT states (its height depending on solvent relaxation), than for intra-CT ones (Fig. 2.4). All these considerations are still valid for alternated AT sequence, but there are two main differences with respect to GC. First, both intra-CT and inter-CT states are higher in energy in the FC region (intra-CT corresponds to S_{10} and inter-CT to S_{12}) and, then, there is a larger energy gap with the bright S_1 state, when compared to GC. Second, due to the pK_a of the involved bases, in AT the PT reaction is not exergonic for intra-CT (Fig. 2.4) (Martinez-Fernandez and Improta 2018). As a consequence, the CT-intra minimum is long living, as confirmed by the comparison between the computed and experimental difference IR spectra of a $d(AT)_9$ single strand (Martinez-Fernandez et al. 2016). DFT/TD-DFT calculations and experiments agree in predicting the decay to a stable CT minimum also for GC duplex at acidic pH, when cytosine is protonated and the two strands associate by Hoogsteen (HG) base pairing (Zhang et al. 2016a). In this case, the protonation of C prevents any PT reaction, and the GC-CT state decays to the ground state via CR.

The indications provided by the analysis of the PES are confirmed by QD calculations on duplex containing consecutive GC steps (Asha et al. 2022). This study, coupling a FrD-LVC approach (the one adopted for GC dimers) (Green et al. 2021a, 2022) with MD simulations, simulates 100 fs of the photoactivated dynamics of (GpC)(GpC) and (CpG)(CpG) fragments, using 11 structures extracted from MD simulations of two different DNA sequences (an oligoGC or an oligoAT). These simulations confirm the effective and fast population of intra-CT states, while inter-CT states do not play a significant role. This population of intra-CT states is more effective for GC than for CG steps, since the former are more closely stacked than the latter (Asha et al. 2022).

As indicated by an interesting semiclassical dynamical study, the photoactivated dynamics of (dG)(dC) DNA homopolymer should be quite different to that just described for the alternated sequences (Francés-Monerris et al. 2018). For the homopolymer, indeed, the inter-strand G → C CT states are more stable than the intra-strand one (C → C and G → G). TD-CAM-B3LYP/MM adiabatic dynamics (performed for 11 structures extracted from MD simulations) on the S_1 PES for a GC WC dimer included in a duplex indicates that the PCET route is the major deactivation channel. On average the first H atom is transferred after ~130 fs and the crossing region with ground state is reached within ~270 fs, which is, however, a lower bound for ground state recovery (Fig. 2.5). The majority of the simulations predicts that PT involves H1 atom, but the transfer of amino hydrogen atom (H21 in Scheme 2.1) can also play a role in the dynamics. Finally, no excited state tautomerization occurs and the ground state recovery occurs via back transfer of H atom to G. This picture does not significantly change when 4 (GG/CC) or 8 bases (GGGG/CCCC) are included in the simulations at the QM level, but these new simulations show than an additional process can be triggered by CT, in which two protons of two adjacent G–C base pairs are transferred to form a biradical zwitterionic intermediate, before decaying back to the ground state. This process is triggered by a G → C CT process not involving two WC paired bases and is activated by the interaction of Na^+ close to the G base (Francés-Monerris et al. 2018).

The relevance of PCET processes in homopolymeric GC sequences is confirmed also by accurate ab initio calculations on a minimal model where a guanine is hydrogen bonded, in a WC arrangement, to one of two stacked cytosine bases (Francés-Monerris et al. 2016).

The involvement of CT states in the photochemistry of bipyrimidine steps has been confirmed also within the duplex. A QM study, integrating TD-DFT and CASPT2 calculations, indicates indeed that the oxetane intermediate, leading to a 64-PP adduct, is formed on a singlet T → T CT state (Fig. 2.5). A small energy barrier (<0.3 eV) is associated to this path, which is favoured by certain conformation of the duplex (a large inter-base angle and slide values). Moreover, this reaction shows a clear directionality, occurring on the PES of 5′-T → 3′-T CT states (Conti et al. 2017).

We end this section by mentioning another interesting family class of excited states with partial CT character, i.e. the so-called HELM states, High-energy Emitting Long-lived Mixed excitons (Fig. 2.4). Identified in alternated polymeric GC and AT sequences (Vayá et al. 2016; Huix-Rotllant et al. 2015), HELM states are delocalized over multiple (at least four) bases and have partial purine→pyrimidine CT character. HELM states are considered responsible for the high-energy nanosecond emission (HENE) present in many NA sequences, a component of the fluorescence (on the blue with respect to the fluorescence of their constituents) decaying on the ns timescale.

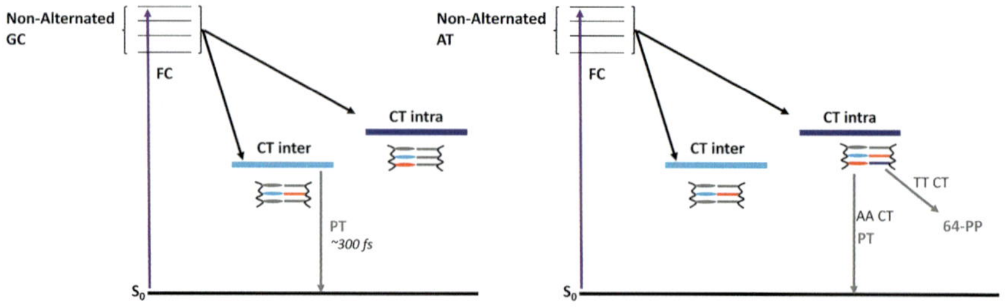

Fig. 2.5 Non-alternated duplexes. Schematic representation of the main decay pathways with partial CT in GC (left) and AT (right) homopolymers

2.5 Non-canonical DNA Structures

2.5.1 Quadruplexes

Well-packed DNA structures as I-motifs (C rich sequences at low pH) or quadruple helices (G-rich sequence, G4) are also good candidates to study the population and evolution of CT states. The characterization of these CT states is crucial since they are thought to be involved in the mechanisms responsible for DNA damages as photodimerization (Martinez-Fernandez and Improta 2021) and low-energy photoionization of G4 (Banyasz et al. 2017; Balanikas et al. 2020).

QM(TD-DFT)/MM calculations on G4 (Martínez-Fernández et al. 2020), using a 21 bases fragment of the human telomeric sequence (Tel21) in the presence of Na^+ as a showcase, revealed important information about the CT excited states at the FC region and upon relaxation. At the FC region, the low-energy (< 4.6 eV) range of the spectra is governed by the combination of the lowest $\pi\pi^*$ excited states localized on guanine bases (usually labelled as L_a), all of them with negligible CT character (< 0.05 a.u.) (Martinez-Fernandez et al. 2019). The first state with significant CT character was predicted to be a few tenths of eV higher in energy (4.80 eV) and it involves CT from a G in the G4 core to a T base in the loop. This study also located two excited state minima in the lowest energy adiabatic PES (S_1) with CT character. On the one hand, a stable minimum was located for the G → T (0.8 a.u.) excited state discussed above. On the other hand, a least stable minimum was optimized also for a G → G CT state (involving the transfer of 0.7 a.u.) between two stacked G bases in two adjacent tetrads (Fig. 2.6). These CT states are likely responsible for the red-tail appearing in the emission spectra of G4 and of the peculiar TR fluorescence anisotropy, whose values decrease upon increasing the emission wavelength (Martinez-Fernandez et al. 2019). Beyond CT states, this study described a complex photophysics for this system, with more excited minima involving monomeric-like decay or even guanine–guanine photodimerization.

These findings indicated that the bases present in the loops or in the dangling edges of the G4s can be involved in CT states and have a very active role in their photoactivated dynamics. With the aim of knowing more about this topic, and

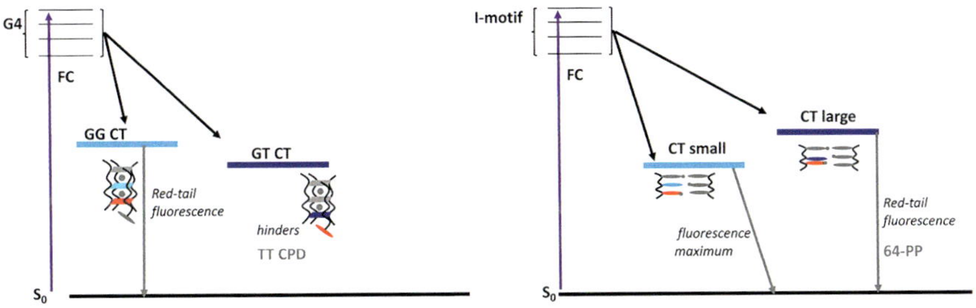

Fig. 2.6 G4 and I-motifs. Schematic representation of the main decay pathways with partial CT character in well-packed DNA structures, G4 (left) and I-motifs (right)

ascertaining if CT to the loops bases can play a role in the low-energy photoionization processes (see below), the authors performed experiments and calculations on model systems (5′-G-G_4-A-3′, 5′-AG-3′, 5′-GA-3′, 5′-TG-3′, and 5′-GT-3′) (Balanikas et al. 2021). Among the examined systems, 5′-GA-3′ showed the larger tendency towards the formation of CT states. In detail, it presents the largest coupling between the bright and CT state at the FC region and a CT minimum (with stability comparable to those of the $\pi\pi^*$ states) that can be reached from the bright states. The population of CT is predicted to be less likely in 5′-AG-3′ or 5′-GT-3′ (less stable minima or more difficult to be reached) and strongly disfavoured for 5′-TG-3′. In summary, the study concluded that CT states' stability/probability follows a clear directionality, i.e. the transfer of an electron towards the 3′ end is more stable than in the opposite direction, in agreement with previous studies and with the ordering of the experimental one-photon ionization quantum yields.

CT states have also been studied using QM(ADC2)/MM calculations in Tel22 (a fragment with 22 bases) in the presence of Na^+ (basket topology) and K^+ (form-3 structure) (Lee and Matsika 2017). The aim of this study was to evaluate how the CT involving T in the loop and G in the G4 core interact with the excited states responsible for T-T CPDs, quenching such reaction. In detail, for some specific isomers, as the *cis*, *syn* isomer, in the FC region the two T bases were found to strongly interact with an adjacent G through π stacking interactions, facilitating CT and then, hindering CPD formation. Through this combination of conformational and electronic effects, the authors reproduce nicely the experimental trends for CPD yields.

We end this section by mentioning low-energy photoionization, an interesting process that could in principle involve CT states. Experiments on duplexes (Banyasz et al. 2018a) and, especially, on G4 show indeed (Balanikas et al. 2020; Balanikas and Markovitsi 2021) one-photon ionization at energies significantly lower than the G vertical ionization potential, eventually triggering several additional photochemical reactions. A possible explanation for this phenomenon, which still waits for a more careful theoretical characterization, involves the participation of G → G and G → T CT states, followed by charge separation (the hole is trapped by the G-rich core and the anion is localized on T base) and electron ejection from the T^- anion.

2.5.2 I-motifs

I-motifs are another non-canonical DNA structure that has recently drawn significant attention, for its biological role, and as potential therapeutic target. It is formed by intercalated hemiprotonated $(C\underset{\cdot}{H}C)^+$ pairs, i.e. formed by a cytosine (C) and a protonated cytosine (CH^+) base, at acidic pH (Abou Assi et al. 2018). A recent computational study focused on the photoactivated dynamics of I-motifs, trying to characterize the main photophysical and photochemical pathways (Improta 2023). The most-populated decay pathway involves two 'stacked' C and CH^+ bases, participating in excited states with a certain degrees of C → CH^+ CT character. The excited states with smaller CT character (involving the transfer of <0.4 a.u.) are not dark and are associated to the maximum of the steady-state fluorescence spectra of I-motifs (Fig. 2.6). The excited states with larger CT character are characterized by red-shift emission and are responsible for the log tail appearing in the low-energy part of the fluorescence spectra. In this minimum, the two bases involved in the CT states get closer and, indeed, a quite accessible path towards the formation of a dimer has been identified, similar to that responsible for the formation of 64-PP in bipyrimidine steps.

2.6 Concluding Remarks

The photoactivated dynamics of a 'simple' oligoadenine single strand is a good example of the large number of interrelated photophysical and photochemical processes that can be triggered by UV irradiation in NA (Improta and Barone 2011). Light absorption leads to the population of 'exciton' states that, depending on the conformation adopted at the moment of the excitation, can involve a different number of bases. Subsequently, experiments and calculations agree that different kinds of states can be populated, 'monomer like', mainly localized on a base, and exciplexes involving two stacked bases. 'Monomer like' excited states can have intrinsically complex dynamics (for adenine it involves the interplay between one dark $n\pi^*$ and 2 bright $\pi\pi^*$states), but can also decay back to ground state on the sub-ps timescale. Exciplexes can have different degrees of CT character, with different spectral features, but after a few ps, excited states with significant $A^+ \rightarrow A^-$ CT character are the most populated ones (Improta and Barone 2011). In addition to this pretty complex photophysics, with a dynamical equilibrium between excited states with different degrees of localization and CT character (Improta and Barone 2011), at least two different photochemical reactions have been shown to occur. For certain syn/anti arrangements of the bases, AA can dimerize (Martinez-Fernandez and Improta 2021; Banyasz et al. 2016) and low-energy photoionization (producing A^+ cations (Banyasz et al. 2018b), which can further react) is possible. When homopolymeric $(dA)_n(dT)_n$ are formed, AA dimerization is not relevant, but CT states can be involved in PT reaction (PCET). This short summary is sufficient to show that it is not possible to predict, in detail, the photoactivated dynamics of a given DNA tract. The particular sequence will obviously determine the relative stability of the different excited state introduced in the previous paragraph. We have reported, for example, the relative stability of the intra-strand CT states with respect to the bright exciton, and inter-strand CT states is larger for an alternated d(GC)d(GC) sequences than for an homopolymeric (dG)(dC) one. The sequence will also dictate the conformational behaviour of the duplex, and, therefore, the average stacking of the bases. This parameter is very important to determine the electronic coupling between bright and CT states, and, therefore, the possibility of populating the latter ones. For example, the stacking geometry of G and C bases is quite different in CpG than in GpC steps, the two bases being more closely stacked (with shorter distance and larger overlap) in the latter. Consequently, not only CT states are relatively more stable in GpC steps, but their coupling with the bright states is larger, and, therefore, their population more effective (Martinez-Fernandez et al. 2024; Asha et al. 2022). Moreover, depending on the sequence and on the conformation of the NA, the degree of CT character of the states we define as CT states can be variable. In the context of closely stacked systems as NA, composed by the bases with similar electrochemical properties (not very strong charge donor or acceptor), the label CT state has to be considered 'cum grano salis', more as a qualitative label, useful for interpretative purpose, than a quantitative one, implying the transfer of an electron from one base to another.

Notwithstanding these caveats, based on the available studies, we can sketch some guidelines which can be useful to predict the involvement of the CT/PCET route in a given sequence, for example involving epigenetic nucleobases.

1. The energy gap between the spectroscopic and the lowest energy CT states in the FC region is only one of the factors determining the population of the CT state.
2. If the CT state can be populated in the FC region, even if at a small extent, we can expect that its population after few ps would be significant. It is indeed likely that most of the population that has not yet decayed to the ground state (usually through the 'monomer' like conical intersections that rule the photophysics of isolated bases) (Improta et al. 2016) has been transferred to CT states. CT states are indeed characterized by a large reorganization energy, i.e. their minima are particularly stable (Green et al. 2022). The B^+ and B^- can reach their optimal geometry 'independently' and are stabilized by the onset of favourable electrostatic interactions between the radical ion pair, which are relatively unaffected by the backbone conformation. The minima of 'neutral' excimers (or of any excited state where both the frontier orbitals are delocalized over two bases) are instead characterized by a close stacking of the bases, which can require significant rearrangements of the backbone (Improta et al. 2016). In other words, any 'transient' population of the CT state, due to the electronic coupling with the spectroscopic states, is likely to be 'irreversible', i.e. these states deactivate only to ground state.
3. The energy of CT states is strongly modulated by environmental effects, such as the interaction with the solvent molecules, with the phosphate groups, counterions, etc. These effects are important to determine the possibility that it is populated in the FC region. The study of GC dimer in chloroform shows that a simple relation exists between the energy of GC-CT in the FC region (and thus its population after 250 fs) and the solute/solvent electrostatic interactions (Green et al. 2022). It is possible that a similar dependence exists also when considering the electrostatic interaction with the (phospho)deoxyribose backbone in NA.
4. Once on the CT surface, depending on the pK_A of the cation and anionic base, PT can occur, leading to diradical species, which, thanks to the proximity with the ground electronic state, can provide an effective non-radiative deactivation route. On the other hand, also this reaction is expected to be modulated by 'environmental' effects. For example, even if the PT process is exergonic, for some

configurations of the solvent (or of the backbone) the system can be trapped in the CT minimum.

Based on these considerations, there are some topics where methodological refinements would be surely extremely useful. For what concerns electronic methods, it is critical to use approaches able to give a balanced description of excited states of different 'nature', both in static calculations (to obtain reliable PES) and in dynamical simulations. For what concerns this latter aspect, effective diabatization procedure has been recently proposed, enabling to describe at a quantum dynamical level, even if in a simple linear vibronic coupling Hamiltonian, the competition between monomeric and inter-monomeric decay channels in multichromophore assemblies as NA (Green et al. 2021a). The same approach can be used also in the parametrization of excitonic Hamiltonians that can consider the role of CT states and of the overlap between the Molecular Orbitals of the different chromophores, and that has been profitably applied to simulate the spectral properties of large NA fragments (Asha et al. 2022; Green et al. 2021b).

Another key point concerns the treatment of the environmental effects, whose accuracy is extremely important. All the processes we are dealing with, i.e. population of CT states from the bright state, PCET, deactivation of CT via CR, involve sudden and large variation of the excited state density. As a consequence, treatments able to explicitly (and accurately) include the coupling between the vibronic degrees of freedom of the bases and those of the solvent (or of the backbone) are critical to provide a reliable description of CT and PCET. More in general, modelling the photoactivated dynamics of NA requires the ability of describing processes which are strongly multiscale both in space and in time. The processes can involve from a single base to several ones, without considering the possible role of large-scale conformational transitions. On the same time, we have to describe the coupling between events occurring in the sub-ps timescale (e.g. the population of CT states or the ground state recovery of 'monomer like' excitation), on a few ps (e.g. local conformational rearrangements of the solvent shell), on dozens of ps (e.g. CR), and beyond, e.g. to include the conformational motion of the NA or to study some photochemical reactions (for example, ionization). It is clear that only the integration of different computational approaches (from Quantum or semiclassical dynamics to classical Molecular Dynamics) and interpretative approaches can provide a solid understanding of these phenomena. For the same reasons, any meaningful advance in this field will also require more sophisticated and accurate experimental approaches, such as, just to make an example, the development and the application of multidimensional TR spectroscopy.

Acknowledgements RI thanks financial support from CNR (progetti@cnr/UCATG4 and Nutrage FOE2022) and from CN3, National Center for GeneTherapy and Drugs based on RNA technology, funded by the European Union-NextGenerationEU-PNRR.

References

Abo-Riziq A, Grace L, Nir E, Kabelac M, Hobza P, de Vries MS (2005) Photochemical selectivity in guanine–cytosine base-pair structures. Proc Natl Acad Sci USA 102(1):20–23

Abou Assi H, Garavís M, González C, Damha MJ (2018) I-motif DNA: structural features and significance to cell biology. Nucleic Acids Res 46(16):8038–8056

Asha H, Green JA, Esposito L, Martinez-Fernandez L, Santoro F, Improta R (2022) Effect of the thermal fluctuations of the photophysics of GC and CG DNA steps: a computational dynamical study. J Phys Chem B 126(50):10608–10621

Balanikas E, Markovitsi D (2021) DNA photoionization: from high to low energies. In: Improta R, Douki T (eds) DNA photodamage: from light absorption to cellular responses and skin cancer. The Royal Society of Chemistry, pp 37–54

Balanikas E, Banyasz A, Douki T, Baldacchino G, Markovitsi D (2020) Guanine radicals induced in DNA by low-energy photoionization. Acc Chem Res 53(8):1511–1519

Balanikas E, Martinez-Fernandez L, Improta R, Podbevšek P, Baldacchino G, Markovitsi D (2021) The structural duality of nucleobases in guanine quadruplexes controls their low-energy photoionization. J Phys Chem Lett 12(34):8309–8313

Banyasz A, Martinez-Fernandez L, Ketola T, Muñoz-Losa A, Esposito L, Markovitsi D, Improta R (2016) Excited state pathways leading to formation of adenine dimers. J Phys Chem Lett 7:2020–2023

Banyasz A, Martínez-Fernández L, Balty C, Perron M, Douki T, Improta R, Markovitsi D (2017) Absorption of low-energy UV radiation by human telomere G-quadruplexes generates long-lived guanine radical cations. J Am Chem Soc 139(30):10561–10568

Banyasz A, Martínez-Fernández L, Improta R, Ketola T-M, Balty C, Markovitsi D (2018a) Radicals generated in alternating guanine–cytosine duplexes by direct absorption of low-energy UV radiation. Phys Chem Chem Phys 20(33):21381–21389

Banyasz A, Ketola T, Martínez-Fernández L, Improta R, Markovitsi D (2018b) Adenine radicals generated in alternating AT duplexes by direct absorption of low-energy UV radiation. Faraday Discuss 207(0):181–197

Barbatti M, Borin AC, Ullrich S (2015) Photoinduced processes in nucleic acids. In: Barbatti M, Borin AC, Ullrich S (eds) Photoinduced phenomena in nucleic acids I: nucleobases in the gas phase and in solvents. Springer International Publishing, Cham, pp 1–32

Bauer B, Sharma R, Chergui M, Oppermann M (2022) Exciton decay mechanism in DNA single strands: back-electron transfer and ultrafast base motions. Chem Sci 13(18):5230–5242

Biemann L, Kovalenko SA, Kleinermanns K, Mahrwald R, Markert M, Improta R (2011) Excited state proton transfer is not involved in the ultrafast deactivation of guanine–cytosine pair in solution. J Am Chem Soc 133(49):19664–19667

Borrego-Varillas R, Cerullo G, Markovitsi D (2019) Exciton trapping dynamics in DNA multimers. J Phys Chem Lett 10(7):1639–1643

Bucher DB, Pilles BM, Carell T, Zinth W (2014a) Charge separation and charge delocalization identified in long-living states of photoexcited DNA. Proc Natl Acad Sci USA 111(12):4369–4374

Bucher DB, Schlueter A, Carell T, Zinth W (2014b) Watson-Crick base pairing controls excited-state decay in natural DNA. Angew Chem Int Ed 53(42):11366–11369

Bucher DB, Kufner CL, Schlueter A, Carell T, Zinth W (2016) UV-induced charge transfer states in DNA promote sequence selective self-repair. J Am Chem Soc 138(1):186–190

Caricato M, Mennucci B, Tomasi J, Ingrosso F, Cammi R, Corni S, Scalmani G (2006) Formation and relaxation of excited states in solution: a new time dependent polarizable continuum model based on time dependent density functional theory. J Chem Phys 124(12):124520

Cerezo J, Martínez-Fernández L, Improta R, Santoro F (2016) Vibronic approach to the calculation of the decay rate of the photoexcited charge-transfer state of Guanine–Cytosine stacked dimer in water solution. Theor Chem Accounts 135(9):221

Conti I, Martínez-Fernández L, Esposito L, Hofinger S, Nenov A, Garavelli M, Improta R (2017) Multiple electronic and structural factors control cyclobutane pyrimidine dimer and 6–4 thymine–thymine photodimerization in a DNA duplex. Chem Eur J 23(60):15177–15188
Corni S, Cammi R, Mennucci B, Tomasi J (2005) Electronic excitation energies of molecules in solution within continuum solvation models: investigating the discrepancy between state-specific and linear-response methods. J Chem Phys 123(13):134512
Crespo-Hernández CE, Cohen B, Hare PM, Kohler B (2004) Ultrafast excited-state dynamics in nucleic acids. Chem Rev 104(4):1977–2020
Crespo-Hernández CE, Cohen B, Kohler B (2005) Base stacking controls excited-state dynamics in A·T DNA. Nature 436(7054):1141–1144
Dreuw A, Head-Gordon M (2005) Single-reference ab initio methods for the calculation of excited states of large molecules. Chem Rev 105(11):4009–4037
Duchi M, O'Hagan MP, Kumar R, Bennie SJ, Galan MC, Curchod BFE, Oliver TAA (2019) Exploring ultraviolet photoinduced charge-transfer dynamics in a model dinucleotide of guanine and thymine. Phys Chem Chem Phys 21(26):14407–14417
Francés-Monerris A, Segarra-Martí J, Merchán M, Roca-Sanjuán D (2016) Theoretical study on the excited-state π-stacking versus intermolecular hydrogen-transfer processes in the guanine–cytosine/cytosine trimer. Theor Chem Accounts 135(2):31
Francés-Monerris A, Gattuso H, Roca-Sanjuán D, Tuñón I, Marazzi M, Dumont E, Monari A (2018) Dynamics of the excited-state hydrogen transfer in a (dG)·(dC) homopolymer: intrinsic photostability of DNA. Chem Sci 9(41):7902–7911
Giussani A, Conti I, Nenov A, Garavelli M (2018) Photoinduced formation mechanism of the thymine–thymine (6–4) adduct in DNA; a QM(CASPT2//CASSCF):MM(AMBER) study. Faraday Discuss 207(0):375–387
González L, Lindh R (2020) Quantum chemistry and dynamics of excited states: methods and applications. Wiley
Green JA, Yaghoubi Jouybari M, Asha H, Santoro F, Improta R (2021a) Fragment diabatization linear vibronic coupling model for quantum dynamics of multichromophoric systems: population of the charge-transfer state in the photoexcited guanine–cytosine pair. J Chem Theory Comput 17(8):4660–4674
Green JA, Asha H, Santoro F, Improta R (2021b) Excitonic model for strongly coupled Multichromophoric systems: the electronic circular dichroism spectra of guanine Quadruplexes as test cases. J Chem Theory Comput 17(1):405–415
Green JA, Gómez S, Worth G, Santoro F, Improta R (2022) Solvent effects on ultrafast charge transfer population: insights from the quantum dynamics of guanine-cytosine in chloroform. Chem Eur J 28(57):e202201731
Gu J, Leszczynski J, Schaefer HF III (2012) Interactions of electrons with bare and hydrated biomolecules: from nucleic acid bases to DNA segments. Chem Rev 112(11):5603–5640
Gustavsson T, Markovitsi D (2021) Fundamentals of the intrinsic DNA fluorescence. Acc Chem Res 54(5):1226–1235
Hendricks JH, Lyapustina SA, de Clercq HL, Bowen KH (1998) The dipole bound-to-covalent anion transformation in uracil. J Chem Phys 108(1):8–11
Huix-Rotllant M, Brazard J, Improta R, Burghardt I, Markovitsi D (2015) Stabilization of mixed Frenkel-charge transfer excitons extended across both strands of guanine–cytosine DNA duplexes. J Phys Chem Lett 6(12):2247–2251
Ibele LM, Sánchez-Murcia PA, Mai S, Nogueira JJ, González L (2020) Excimer intermediates en route to long-lived charge-transfer states in single-stranded adenine DNA as revealed by nonadiabatic dynamics. J Phys Chem Lett 11(18):7483–7488
Improta R (2008) The excited states of π-stacked 9-methyladenine oligomers: a TD-DFT study in aqueous solution. Phys Chem Chem Phys 10(19):2656–2664
Improta R (2012) Photophysics and photochemistry of thymine deoxy-dinucleotide in water: a PCM/TD-DFT quantum mechanical study. J Phys Chem B 116(49):14261–14274

Improta R (2023) Shedding light on the photophysics and photochemistry of I-motifs using quantum mechanical calculations *Int*. Int J Mol Sci [Online] 24:12614
Improta R, Barone V (2011) Interplay between "neutral" and "charge-transfer" excimers rules the excited state decay in adenine-rich polynucleotides. Angew Chem Int Ed 50(50):12016–12019
Improta R, Barone V, Scalmani G, Frisch MJ (2006) A state-specific polarizable continuum model time dependent density functional theory method for excited state calculations in solution. J Chem Phys 125(5):054103
Improta R, Scalmani G, Frisch MJ, Barone V (2007) Toward effective and reliable fluorescence energies in solution by a new state specific polarizable continuum model time dependent density functional theory approach. J Chem Phys 127(7):074504
Improta R, Santoro F, Blancafort L (2016) Quantum mechanical studies on the photophysics and the photochemistry of nucleic acids and nucleobases. Chem Rev 116:3540–3593
Jian Y, Maximowitsch E, Liu D, Adhikari S, Li L, Domratcheva T (2017) Indications of 5′ to 3′ interbase electron transfer as the first step of pyrimidine dimer formation probed by a dinucleotide analog. Chem Eur J 23(31):7526–7537
Jouybari MY, Green JA, Improta R, Santoro F (2021) The ultrafast quantum dynamics of photoexcited adenine–thymine basepair investigated with a fragment-based diabatization and a linear vibronic coupling model. J Phys Chem A 125(40):8912–8924
Ko C, Hammes-Schiffer S (2013) Charge-transfer excited states and proton transfer in model guanine-cytosine DNA duplexes in water. J Phys Chem Lett 4(15):2540–2545
Lee W, Matsika S (2017) Conformational and electronic effects on the formation of anti cyclobutane pyrimidine dimers in G-quadruplex structures. Phys Chem Chem Phys 19(4):3325–3336
Lee W, Matsika S (2023) Mechanistic aspects of the effect of flanking nucleotide sequence on CPD formation and CPD self-repair in DNA. J Phys Chem B 127(1):18–25
Luo J, Liu Y, Yang S (2017) Role of base arrangements and intermolecular hydrogen bonding in charge-transfer states of thymine-adenine dinucleotide in aqueous solution. J Photochem Photobiol, A 337:1–5
Mai S, González L (2020) Molecular photochemistry: recent developments in theory. Angew Chem Int Ed 59(39):16832–16846
Martínez Fernández L, Santoro F, Improta R (2022) Nucleic acids as a playground for the computational study of the photophysics and photochemistry of multichromophore assemblies. Acc Chem Res 55(15):2077–2087
Martinez-Fernandez L, Improta R (2017) Novel adenine/thymine photodimerization channels mapped by PCM/TD-DFT calculations on dApT and TpdA dinucleotides. Photochem Photobiol Sci 16:1277–1283
Martínez-Fernández L, Improta R (2018) Sequence dependence on DNA photochemistry: a computational study of photodimerization pathways in TpdC and dCpT dinucleotides. Photochem Photobiol Sci 17(5):586–591
Martinez-Fernandez L, Improta R (2018) Photoactivated proton coupled electron transfer in DNA: insights from quantum mechanical calculations. Faraday Discuss 207(0):199–216
Martinez-Fernandez L, Improta R (2021) Photoinduced dimerization in DNA: singlet excited state reaction paths. In: Improta R, Douki T (eds) DNA photodamage: from light absorption to cellular responses and skin cancer. The Royal Society of Chemistry, pp 17–36
Martinez-Fernandez L, Zhang Y, de La Harpe K, Beckstead AA, Kohler B, Improta R (2016) Photoinduced long-lived charge transfer excited states in AT-DNA strands. Phys Chem Chem Phys 18(31):21241–21245
Martinez-Fernandez L, Prampolini G, Cerezo J, Liu Y, Santoro F, Improta R (2018) Solvent effect on the energetics of proton coupled electron transfer in guanine-cytosine pair in chloroform by mixed explicit and implicit solvation models. Chem Phys 515:493–501
Martinez-Fernandez L, Changenet P, Banyasz A, Gustavsson T, Markovitsi D, Improta R (2019) Comprehensive study of guanine excited state relaxation and photoreactivity in G-quadruplexes. J Phys Chem Lett 10(21):6873–6877

Martínez-Fernández L, Esposito L, Improta R (2020) Studying the excited electronic states of guanine rich DNA quadruplexes by quantum mechanical methods: main achievements and perspectives. Photochem Photobiol Sci 19(4):436–444

Martinez-Fernandez L, Green JA, Esposito, Yaghoubi Jouybari M, Zhang Y, Santoro F, Kohler B, Improta R (2024) The photoactivated dynamics of dGpdC and dCpdG sequences in DNA: a comprehensive quantum mechanical study Chemical Science 15(25):9676–9693. https://doi.org/10.1039/D4SC00910J

Middleton CT, de La Harpe K, Su C, Law YK, Crespo-Hernández CE, Kohler B (2009) DNA excited-state dynamics: from single bases to the double helix. Annu Rev Phys Chem 60(1):217–239

Muñoz-Losa A, Markovitsi D, Improta R (2015) A state-specific PCM–DFT method to include dynamic solvent effects in the calculation of ionization energies: application to DNA bases. Chem Phys Lett 634:20–24

Nogueira JJ, Plasser F, González L (2017) Electronic delocalization, charge transfer and hypochromism in the UV absorption spectrum of polyadenine unravelled by multiscale computations and quantitative wavefunction analysis. Chem Sci 8(8):5682–5691

Röttger K, Marroux HJB, Grubb MP, Coulter PM, Böhnke H, Henderson AS, Galan MC, Temps F, Orr-Ewing AJ, Roberts GM (2015) Ultraviolet absorption induces hydrogen-atom transfer in G·C Watson–Crick DNA base pairs in solution. Angew Chem Int Ed 54(49):14719–14722

Röttger K, Marroux HJB, Chemin AFM, Elsdon E, Oliver TAA, Street STG, Henderson AS, Galan MC, Orr-Ewing AJ, Roberts GM (2017) Is UV-induced electron-driven proton transfer active in a chemically modified A·T DNA base pair? J Phys Chem B 121(17):4448–4455

Saha S, Quiney HM (2017) Solvent effects on the excited state characteristics of adenine–thymine base pairs. RSC Adv 7(53):33426–33440

Santoro F, Green JA, Martinez-Fernandez L, Cerezo J, Improta R (2021) Quantum and semiclassical dynamical studies of nonadiabatic processes in solution: achievements and perspectives. Phys Chem Chem Phys 23(14):8181–8199

Schreier WJ, Gilch P, Zinth W (2015) Early events of DNA photodamage. Annu Rev Phys Chem 66(1):497–519

Segatta F, Cupellini L, Garavelli M, Mennucci B (2019) Quantum chemical modeling of the photoinduced activity of multichromophoric biosystems. Chem Rev 119(16):9361–9380

Slavíček P, Winter B, Faubel M, Bradforth SE, Jungwirth P (2009) Ionization energies of aqueous nucleic acids: photoelectron spectroscopy of pyrimidine nucleosides and ab initio calculations. J Am Chem Soc 131(18):6460–6467

Sobolewski AL, Domcke W (2004) Ab initio studies on the photophysics of the guanine–cytosine base pair. Phys Chem Chem Phys 6(10):2763–2771

Sobolewski AL, Domcke W, Hättig C (2005) Tautomeric selectivity of the excited-state lifetime of guanine/cytosine base pairs: the role of electron-driven proton-transfer processes. Proc Natl Acad Sci USA 102(50):17903–17906

Szabla R, Kruse H, Stadlbauer P, Šponer J, Sobolewski AL (2018) Sequential electron transfer governs the UV-induced self-repair of DNA photolesions. Chem Sci 9(12):3131–3140

Szkaradek KE, Stadlbauer P, Šponer J, Góra RW, Szabla R (2020) UV-induced hydrogen transfer in DNA base pairs promoted by dark $n\pi^*$ states. Chem Commun 56(2):201–204

Toldo JM, do Casal MT, Ventura E, do Monte SA, Barbatti M (2023) Surface hopping modeling of charge and energy transfer in active environments. Phys Chem Chem Phys 25(12):8293–8316

Vayá I, Brazard J, Huix-Rotllant M, Thazhathveetil AK, Lewis FD, Gustavsson T, Burghardt I, Improta R, Markovitsi D (2016) High-energy long-lived mixed Frenkel–charge-transfer excitons: from double stranded (AT)n to natural DNA. Chem Eur J 22(14):4904–4914

Wohlgemuth M, Mitrić R (2020) Excitation energy transport in DNA modelled by multichromophoric field-induced surface hopping. Phys Chem Chem Phys 22(29):16536–16551

Zhang Y, Dood J, Beckstead AA, Li XB, Nguyen KV, Burrows CJ, Improta R, Kohler B (2014) Efficient UV-induced charge separation and recombination in an 8-oxoguanine-containing dinucleotide. Proc Natl Acad Sci USA 111(32):11612–11617

Zhang Y, de La Harpe K, Beckstead AA, Improta R, Kohler B (2015a) UV-induced proton transfer between DNA strands. J Am Chem Soc 137(22):7059–7062
Zhang Y, Dood J, Beckstead AA, Li XB, Nguyen KV, Burrows CJ, Improta R, Kohler B (2015b) Photoinduced electron transfer in DNA: charge shift dynamics between 8-Oxo-guanine anion and adenine. J Phys Chem B 119:7491–7502
Zhang Y, de La Harpe K, Beckstead AA, Martínez-Fernández L, Improta R, Kohler B (2016a) Excited-state dynamics of DNA duplexes with different H-bonding motifs. J Phys Chem Lett 7(6):950–954
Zhang Y, Li X-B, Fleming AM, Dood J, Beckstead AA, Orendt AM, Burrows CJ, Kohler B (2016b) UV-induced proton-coupled electron transfer in cyclic DNA miniduplexes. J Am Chem Soc 138(23):7395–7401
Zhu L, Zhou Q, Wan Y, Li Q, Wan Y, Yin H, Shi Y (2023) The dynamical temporal behaviors of guanine–cytosine coherent charge transfer. Phys Chem Chem Phys 25(15):10661–10670
Zuluaga C, Spata VA, Matsika S (2021) Benchmarking quantum mechanical methods for the description of charge-transfer states in π-stacked nucleobases. J Chem Theory Comput 17(1):376–387

Chapter 3
The Photodynamics of Thionated Pyrimidine Nucleobases: Using Time-Resolved Photoelectron Spectroscopy to Explore Characteristics of Excited-State Topographies

Susanne Ullrich

Abstract The UV photophysics of the pyrimidine nucleobases are surprisingly sensitive to single atom structural modifications. The canonical nucleobases undergo efficient ultrafast internal conversion, and intersystem crossing only becomes relevant on sufficiently long timescales. Thionation, on the other hand, promotes intersystem crossing from the lowest singlet excited state into the triplet manifold with high quantum yields. The subsequent triplet state dynamics uniquely depend on the degree and position of thionation and tautomeric forms of the nucleobases. Here, the electronic population dynamics as observed with gas-phase time-resolved photoelectron spectroscopy are reviewed and placed in context of the underlying potential energy topographies and mechanistic details of the deactivation process.

Keywords Nucleobase · Thiobase · Pyrimidine · Tautomer · Isomer · Time-resolved photoelectron spectroscopy · Excited-state dynamics · Internal conversion · Intersystem crossing · Triplet state potentials

3.1 Motivation

The canonical nucleobases respond to UV-B photoexcitation with ultrafast internal conversion (IC) from singlet excited states back to the ground state. (Crespo-Hernández et al. 2004; Middleton et al. 2009; Yu et al. 2016; Bisgaard et al. 2009; Evans and Ullrich 2010; Hudock et al. 2007; Barbatti et al. 2010, 2015a, b) These processes dominate over slower intersystem crossing (ISC) or radiative decay

S. Ullrich (✉)
Department of Physics and Astronomy, University of Georgia, Athens, GA, USA
e-mail: ullrich@uga.edu

S. Matsika, A. H. Marcus (eds.), *Nucleic Acid Photophysics and Photochemistry*, Nucleic Acids and Molecular Biology 36,
https://doi.org/10.1007/978-3-031-68807-2_3

mechanisms and provide the nucleobases with an inherent protection against UV photodamage. When one or two oxygen atoms are replaced with a sulfur atom, these nucleobase analogs (Saenger 1984) not only exhibit highly efficient ISC but also behave significantly different depending on the degree and position of thionation (Teles-Ferreira et al. 2020; Mohamadzade and Ullrich 2020; Manae and Hazra 2019; Ashwood et al. 2019; Mohamadzade 2022). Their unique photophysical and photochemical properties have brought a surge of experimental (Yu et al. 2016; Teles-Ferreira et al. 2020; Mohamadzade and Ullrich 2020; Pollum et al. 2014a, b; 2015; Mai et al. 2016a, b; Martínez-Fernández et al. 2017; Sanchez-Rodriguez et al. 2017; Mohamadzade et al. 2018; Borrego-Varillas et al. 2018) and theoretical (Mai et al. 2015, 2016b, 2018; Xie et al. 2017; Ruckenbauer et al. 2016a; Jiang et al. 2015; Martinez-Fernandez et al. 2017; Cui and Thiel 2014; Nenov et al. 2018; Cao and Chen 2020; Arslancan et al. 2017) studies to establish the underlying mechanistic details.

Beyond fundamental interest in photochemical principles, the potential of thiobases for use in cancer treatments and phototherapeutic applications has also been investigated (Pollum et al. 2014a; Cui and Thiel 2014; Massey et al. 2001; Harada et al. 2007; Karran and Attard 2008; Pridgeon et al. 2011; Gemenetzidis et al. 2013; Trigiante and Xu 2015; Crespo-Hernandez and Pollum 2016; Ortiz-Rodríguez and Crespo-Hernández 2020). Photosensitizers make use of long-lived, reactive triplet states (Mai et al. 2016a; Ortiz-Rodríguez and Crespo-Hernández 2020; Baptista et al. 2017) which interact with their surroundings by either direct photo-crosslinking with canonical nucleic acids and amino acids, or electron or energy transfer processes from the thiobase triplet state to molecular oxygen generating superoxide or singlet reactive oxygen species, respectively. Ultimately, both types will react with the intracellular environment to kill the cell (Ashwood et al. 2019; Pollum et al. 2014a; Ortiz-Rodríguez and Crespo-Hernández 2020). The triplet state dynamics of thiobases, therefore, play a crucial role in the therapeutic efficacy. The ISC dynamics are affected by several factors, including spin–orbit coupling (SOC) between the triplet (T) and the ground state (GS), the barrier height from the triplet minimum to T/GS crossing points, vibrational energy (re)distribution among active and inactive modes, and the environment that surrounds the molecule. These fundamental photoproperties are effectively investigated by a synergistic approach involving theory and experiment. Furthermore, environmental effects can be removed by looking at the isolated gas-phase molecules. The focus, here, is on the use of time-resolved photoelectron spectroscopy (TRPES) to characterize the electronic population dynamics and the excited-state potential energy topographies that govern the photodynamics of thionated pyrimidine nucleobases.

3.2 Methodology

Femtosecond time-resolved photoionization measurements employ a pump–probe type two- (or multi-) photon excitation–ionization scheme with subsequent detection of the cation or photoelectron. The probe process of ionization can prove advantageous over absorption/emission-based techniques with more stringent selection rules. Since the electron comes free, ionization is in principle always allowed as long as the photon energy is sufficient for ionization. Thus, time-resolved photoionization measurements can be used to characterize dark states commonly encountered in non-radiative decays.

In time-resolved ion yield experiments (TRIY), mass spectra are recorded for a series of different pump–probe delays and contain information on the molecular evolution through time-dependent changes of parent or fragment ion yields. This technique, in its simplest adaptation, is mass specific but provides no direct spectroscopic information on relaxation processes. However, it becomes particularly powerful when extended to coincidence measurements with photoelectron spectroscopy or multi-fragment imaging approaches (Davies et al. 1999; Forbes et al. 2017; Stavros and Verlet 2016; Continetti 2001; Crane et al. 2023; Suits 2018; Chandler and Houston 1987; Eppink and Parker 1997).

Femtosecond TRPES (Stolow et al. 2004; Stolow and Underwood 2008; Fielding and Worth 2018; Kotsina and Townsend 2021) employs a differential detection scheme that monitors changes in the kinetic energies of photoemitted electrons. Partial ionization probabilities for ionization into cationic states of specific electronic character can differ drastically with respect to the molecular orbital nature of the neutral excited state. The series of photoelectron spectra recorded at different pump–probe delays therefore contains *both dynamical and spectral information* associated with the relaxation process. Electronic configurational changes can often be assessed using a simplistic approach of Koopman's-like ionization correlations to extract the population dynamics of the neutral electronic states (see Fig. 3.1) (Blanchet et al. 1999, 2001; Schmitt et al. 2001). When extended to imaging techniques, e.g. velocity map imaging, the angular distribution of the emitted photoelectrons is obtained which reflects orbital symmetry chararacteristics (Mabbs et al. 2009; Suzuki 2006). The necessary information for the analysis of the photoelectron spectra and images is available through ab initio calculations and simulations. Particularly useful are calculations of the changes in ionization potential and the orbital nature for critical points along the relaxation pathway. This synergistic approach between experiment and theory has been demonstrated throughout the literature and, here, is highlighted in the thiobase studies discussed below.

Experimentally this can be realized with a setup that consists of an amplified ultrafast laser system and molecular beam photoelectron spectrometer as shown in Fig. 3.2. The molecular sample of interest is prepared in a molecular beam and transported into the ionization region of a photoelectron spectrometer where it interacts with the femtosecond pump and probe pulses. The molecules are photoexcited to an electronically excited state with a UV pump pulse and subsequently ionized

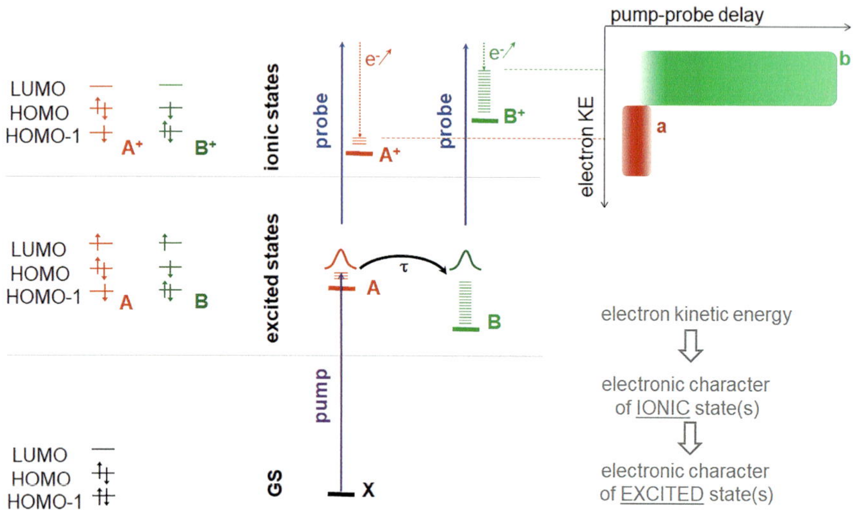

Fig. 3.1 Schematic visualizing the principle of time-resolved photoelectron spectroscopy. Molecules in the electronic ground state (GS) are photoexcited by a UV pump pulse to an electronically excited state A, which is bright in absorption and subsequently decays to the lower excited state B. The electronic relaxation process is associated with a change in electron configuration (left). This translates into preferential ionization to different cationic states, A^+ and B^+, respectively (middle), and is observed as distinct peaks in the photoelectron spectrum that rise and decay according to the population dynamics of the neutral excited states (right, top)

with a time-delayed probe pulse. The kinetic energies of the emitted photoelectrons are measured at various pump–probe delays to obtain a TRPES spectrum using either time-of-flight or velocity map imaging techniques (Eppink and Parker 1997). The latter additionally provides information on the angular distribution of the emitted photoelectrons. The schematic in Fig. 3.2 shows a magnetic bottle type photoelectron photoion coincidence spectrometer. The bottle shaped magnetic field is produced by a ring magnet and solenoid below and above the ionization region, respectively, and redirects and guides electrons to the microchannel plate detector. Ions can be extracted from the ionization region and into the linear mass spectrometer through the center aperture of the ring magnet by applying a high voltage repelling pulse. Both the photoelectron kinetic energy and ion mass spectra are derived from time-of-flight measurements using calibration gases with known ionization potentials/cationic states and masses. Timing calibrations (i.e., instrument response function) are performed in a cross-correlation measurement of the pump and probe pulses using a calibration gas of negligible excited-state lifetime.

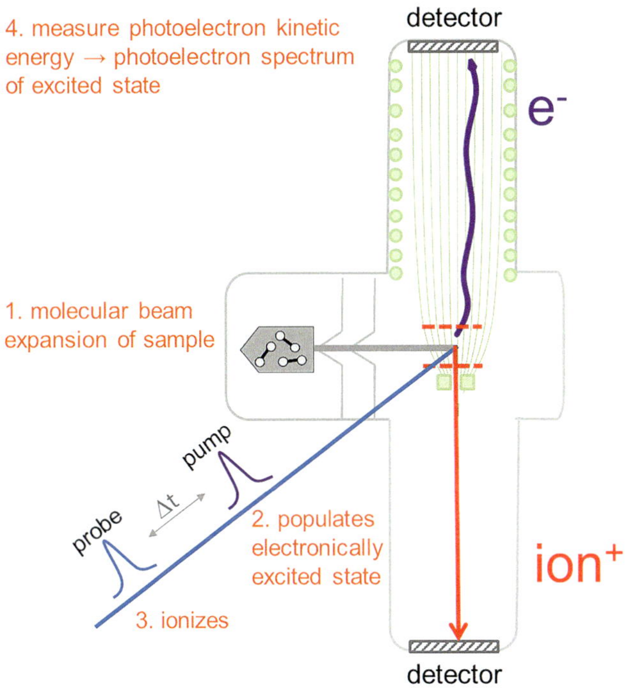

Fig. 3.2 A typical experimental setup for time-resolved photoionization spectroscopy. Here, a magnetic bottle-based photoelectron photoion coincidence spectrometer is shown, but velocity map imaging spectrometers are similarly common

3.3 General Deactivation Mechanism: Canonical Versus Thionated Nucleobases

The excited-state relaxation dynamics of the canonical nucleobase uracil (U) and modified nucleobases 2-thiouracil (2TU), 4-thiouracil (4TU), and 2,4-dithiouracil (DTU) have been studied in solution and gas-phase environments, both experimentally and theoretically (Hudock et al. 2007; Teles-Ferreira et al. 2020; Mohamadzade and Ullrich 2020; Mai et al. 2016b, 2018; Borrego-Varillas et al. 2018; Jiang et al. 2015; Nenov et al. 2018; Ghafur et al. 2018; Koyama et al. 2017; Jin and Zheng 2017; Zou et al. 2014; Nachtigallová et al. 2011; Richter et al. 2014; Mohamadzade et al. 2023). While IC back to the GS is efficient in U, thiouracils undergo ISC to the triplet state with quantum yields close to unity (Yu et al. 2016; Teles-Ferreira et al. 2020; Mohamadzade and Ullrich 2020; Pollum et al. 2015; Sanchez-Rodriguez et al. 2017; Mohamadzade et al. 2018; Borrego-Varillas et al. 2018; Ullrich et al. 2004; Gobbo and Borin 2014). Nonetheless, some general characteristics of their relaxation mechanisms can be identified which are visualized in Fig. 3.3. The lowest singlet excited state, $1^1n\pi^*$, takes the center stage in accessing the triplet manifold but is dark in absorption. Instead, UV photoexcitation populates a higher-lying,

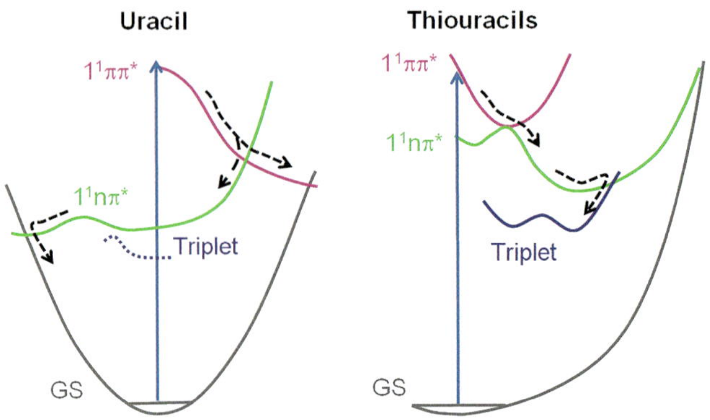

Fig. 3.3 Generalized picture of the relaxation mechanisms in uracil and thiouracils. The accessibility of CIs between singlet excited states plays a central role in the competition between IC back to the ground state versus ISC into the triplet manifold. A key player in the relaxation dynamics is the lowest singlet excited state, $1^1n\pi^*$, which is dark in absorption and populated indirectly from the $1^1\pi\pi^*$ or higher states of $^1\pi\pi^*$ character. ISC close to the $1^1n\pi^*$ minimum is controlled by subtle changes in electronic and structural factors that also affect the subsequent triplet state dynamics

singlet $^1\pi\pi^*$ state with non-negligible oscillator strength. In the case of uracil, conical intersections (CI) with the GS and $1^1n\pi^*$ states are easily accessible and provide pathways for direct or indirect ground state repopulation. IC is efficient and dominates at early times, while picosecond trapping due to a barrier on $1^1n\pi^*$ facilitates ISC on sufficiently long timescales (Richter et al. 2014). Thionation, however, alters the topography of the excited-state potential and restricts IC from the singlet excited states to the GS. While CI energies remain mostly unaffected by the substitution, excited-state minima shift toward lower energies resulting in a significant barrier that hinders ground state access (Mai et al. 2016a). Population is funneled into the triplet manifold along a barrierless pathway via the lowest $1^1n\pi^*$. Besides this common deactivation mechanism (Arslancan et al. 2017), which is shared by other pyrimidine thiobases, their triplet state dynamics are surprisingly sensitive to the position and degree of substitution. In cytosine derivatives, the mechanistic details are tautomer specific and more complex due to multiple, competing pathways.

3.4 Position, Degree, and Type of Substitution: Thiouracils

3.4.1 How Does the Position of Thionation Affect the Photodynamics of Uracil?

Replacing one of the oxygen atoms in U with sulfur causes a significant redshift of the first absorption band which depends on the substituent's position. The UV absorption spectra of U, 2TU, and 4TU are shown in Fig. 3.4, left. The low-lying bright $^1\pi\pi^*$ transitions that contribute to the lowest absorption band of each

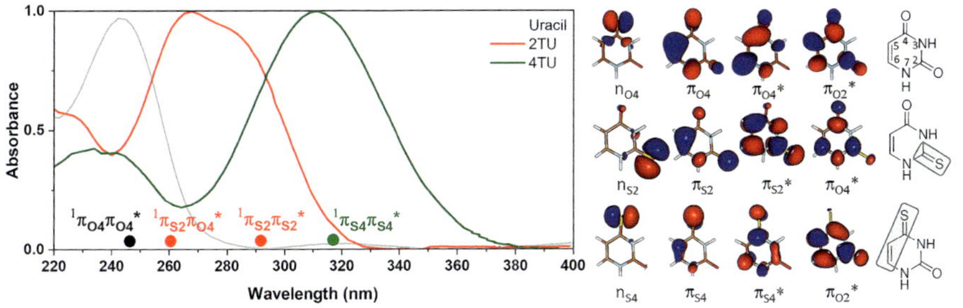

Fig. 3.4 Gas-phase UV absorption spectra (Gas-phase absorption spectra replotted from (Mayer et al. 2022a) Mayer, D.; Picconi, D.; Robinson, M. S.; Gühr, M. Experimental and theoretical gas-phase absorption spectra of thionated uracils. *Chemical Physics* **2022**, *558*, 111,500. DOI: https://doi.org/10.1016/j.chemphys.2022.111500. with permission from Elsevier) (left) of uracil, 2-thiouracil, and 4-thiouracils adapted from (Mayer et al. 2022a) and the orbital transitions contributing to the first absorption band are indicated with dots (Nenov et al. 2018; Chakraborty et al. 2021). The relevant orbitals (Uracil orbitals are replotted from (Chakraborty et al. 2021) Chakraborty, P.; Liu, Y.; Weinacht, T.; Matsika, S. Effect of dynamic correlation on the ultrafast relaxation of uracil in the gas phase. *Faraday Discussions* **2021**, *228* (0), 266–285, 10.1039/D0FD00110D. DOI: 10.1039/D0FD00110D. with permission from Royal Society of Chemistry; 2-Thiouracil and 4-Thiouracil orbitals reproduced from (Mohamadzade 2022) Mohamadzade, A. TIME-RESOLVED PHOTOELECTRON SPECTROSCOPY OF URACIL DERIVATIVES: FEMTOSECOND-NANOSECOND EXCITED STATE DYNAMICS. University of Georgia, 2022) are displayed in the center. The extent of delocalization and conjugation in the vicinity of the sulfur substitution is visualized on the right. The same atom numbering scheme as indicated for uracil is used for all thiobases

molecule are indicated as dots and orbital localizations of these transitions are visualized (Fig. 3.4, center). While n orbitals are almost entirely localized on oxygen or sulfur lone pairs, π and π* orbitals also have significant ring components. For simplicity, the labeling of transitions is chosen to reflect the position 4 and position 2 contributions, i.e. it indicates the oxygen or sulfur atom that participates most in the orbital transition. The prefix indicates the energy ordering of states, counted independently for nπ* and ππ* character, and the superscript describes the multiplicity. For example, the lowest $1^1\pi\pi^*$ transitions correspond to $^1\pi_{O4}\pi_{O4}{}^*$ in uracil, $^1\pi_{S2}\pi_{S2}{}^*$ in 2-thiouracil, and $^1\pi_{S4}\pi_{S4}{}^*$ in 4-thiouracil. The second $2^1\pi\pi^*$ transition originates from the same bonding orbital but extends to a higher antibonding orbital ($^1\pi_{S2}\pi_{O4}{}^*$). The $2^1\pi\pi^*$ only contributes to the first absorption band in 2-thiouracil.

Thionation destabilizes the π orbital while stabilizing the π* orbital which effectively shifts the transition toward longer wavelengths. The extent of the redshift is determined by factors such as electron delocalization and conjugation with neighboring bonds. Electrons are more delocalized over a C=S bond compared to a C=O bond due to the larger size of the sulfur atom. In the case of 4TU, the conjugation of the C4=S and C5=C6 bonds (see Fig. 3.4, right) further increases the redshift. The lowest transition of $1^1n\pi^*$ character is also sulfur localized ($n_{S2} \rightarrow \pi_{S2}{}^*$ for 2TU, $n_{S4} \rightarrow \pi_{S4}{}^*$ for 4TU) but its oscillator strength and contribution to the absorption spectrum is negligible. These states, however, play a key role in the electronic relaxation dynamics of the thiobases.

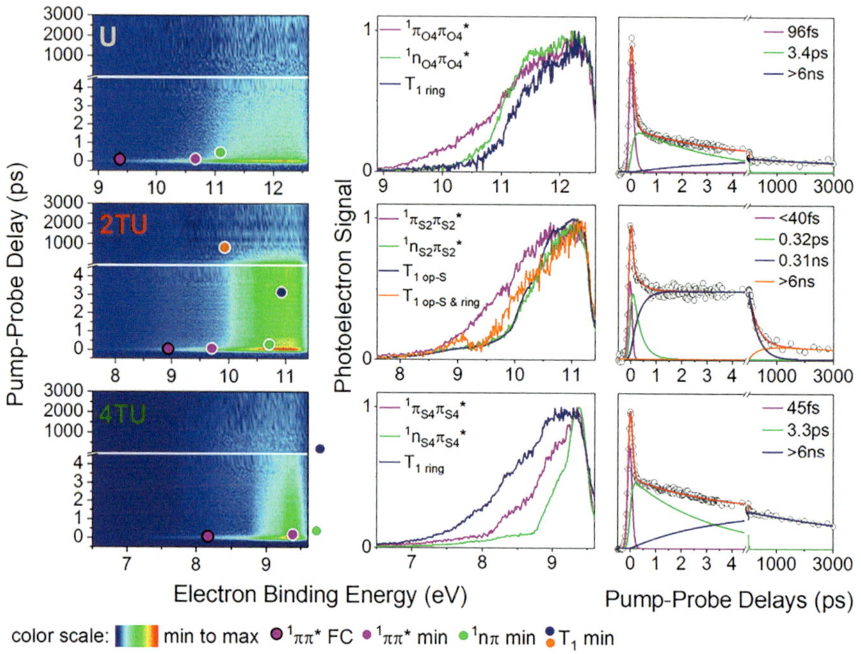

Fig. 3.5 TRPES of U, 2TU, and 4TU (left column) (TRPES data replotted from (Mohamadzade 2022) ibid) following photoexcitation on the rising edge of their absorption spectrum. The superimposed dots indicate the electron binding energies where photoelectron signals for ionization from specific electronically excited states are expected. The evolution associated spectra (middle column) and timetraces (right column) are extracted using global analysis techniques assuming sequential, exponential decay dynamics with three (U, 4TU) or four (2TU) components (Yu et al. 2016; Sanchez-Rodriguez et al. 2017; Mohamadzade et al. 2018; Ullrich et al. 2022)

The TRPES of U, 2TU, and 4TU, evolution associated spectra (EAS) and their timetraces are shown in Fig. 3.5. All spectra were recorded with a one-photon excitation and two-photon ionization scheme at wavelengths on the rising edge (270 nm—U, 304 nm—2TU, 342 nm—4TU) of the absorption spectra and below the onset, respectively. This prepares all molecules with similar vibrational excess energy on the lowest singlet $1^1\pi\pi^*$ and allows a direct comparison of their photodynamics. According to theory, the dominant deactivation pathway proceeds from the $1^1\pi\pi^*$ to the $1^1n\pi^*$ and eventually to the lowest $1^3\pi\pi^*$. The colored dots superimposed onto the TRPES colormaps in Fig. 3.5 indicate the energy range where photoelectron bands for ionization from critical points along the pathway are expected, e.g. ionization in the Franck-Condon (FC) region and from the minima of the excited states. The estimates assume vertical ionization into a vibrationally excited cationic state with $\Delta\nu = 0$ propensity and subject to Koopman's-like ionization correlations. There is generally good agreement between the observed and predicted shifts in the photoelectron bands which confirms involvement of the $1^1\pi\pi^*$, $1^1n\pi^*$, and T_1 states in the population dynamics. However, stark differences in the

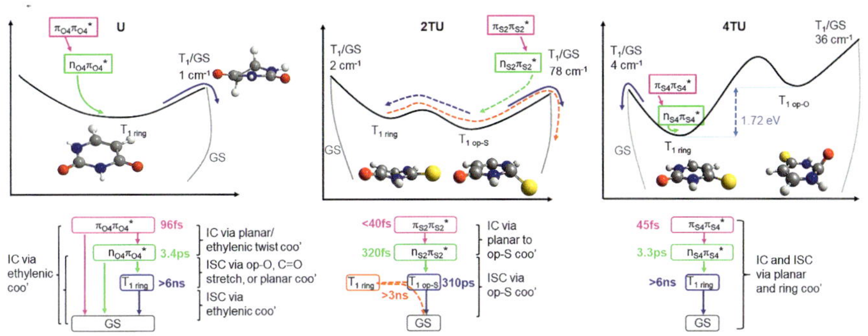

Fig. 3.6 Relaxation pathways and triplet state topographies of U (left), 2TU (middle), and 4TU (right) adapted from (Richter et al. 2014) (U) (Ullrich et al. 2022), (2TU) (Mohamadzade et al. 2018), (4TU), respectively (The triplet state potentials are adapted from (Mohamadzade 2022) ibid). The time constants represent the photodynamics for excitation to the lowest singlet state of $^1\pi\pi^*$ character on the rising edge of absorption spectrum

photodynamics are observed dependent on the position of the substituent. Overall, 2TU undergoes much faster IC and ISC processes than 4TU, but the 2TU dynamics on the lowest triplet state cannot be described by a single exponential decay back to the ground state. Instead, the 2TU EAS (Fig. 3.5, middle) and timetraces (Fig. 3.5, right) show contributions from much slower dynamics. In the case of U, the depopulation dynamics of these excited states can be described with three exponentials despite a complex decay mechanism along competing pathways. In the absence of characteristic spectral features associated with ground state repopulation, the branching along individual IC and ISC pathways cannot be disentangled from the TRPES. These distinctly substituent-dependent dynamics are best discussed in conjunction with the excited-state topographies of these molecules. Figure 3.6 illustrates the mechanistic details leading to population of the lowest triplet states (T_1), the T_1 potential energy profiles, and crossing points back to the GS. Only the dominant pathways are included in the schematics.

Uracil

IC dominates the relaxation dynamics of uracil. Following excitation to the $1^1\pi_{O4}\pi_{O4}^*$ state multiple, distinctly different ethylenic CIs connect the singlet excited states and the ground state. The initially excited state depopulates along two competing pathways to the ground state either directly on ultrafast timescales (<100 fs) or via the $1^1n_{O4}\pi_{O4}^*$ state within a few picoseconds. Along the latter route, carbonyl stretching or out-of-plane bending of the oxygen as well as motion along some planar coordinates promotes ISC into the triplet manifold. Population then remains trapped on the lowest triplet state, T_1, over extended times (>6 ns) due to a barrier and low SOC at the ethylenic crossing point to the ground state.

2-Thiouracil

Thionation in the 2-position drastically changes the decay mechanism (Ullrich et al. 2022). While the ethylenic CIs between singlet excited states become inaccessible, ultrafast IC proceeds from the planar $1^1\pi_{S2}\pi_{S2}{}^*$ minimum to the $1^1n_{S2}\pi_{S2}{}^*$ along the sulfur-out-of-plane (op-S) coordinate. Continuation along the same coordinate also drives highly efficient ISC that populates the $T_{1\ op\text{-}S}$ minimum on timescales of a few hundred femtoseconds. The observed photoelectron band in the ps $T_{1\ op\text{-}S}$ EAS (10-11 eV range) is consistent with the calculated eBE for ionization from the $T_{1\ op\text{-}S}$ minimum, confirming this assignment. In stark contrast to other thiouracils and U, the subsequent triplet relaxation dynamics cannot adequately be described by a single exponential decay time constant. This can be attributed to its unique triplet state topography. The lowest triplet state, $T_{1,}$ is a double-well potential with two almost isoenergetic minima, $T_{1\ op\text{-}S}$ and $T_{1\ ring,}$ that are separated by a ~0.1 eV barrier. A $T_{1\ op\text{-}S}$/GS crossing point provides access to the ground state with high SOC at a sulfur-out-of-plane geometry, whereas the $T_{1\ ring}$/GS crossing is higher in energy and has low SOC. Two processes, i.e. equilibration via barrier crossing into the $T_{1\ ring}$ minimum and $T_{1\ op\text{-}S}$/GS crossing to the ground state, contribute to a depopulation of the $T_{1\ op\text{-}S}$ minimum and a decay of the 10–11 eV band with an average time constant of a few hundred picoseconds. The accumulation of population in the $T_{1\ ring}$ minimum is reflected in the appearance of an additional small feature in the photoelectron spectrum around 9 eV eBE. Insight into the fate of the $T_{1\ ring}$ population can be obtained from the ns $T_{1\ op\text{-}S}$ and $T_{1\ ring}$ EAS. Generally, ISC through the $T_{1\ ring}$/GS and $T_{1\ op\text{-}S}$/GS are possible avenues for non-radiative decay to the ground state. However, the presence of $T_{1\ op\text{-}S}$ and $T_{1\ ring}$ associated photoelectron bands in the ns EAS indicates that both T_1 minima are populated and are involved in the slower dynamics. A deactivation mechanism involving reverse barrier crossing back to $T_{1\ op\text{-}S}$ followed by ISC through $T_{1\ op\text{-}S}$/GS with high SOC seems favored over direct ground state repopulation via $T_{1\ ring}$/GS. Radiative decay to the GS, e.g. through phosphorescence, may be another avenue.

4-Thiouracil

Sulfur substitution of the oxygen in the 4 position alters the deactivation mechanism significantly (Mohamadzade et al. 2018) due to electron delocalization and conjugation across the substitution site and ethylenic bond. IC from $1^1\pi_{S4}\pi_{S4}{}^*$ to the $1^1n_{S4}\pi_{S4}{}^*$ minimum proceeds along planar and ring coordinates and leads to intersystem crossing into the $T_{1\ ring}$ minimum. In 4TU, the $T_{1\ op\text{-}O}$ minimum is significantly higher in energy which renders the T_1 topography effectively a single-well potential. Consequently, the ISC dynamics more closely resemble those observed in U. The $T_{1\ ring}$ is populated on timescales of a few picoseconds but then remains trapped (>6 ns) due to low SOC and a high barrier to access the $T_{1\ ring}$/GS crossing point.

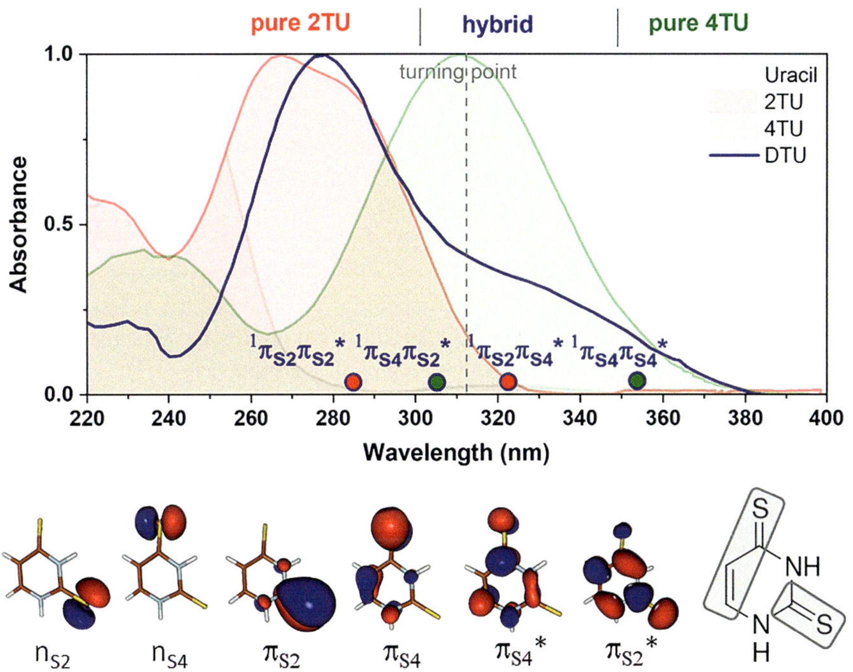

Fig. 3.7 Gas-phase absorption spectrum of DTU (blue line) from (Mayer et al. 2022a) (Gas-phase absorption spectra replotted from (Mayer et al. 2022a) Mayer, D.; Picconi, D.; Robinson, M. S.; Gühr, M. Experimental and theoretical gas-phase absorption spectra of thionated uracils. *Chemical Physics* **2022**, *558*, 111,500. DOI: https://doi.org/10.1016/j.chemphys.2022.111500. with permission from Elsevier), and its underlying orbital transitions (bottom row) (DTU orbitals are adapted from (Mohamadzade 2022) Mohamadzade, A. TIME-RESOLVED PHOTOELECTRON SPECTROSCOPY OF URACIL DERIVATIVES: FEMTOSECOND-NANOSECOND EXCITED STATE DYNAMICS. University of Georgia, 2022). The incorporation of two sulfur atoms in DTU increases the delocalization and conjugation (bottom right) which causes an additional red shift compared to 2TU and 4TU. The absorption spectrum is categorized into three regions (top): pure 2TU or 4TU paths following excitation of the $^1\pi_{S2}\pi_{S2}{}^*$ and $^1\pi_{S4}\pi_{S4}{}^*$ transitions; excitation of mixed orbital transitions launches the dynamics along hybrid paths. The turning point (gray dashed line) indicates a switch in the terminal triplet state, i.e. $T_{1\ ring}$ versus $T_{1\ op\text{-}S,}$ and distinguishes 2TU- and 4TU-like triplet state dynamics (Mohamadzade et al. 2023)

3.4.2 How Does Doubling the Thionation Affect the Photodynamics of Uracil?

In the case of DTU, when both oxygen atoms are replaced by sulfur, the photophysics and dynamics become complex due to a large number of low-lying excited states (Mohamadzade and Ullrich 2020; Mohamadzade et al. 2023). The absorption spectrum of DTU is shown in Fig. 3.7 (blue line) and compared to the spectra of U (gray), 2TU (red), and 4TU (green). The UV absorption spectrum of DTU shows two bands, which roughly fall within the regions of 4TU and 2TU absorption except for an additional redshift due to increased electron delocalization. The lowest singlet $^1\pi\pi^*$ transitions including their orbital contributions are indicated in the absorption spectrum. On the basis of these orbital transitions, the absorption spectrum can be categorized into three regions. In the UV-C and UV-A, pure 2TU and 4TU regions are associated with pure $^1\pi_{S2}\pi_{S2}^*$ and $^1\pi_{S4}\pi_{S4}^*$ transitions, respectively, whereas the intermediate region involves hybrid orbital transitions, $\pi_{S4}\pi_{S2}^*$ and $\pi_{S2}\pi_{S4}^*$. Photoexcitation within these regions launches the dynamics along either pure or hybrid paths which all involve a dark singlet $^1n\pi^*$ state and triplet $^3\pi\pi^*$ with orbital localizations on either the 2- or 4-position. The photodynamics can further be distinguished by their triplet state dynamics, i.e. whether the relaxation path terminates in the $T_{1\ ring}$ (4TU-like) or $T_{1\ op\text{-}S}$ (2TU-like) minimum. This turning point in the triplet state dynamics is marked by a gray dashed line (Fig. 3.7).

The TRPES colormaps of DTU at two excitation wavelengths in the UV-A and UV-C regions of the absorption spectrum are shown in Fig. 3.8. The colored dots represent the estimated photoelectron binding energies for ionization in the FC region and state minima along the predicted pure 4TU- and 2TU-like relaxation paths. For the 351 nm and 267 nm TRPES, these correspond to FC-$^1\pi_{S4}\pi_{S4}^*$ → min-$^1n_{S4}\pi_{S4}^*$ → min-$^3\pi_{S4}\pi_{S4}^*$ ($T_{1\ ring}$) and FC-$^1\pi_{S2}\pi_{S2}^*$ → min-$^1n_{S2}\pi_{S2}^*$ → min-$^3\pi_{S2}\pi_{S2}^*$ ($T_{1\ op\text{-}S}$) → GS and min-$^3\pi_{S4}\pi_{S4}^*$ ($T_{1\ ring}$), respectively. Details of the wavelengths-dependent relaxation pathways were mapped out in joint theoretical and experimental work by Mohamadzade et al. (2023), which also includes a thorough analysis of the EAS band shapes, but Fig. 3.8 conveys the general agreement between predicted and observed shifts. The resemblance of the DTU photodynamics to the singly

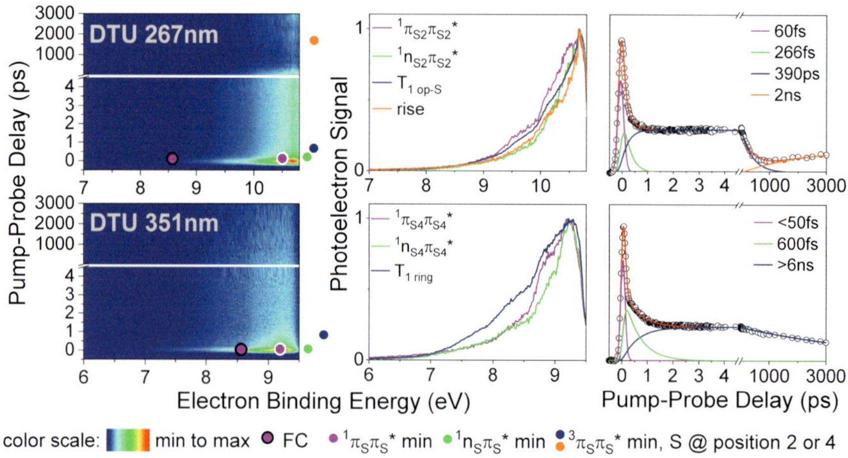

Fig. 3.8 TRPES of 2,4-Dithiouracil photoexcited in the pure 4TU- and 2TU-like regions of the absorption spectrum at 351 nm (colormap bottom) and 267 nm (colormap top) (TRPES data replotted from (Mohamadzade 2022) ibid). The colored dots indicate the estimated electron binding energy for ionization from critical points along the relaxation pathway. The EAS (middle) and timetraces (right) for both excitation wavelengths are extracted using a sequential model with three exponentials, and an additional rise for 267 nm only

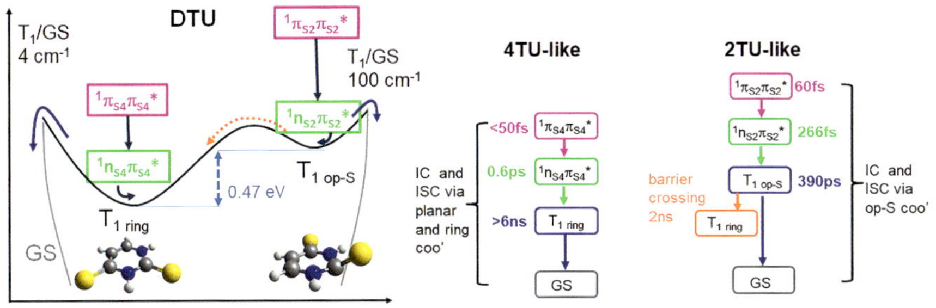

Fig. 3.9 DTU triplet state topography (left) (The triplet state potentials are adapted from (Mohamadzade 2022) ibid) and pure 2TU-like and 4TU-like relaxation pathways (right) for photoexcitation in the pure UV-A (351 nm) and UV-C (267 nm) regions of the DTU absorption spectrum

substituted thiouracils is also apparent in the timetraces (Fig. 3.8, right) which are summarized and related to the underlying potential energy curves in Fig. 3.9.

Photoexcitation in the UV-A at 351 nm exclusively populates the lowest $^1\pi_{S4}\pi_{S4}{}^*$ state which launches the photodynamics along relaxation coordinates involving ring distortions and electronic states with orbital localizations primarily on S4. Ultrafast IC to $^1n_{S4}\pi_{S4}{}^*$ is followed by ISC on hundreds of femtosecond timescales and populates a T_1 minimum of $^3\pi_{S4}\pi_{S4}{}^*$ character. While ISC into the triplet state is barrierless, the population remains trapped in the $T_{1\ ring}$ minimum due to the high access energy of the ring puckered $T_{1\ ring}$/GS crossing and low SOC. This behavior is similar to the 4TU dynamics depicted in Fig. 3.6, right.

In the UV-C at 267 nm, in contrast, the photoexcitation energy is high enough to reach the 2TU-like $^1\pi_{S2}\pi_{S2}{}^*$, with orbital localization on the S2. Along an increasing sulfur-out-of-plane bend, DTU internally converts to the $^1n_{S2}\pi_{S2}{}^*$ and intersystem crosses into the $T_{1\ op\text{-}S}$ ($^3\pi_{S2}\pi_{S2}{}^*$) within a few hundred femtoseconds. The topography of the T_1 state of DTU differs from 2TU and 4TU in various aspects that control the triplet state dynamics. The $T_{1\ op\text{-}S}$ minimum is located 0.75 eV above the $T_{1\ ring}$ minimum and can depopulate either through ISC back to the ground state or by crossing a 0.21 eV barrier into $T_{1\ ring}$. Although $T_{1\ op\text{-}S}$/GS crossing point is located 0.37 eV above the $T_{1\ op\text{-}S}$ minimum this route continues along the sulfur-out-of-plane coordinate and dynamically favored. In analogy with 2TU, the few hundred picosecond time constant is associated with $T_{1\ op\text{-}S} \rightarrow$ GS decay whereas the slow signal rise at longer delays (nanoseconds) is tentatively attributed to population equilibration into $T_{1\ ring}$. However, in DTU, reverse barrier crossing back to the $T_{1\ op\text{-}S}$ minimum can be excluded on an energetic basis.

Photoexcitation in the intermediate UV-B region leads to hybrid relaxation pathways. For example, from the FC-$^1\pi_{S2}\pi_{S4}{}^*$ region dynamics evolve through an entropic trap ($^1\pi_{S2}\pi_{S4}{}^*/^1n_{S2}\pi_{S4}{}^* \leftrightarrow {}^1\pi_{S2}\pi_{S4}{}^*/^1n_{S4}\pi_{S4}{}^*$ CIs) to $^1n_{S4}\pi_{S4}{}^*$ and merge onto the 4TU-like path which ends in the long-lived $T_{1\ ring}$ ($^3\pi_{S4}\pi_{S4}{}^*$) minimum. Conversely, photodynamics that launch from the FC-$^1\pi_{S4}\pi_{S2}{}^*$ region proceed via a $^1\pi_{S4}\pi_{S2}{}^*/^1n_{S2}\pi_{S2}{}^*$ CI onto the 2TU-like path and $T_{1\ op\text{-}S}$ ($^3\pi_{S4}\pi_{S2}{}^*$) minimum. The latter depopulates to the ground state and/or $T_{1\ ring}$ minimum with hundreds of picosecond lifetime that decreases with increasing photon energy.

3.4.3 How Does Methylation Affect the Photodynamics of Thiouracils?

To date, gas-phase TRPES studies on thiothymines that look into the effects of methylation have been limited. A direct comparison between 2TU and 2-Thiothymine (2TT) showed that the general deactivation mechanism is preserved (Ullrich et al. 2022) and is promoted by a sulfur-out-of-plane motion that leads from the $^1\pi_{S2}\pi_{S2}^*$ via the $^1n_{S2}\pi_{S2}^*$ into the $T_{1\ op\text{-}S}$ ($^3\pi_{S2}\pi_{S2}^*$). These initial steps are only minimally affected by the presence of the distant methyl group on the other side of the molecule. The triplet state topography resembles a double-well potential similar to 2TU, but barriers to T_1/GS crossing points are higher in energy and SOC constants are lower. Furthermore, the barrier for equilibration into the $T_{1\ ring}$ ($^3\pi_{S4}\pi_{S4}^*$) minimum increased. Overall, this leads to slower triplet state dynamics in 2TT compared to 2TU. For example, ISC crossing from the $T_{1\ op\text{-}S}$ ($^3\pi_{S2}\pi_{S2}^*$) to the ground state takes a few ns in 2TT instead of a few hundred picoseconds in 2TU and a smaller fraction of the $T_{1\ op\text{-}S}$ population equilibrates into the $T_{1\ ring}$ minimum.

In 4-Thiothymine (4TT) and 2,4-Dithiothymine (DTT), where the methylation occurs in closer proximity to the sulfur substituent more significant effects on the photodynamics are to be expected. For example, theoretical studies on DTT have predicted a different ordering of the electronically excited states compared to DTU (Xie et al. 2017; Mohamadzade et al. 2023). The lowest bright state, $^1\pi_{S2}\pi_{S4}^*$, is of mixed orbital localization and launches the dynamics toward a four-state near-degeneracy ($^1n_{S2}\pi_{S2}^*/^1n_{S4}\pi_{S4}/^3\pi_{S2}\pi_{S2}^*/^3n_{S4}\pi_{S4}$). Among various competing pathways, $^1\pi_{S2}\pi_{S4}^* \rightarrow {}^1n_{S2}\pi_{S2}^* \rightarrow {}^1n_{S4}\pi_{S4}$ IC is expected to dominate and populate the long-lived $T_{1\ ring}$ ($^3\pi_{S4}\pi_{S4}^*$). The relaxation proceeds along near-planar coordinates except for a slight S2-out-of-plane bend during ISC. However, these predictions have yet to be verified with gas-phase TRPES. In the case of the 4-thionated compounds (Cui and Thiel 2014; Cao and Chen 2020; Teles-Ferreira et al. 2022), the dominant relaxation pathway originates from the FC region of the photoexcited $^1\pi_{S4}\pi_{S4}^*$ and leads to population of the near-planar $T_{1\ ring}$ minimum along mostly planar coordinates with comparable in-plane deformations in 4TT and 4TU. These involve bending and stretching motions of the C=S bond and C5-C6 elongation. While 4TU maintains its near-planarity throughout this entire process, 4TT additionally undergoes pronounced out-of-plane bending of the methyl group at the CI to the $^1n_{S4}\pi_{S4}^*$ state and subsequently intersystem crosses with C=S pyramidalization through the crossing into $T_{1\ ring}$. When comparing the photodynamics of 4TT and 4TU, the out-of-plane motion of the methyl group is expected to slow down internal conversion in 4TT. On the other hand, sulfur-out-of-plane bending, similar to 2TU/2TT, may promote ISC. Gas-phase TRPES studies on these thiothymines have not yet been reported, but this intuitive picture is consistent with dynamics simulations (Cao and Chen 2020).

3.5 Tautomer Aspects: (Thio)Cytosine

In thiopyrimidines, the thiocarbonyl group plays an instrumental role in facilitating IC and ISC processes. Depending on the substituent position, it is central to in-plane deformations such as C=S elongation or bending as well as out-of-plane pyramidalizations that lead to crossing points between the lowest singlet and triplet excited states. Thiouracils are most stable in their thione forms and all undergo efficient ISC along a $^1\pi\pi^* \rightarrow {}^1n\pi^* \rightarrow$ Triplet pathway. Cytosines and their thionated analogs however exist in their keto/thione and enol/thiol tautomeric forms and thus provide a platform to investigate the role of the thiocarbonyl group in more detail.

Figure 3.10 (left) shows the experimental absorption spectrum of cytosine (black line) in an argon matrix (Bazsó et al. 2011). The contributions of the individual tautomers (right), which are present in the keto (red), enol (green), and imino (blue) forms at a ratio of 1:1:0.25, were evaluated theoretically. For tautomer-specific studies, the first absorption band, which corresponds to the lowest bright $1^1\pi\pi^*$ state, is of particular interest. The origin of the $\pi \rightarrow \pi^*$ transition of the enol tautomer is higher in energy (~4.46 eV) than in the keto form (3.95 eV) and therefore allows selective photoexcitation of the latter in the wavelength region close to the absorption onset (marked by red arrow) (Kosma et al. 2009; Nir et al. 2002). Conversely, mostly the enol form contributes to the absorption band maximum (marked by green arrow) and therefore also the observed photodynamics are dominated by this tautomer. Theoretical studies have predicted distinct differences in the photophysics of the keto and enol tautomers (Mai et al. 2013; Ruckenbauer et al. 2016b; Nakayama et al. 2013). For example, keto cytosine generally undergoes faster IC dynamics than the enol form but a small fraction intersystem crosses into the triplet manifold with longer lifetimes. The mechanistic details however are complex due to multiple parallel pathways with distinctly different crossing points that lead back to the ground state (Fig. 3.11, right). The initial $1^1\pi\pi^*$ state corresponds to the HOMO→LUMO transition with mostly ring localized orbitals. The character of the $1^1n\pi^*$ is of mixed n_N and n_O origin unless otherwise noted by a subscript that indicates an orbital localization (Mai et al. 2013). Although the $1^1n_N\pi^*$ state in the FC region is lower in energy than $1^1n_O\pi^*$, this ordering reverses at the minimum of the lowest singlet excited state, which is dominantly $n_O \rightarrow \pi^*$ character. In keto cytosine, initial population in the $1^1\pi\pi^*$ state internally converts back to the ground state either directly through a three-state CI (3-ST) or indirectly via the $1^1n\pi^*$ through CIs along a crossing seam with either semi-planar (SP) or NH_2 out-of-plane (oop) structures that connect to the GS (Barbatti et al. 2011). ISC along a C=O stretching coordinate and shortening of C2-N3 proceeds via a three-state $1^1n_O\pi^*/2^3n_O\pi^*/1^3\pi\pi^*$ near-degeneracy with SOC constants of ~30–40 cm^{-1} (Mai et al. 2013). Although in violation of El-Sayed's rule, the contribution from the $1^3\pi\pi^*$ enhances the SOC. In

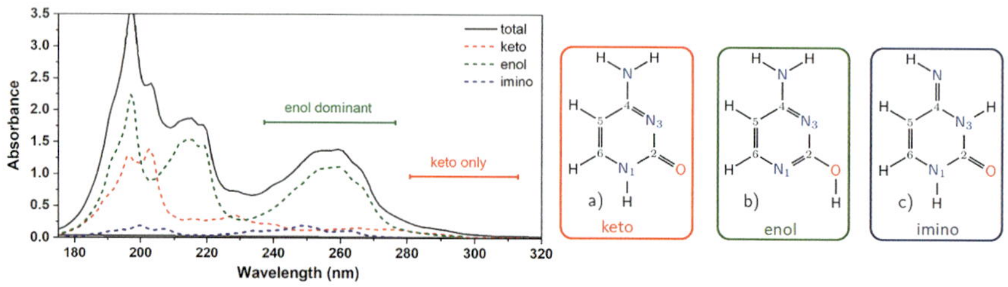

Fig. 3.10 Absorption spectrum of cytosine (left) in an Ar matrix and contribution from different tautomers (right) based on (Bazsó et al. 2011) (Absorption spectra replotted from (Bazsó et al. 2011) Bazsó, G.; Tarczay, G.; Fogarasi, G.; Szalay, P. G. Tautomers of cytosine and their excited electronic states: a matrix isolation spectroscopic and quantum chemical study. *Physical Chemistry Chemical Physics* **2011**, *13* (15), 6799–6807, 10.1039/C0CP02354J. DOI: 10.1039/C0CP02354J. with permission from the Royal Society of Chemistry), The red and green arrows indicate regions where the keto and enol forms dominate, respectively, and can be selectively photoexcited

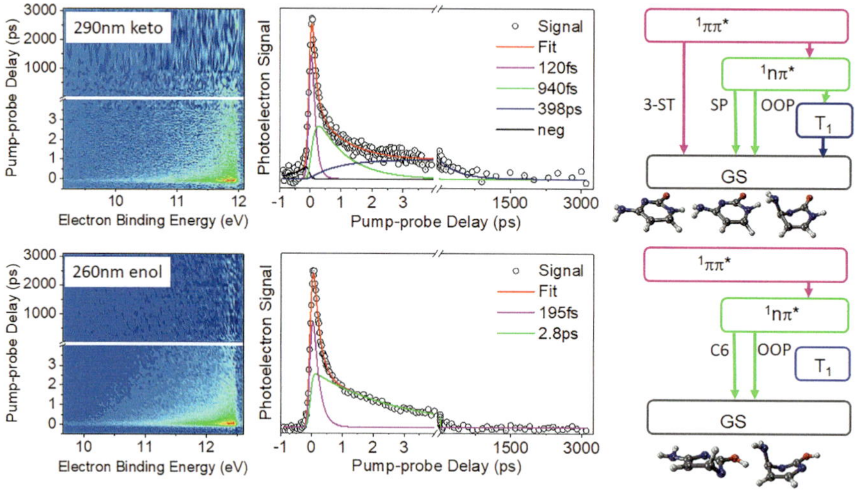

Fig. 3.11 TRPES spectra (left) of the keto (top) and enol (bottom) forms of cytosine. Timetraces and components (middle) are extracted using global analysis techniques with a sequential decay model. The time constants for depopulation of the electronic excited states are correlated with the theoretical pathways (right) (Ruckenbauer et al. 2016b)

the enol tautomer, in contrast, the dominant pathway from the $1^1\pi\pi^*$ to GS repopulation is indirectly via $1^1n_O\pi^*$, with the first IC step along planar and the second step along C6 puckering and out-of-plane coordinates (Mai et al. 2013). In the absence of the carbonyl group, enol cytosine lacks the C=O-stretching ISC pathway of the keto form. No extended areas of the potential energy surfaces were identified where singlets and triplets remain in close proximity and SOC constants are low (~10 cm^{-1}). ISC is therefore expected to only play a minor role.

The keto and enol TRPES (Fig. 3.11, left) generally support these theoretical models (Fig. 3.11, right). A sequential exponential decay model provides a good description of the measured electronic population dynamics. Parallel pathways occur on similar timescales and lack distinguishable, characteristic spectral features which prevents closer insight into the branching dynamics from an experimental perspective using TRPES (Ruckenbauer et al. 2016b). Instead, the extracted time constants represent an average of the deactivation times along the different available pathways. Keto cytosine requires a sequential decay model with three exponentials: an ultrafast depopulation of the $^1\pi\pi^*$ to the ground state or the $^1n\pi^*$. Population in this intermediate state then decays on timescales of several hundred femtoseconds back to the ground state or into a long-lived (hundreds of picoseconds) triplet. The complex dynamics are in stark contrast to the enol form which predominantly undergoes a two-step sequential exponential decay. Internal conversion from the $^1\pi\pi^*$ to the $^1n\pi^*$ occurs ultrafast within a few hundreds of femtoseconds, but the population remains trapped for a few picoseconds before repopulating the ground state via two conical intersections.

These distinct tautomer-specific deactivation mechanisms in cytosine raise questions about how thionation will affect their photodynamics and whether this tautomer specificity is preserved. In analogy with thiouracils, one might anticipate an enhancement of ISC in thione thiocytosine due to inaccessibility of singlet-only pathways and higher SOC at singlet–triplet crossing points. On the other hand, thiol thiocytosine, which lacks a thiocarbonyl group, should exhibit photophysics that resemble enol cytosine, i.e. only minor ISC. This hypothesis can be evaluated experimentally but only with limitations. While thione thiocytosine is the dominant tautomer in solution, exclusively the thiol form is present in the gas phase. This hampers a direct comparison of the tautomer-specific photodynamics due to potential perturbations by the solvent environment.

To date, only thione thiocytosine has been investigated thoroughly in solution-phase experiments and by theory and a general pathway similar to other thiouracils has been confirmed. Specifically, ISC via a three-state near-degeneracy of the $^1n\pi^*$, T_2 and T_1 states is efficient and promoted by a high SOC constant of ~160 cm^{-1}. The topography of the lowest triplet state T_1, in analogy to 2TU, resembles a double-well potential with $T_{1\ ring}$ and $T_{1\ op\text{-}S}$ minima (Mai et al. 2016a; Bai and Barbatti 2017).

For thiol thiocytosine, the information is currently sparse. The gas-phase absorption spectrum is shown in Fig. 3.12 including a simulated spectrum of enol cytosine. Compared to thiocarbonyl bases, such as thiouracils, the redshift upon thionation is only moderate. This observation supports a hypothesis that the electronic structure remains similar and therefore the photodynamics of thiol thiocytosine should mirror those of enol cytosine.

The TRPES colormaps of thiocytosine for excitation on the onset and maximum of the first absorption band are displayed in Fig. 3.13 (left). In contrast to cytosine, both spectra are associated with thiol thiocytosine which is the only tautomer present in the gas phase. Thione thiocytosine is predicted to be higher in energy by ~5 kcal/mol and therefore its ground state population under gas-phase molecular beam conditions can be considered negligible (Mai et al. 2017). The population

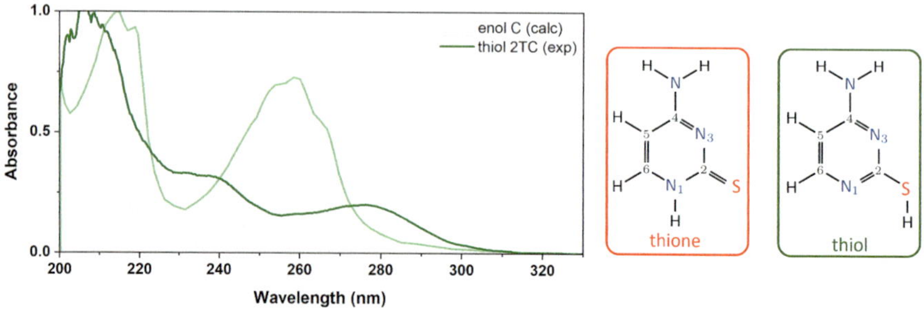

Fig. 3.12 Experimental gas-phase absorption spectrum of thiocytosine (green) in its thiol form and theoretical absorption spectrum of enol cytosine (green-gray) (Bazsó et al. 2011) (Enol Cytosine spectrum replotted from (Bazsó et al. 2011) ibid. with permission from the Royal Society of Chemistry), for comparison. Thiol thiocytosine is the only tautomer present in the gas phase, whereas the thione form dominates in solution

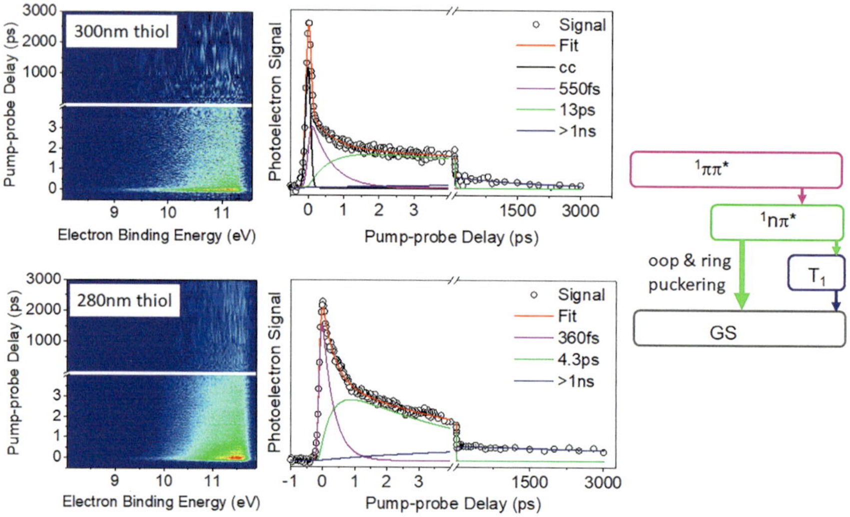

Fig. 3.13 TRPES of thiol thiocytosine with photoexcitation on the onset and maximum of the absorption band (left). The timetrace and components were extracted from global analysis assuming a sequential decay (middle). A hypothetical model for the deactivation mechanism which is derived from enol cytosine (right)

dynamics observed in both TRPES can be described by a sequential model with three exponential decays. An additional Gaussian (cc) is required to account for the initial spike in the cross-correlation region, which is likely caused by significant probe–pump contributions when exciting at the onset where the pump absorption is low. However, only the pump-induced population dynamics are of interest and associated with the photophysical model in Fig. 3.13 (right). In the absence of theoretical studies on thiol thiocytosine, the following proposed model is hypothesized from enol cytosine and modified to align with the TRPES observations. Ultrafast IC within a few hundred femtoseconds transfers population from the initially excited $^1\pi\pi^*$ state to the $^1n\pi^*$. From there, ring puckering and oop deformations are expected to lead to multiple conical intersections with the GS. The mechanistic details may differ, in particular, for molecular distortions in the vicinity of the substituent. An alternative path with a crossing from $^1n\pi^*$ into the triplet manifold is also proposed. All these pathways contribute to the depopulation of the $^1n\pi^*$ state which takes place on timescales of a few picoseconds. In contrast to enol cytosine, a long-lived (>1 ns) signal is present which suggests non-negligible ISC and subsequent trapping in the lowest triplet state, T_1. Despite the absence of a thiocarbonyl group, thionation of enol cytosine enhances the SOC or accessibility of the S_1/Triplet crossing and appreciable contributions from the ISC pathway are observed. Additional excess energy from photoexcitation at the absorption band maximum as opposed to the onset speeds up the electronic population dynamics but the general decay mechanism seems preserved.

3.6 Summary

Thionation of the pyrimidine nucleobases induces changes to the excited-state potential energy topographies and, although these thiobases share a common deactivation pathway, their electronic relaxation mechanisms are uniquely dependent on position, degree, and type of substitution. The dominant deactivation pathway, $1^1\pi\pi^* \rightarrow 1^1n\pi^* \rightarrow$ Triplet, involves three sequential steps (internal conversion, intersystem crossing, and slower triplet state dynamics) that are promoted by characteristic relaxation coordinates that either result in population of the $T_{1\ \mathrm{ring}}$ or $T_{1\ \mathrm{op\text{-}S}}$ minimum. In (thio)cytosines, the photodynamics are tautomer specific and ISC competes with multiple IC pathways. Timescales of these processes are determined by the accessibility of conical intersections and crossing points, and non-adiabatic and spin–orbit couplings between the states. The following take-home message summarizes the unique photoproperties of each of the (thio)bases.

Uracil

- The ultrafast photodynamics of uracil are dominated by singlet-only pathways. For example, several distinct ethylenic conical intersections connect the $1^1\pi\pi^*$, directly or indirectly via $1^1n\pi^*$, to the ground state. Intersystem crossing from the $1^1n\pi^*$ to the triplet plays a role on sufficiently long timescales.
- The same general electronic relaxation pathways are preserved upon methylation, but the photodynamics of thymine are slowed down.

Substituted uracils

- Photoexcitation populates the lowest singlet $1^1\pi\pi^*$ state with near-planar geometry. In the case of the thionated analogs, the transition includes sulfur-localized orbital contributions.
- Evolution from the Franck–Condon region toward the $1^1\pi\pi^*$ minimum is driven by molecular deformations that are dependent on the substituent's position. For

example, in 2TU the motion evolves from planarity into a sulfur-out-of-plane distortion, whereas in 4TU the relaxation involves in-plane bending and bond elongations.

- In accordance with Kasha's rule, the dominant pathway involves internal conversion to the lowest $1^1n\pi^*$. Subsequent intersystem crossing populates the lowest triplet $1^3\pi\pi^*$ state and thereby undergoes a change in molecular orbital nature (El-Sayed's rule). The $1^3\pi\pi^*$ topography resembles a double-well potential with minima of a near-planar geometry and a pyramidalization at the 2 positions (i.e., sulfur or oxygen out-of-plane). In 2TU the minima are isoenergetic, but in the other analogs the near-planar minimum is significantly lower in energy.
- The deactivation dynamics in 2TU are promoted by an increasing sulfur-out-of-plane distortion to populate the pyramidalized sulfur-out-of-plane triplet minimum. The other uracil analogs maintain near-planarity during electronic relaxation and populate the near-planar triplet minimum.
- If methylation occurs in close proximity of the sulfur substituent, it can profoundly alter the deactivation mechanism. Methylation on the opposite side of the molecule as is the case in 2TU/2TT maintains the pathway but slows down intersystem crossing and triplet state dynamics due to higher barriers and smaller spin–orbit couplings. In 4TT out-of-plane distortion of the methyl group and sulfur become important during the internal conversion and intersystem crossing steps, respectively.
- Doubling thionation causes a large number of low-lying electronic states with sulfur-localized orbital transitions. Multiple crossings between these states lead to a complex network of electronic relaxation pathways. Photoexcitation of DTU triggers 4TU- or 2TU-like behavior for excitation wavelengths in the UV-A and UV-C, respectively, because these regions are associated with sulfur-localized orbital transitions in the 2 or 4 position. In the intermediate UV-B region mixed orbital transitions activate hybrid paths that initially launch with 4TU-like behavior but then merge and end on the 2TU-like path, and vice versa.
- Methylation of DTU perturbs the ordering of the low-lying excited states and significantly alters the internal conversion and intersystem crossing mechanism. Some general characteristics of DTT remain similar to DTU, such as the importance of the lowest singlet $1^1n\pi^*$ state in promoting intersystem crossing and a long-lived, near-planar $T_{1\ ring}$ minimum.

Cytosine

- Tautomer aspects need to be taken into consideration to explain the distinctly different photodynamics of keto and enol cytosine.
- The keto form can be selectively excited at the onset of the cytosine absorption spectrum. Three distinct internal conversion pathways (3-state intersection, semi-planar, and out-of-plane) lead back to the ground state. Intersystem crossing from the $1^1n\pi^*$ to the triplet plays a minor role.
- The enol form dominates the absorption spectrum at the first band maximum. The photodynamics are characterized by internal conversion to the $1^1n\pi^*$ and C6 puckering and out-of-plane pathways that repopulate the ground state.

Substituted cytosines

- In the gas phase, thiocytosine exists exclusively in the thiol form, and information on the photodynamics is currently sparse.
- Efficient internal conversion pathways similar to enol cytosine can be expected. However, despite the absence of a thiocarbonyl group, intersystem crossing into a long-lived triplet state has been observed.

3.7 Outlook

TRPES is a powerful technique that provides insight into the electronic population dynamics of molecules but lacks sensitivity to the structural changes along the deactivation pathway. The latter information is desirable to obtain a detailed mechanistic picture of the relaxation process. TRPES experiments with sufficient time resolution (e.g. ~10 fs) have captured the motion of vibrational wave packets as coherent oscillations that are superimposed onto the excited-state time-resolved photoelectron spectra or ion yields (Teles-Ferreira et al. 2022; Solling 2021). The frequencies of the vibrational modes are obtained from Fourier analysis and identified with the help of calculations. Such experiments provide information on the vibrational modes that drive the electronic relaxation.

In the case of complex organic molecules, multiple competing electronic deactivation pathways are often indistinguishable on the basis of their broad, featureless photoelectron spectra alone. However, critical points along the electronic relaxation pathways, such as state minima, conical intersections, and crossings, often have unique geometric characteristics and can therefore be identified with time-resolved structural spectroscopies (Crane et al. 2023; Young et al. 2018; Xu et al. 2016; Schnappinger et al. 2022; Schuurman and Blanchet 2022; Boll et al. 2022; Yang et al. 2020; Cavaletto et al. 2023). Ultrafast electron diffraction and Coulomb explosion experiments are complementary techniques to TRPES that can provide such structural information. Time-resolved core level photoelectron spectroscopy offers another avenue for atom specific, localized probing of the deactivation mechanisms.

Regarding the photodynamics of thiobases, these techniques are expected to provide further clarification on the ultrafast electronic relaxation processes that are driven by sulfur-out-of-plane pyramidalization versus ring deformations. They can also yield better insight into the slower triplet state dynamics such as equilibration between the isoenergetic $T_{1\ op\text{-}S}$ and $T_{1\ ring}$ minima. First studies demonstrating the full potential of structural/site-specific techniques to address these open questions about the thiobase photodynamics are currently emerging (Robinson et al. 2023; Mayer et al. 2022b; Lever et al. 2020, 2021).

Acknowledgments The author thanks Dr. Abed Mohamadzade, Yingqi Qu, and Bijay Duwal who performed the thiobase TRPES studies that are reviewed in this chapter as part of their PhD research. This work was supported by the National Science Foundation grants CHE-1800050, 2106353, 2135351, and 2154852.

References

Arslancan S, Martínez-Fernández L, Corral I (2017) Photophysics and photochemistry of canonical nucleobases' Thioanalogs: from quantum mechanical studies to time resolved experiments. Molecules 22

Ashwood B, Pollum M, Crespo-Hernández CE (2019) Photochemical and photodynamical properties of sulfur-substituted nucleic acid bases. Photochem Photobiol 95(1):33–58

Bai S, Barbatti M (2017) On the decay of the triplet state of thionucleobases. Phys Chem Chem Phys 19(20):12674–12682. https://doi.org/10.1039/C7CP02050C

Baptista MS, Cadet J, Di Mascio P, Ghogare AA, Greer A, Hamblin MR, Lorente C, Nunez SC, Ribeiro MS, Thomas AH (2017) Type I and type II photosensitized oxidation reactions: guidelines and mechanistic pathways. Photochem Photobiol 93(4):912–919

Barbatti M, Aquino AJ, Lischka H (2010) The UV absorption of nucleobases: semi-classical ab initio spectra simulations. Phys Chem Chem Phys 12(19):4959–4967

Barbatti M, Aquino AJA, Szymczak JJ, Nachtigallová D, Lischka H (2011) Photodynamical simulations of cytosine: characterization of the ultrafast bi-exponential UV deactivation. Phys Chem Chem Phys 13(13):6145–6155. https://doi.org/10.1039/C0CP01327G

Barbatti M, Borin AC, Ullrich S (2015a) Photoinduced phenomena in nucleic acids I nucleobases in the gas phase and in solvents. Springer International Publishing, Cham. https://doi.org/10.1007/978-3-319-13371-3

Barbatti M, Borin AC, Ullrich S, Aquino A (2015b) Photoinduced phenomena in nucleic acids. II, DNA fragments and phenomenological aspects. Springer, Cham. https://doi.org/10.1007/978-3-319-13272-3

Bazsó G, Tarczay G, Fogarasi G, Szalay PG (2011) Tautomers of cytosine and their excited electronic states: a matrix isolation spectroscopic and quantum chemical study. Phys Chem Chem Phys 13(15):6799–6807. https://doi.org/10.1039/C0CP02354J

Bisgaard CZ, Satzger H, Ullrich S, Stolow A (2009) Excited-state dynamics of isolated DNA bases: a case study of adenine. ChemPhysChem 10(1):101–110

Blanchet V, Zgierski MZ, Seideman T, Stolow A (1999) Discerning vibronic molecular dynamics using time-resolved photoelectron spectroscopy. Nature 401(6748):52–54. https://doi.org/10.1038/43410

Blanchet V, Zgierski MZ, Stolow A (2001) Electronic continua in time-resolved photoelectron spectroscopy. I. Complementary ionization correlations. J Chem Phys 114(3):1194–1205. https://doi.org/10.1063/1.1331636

Boll R, Schäfer JM, Richard B, Fehre K, Kastirke G, Jurek Z, Schöffler MS, Abdullah MM, Anders N, Baumann TM et al (2022) X-ray multiphoton-induced Coulomb explosion images complex single molecules. Nat Phys 18(4):423–428. https://doi.org/10.1038/s41567-022-01507-0

Borrego-Varillas R, Teles-Ferreira DC, Nenov A, Conti I, Ganzer L, Manzoni C, Garavelli M, Maria de Paula A, Cerullo G (2018) Observation of the sub-100 femtosecond population of a dark state in a thiobase mediating intersystem crossing. J Am Chem Soc 140(47):16087–16093

Cao J, Chen D-C (2020) The excited-state relaxation mechanism of potential UVA-activated phototherapeutic molecules: trajectory surface hopping simulations of both 4-thiothymine and 2, 4-dithiothymine. Phys Chem Chem Phys 22(19):10924–10933

Cavaletto SM, Nam Y, Rouxel JR, Keefer D, Yong H, Mukamel S (2023) Attosecond monitoring of nonadiabatic molecular dynamics by transient X-ray transmission spectroscopy. J Chem Theory Comput 19(8):2327–2339. https://doi.org/10.1021/acs.jctc.3c00062

Chakraborty P, Liu Y, Weinacht T, Matsika S (2021) Effect of dynamic correlation on the ultrafast relaxation of uracil in the gas phase. Faraday Discuss 228(0):266–285. https://doi.org/10.1039/D0FD00110D

Chandler DW, Houston PL (1987) Two-dimensional imaging of state-selected photodissociation products detected by multiphoton ionization. J Chem Phys 87(2):1445–1447. https://doi.org/10.1063/1.453276

Continetti RE (2001) Coincidence spectroscopy. Annu Rev Phys Chem 52(1):165–192. https://doi.org/10.1146/annurev.physchem.52.1.165
Crane SW, Lee JWL, Ashfold MNR, Rolles D (2023) Molecular photodissociation dynamics revealed by Coulomb explosion imaging. Phys Chem Chem Phys 25(25):16672–16698. https://doi.org/10.1039/D3CP01740K
Crespo-Hernandez CE, Pollum M (2016) Thiobase compounds for photodynamic therapy. Google Patents
Crespo-Hernández CE, Cohen B, Hare PM, Kohler B (2004) Ultrafast excited-state dynamics in nucleic acids. Chem Rev 104(4):1977–2020
Cui G, Thiel W (2014) Intersystem crossing enables 4-thiothymidine to act as a photosensitizer in photodynamic therapy: an ab initio QM/MM study. J Phys Chem Lett 5(15):2682–2687
Davies JA, LeClaire JE, Continetti RE, Hayden CC (1999) Femtosecond time-resolved photoelectron–photoion coincidence imaging studies of dissociation dynamics. J Chem Phys 111(1):1–4. https://doi.org/10.1063/1.479248
Eppink ATJB, Parker DH (1997) Velocity map imaging of ions and electrons using electrostatic lenses: application in photoelectron and photofragment ion imaging of molecular oxygen. Rev Sci Instrum 68(9):3477–3484. https://doi.org/10.1063/1.1148310
Evans NL, Ullrich S (2010) Wavelength dependence of electronic relaxation in isolated adenine using UV femtosecond time-resolved photoelectron spectroscopy. Chem A Eur J 114(42):11225–11230
Fielding HH, Worth GA (2018) Using time-resolved photoelectron spectroscopy to unravel the electronic relaxation dynamics of photoexcited molecules. Chem Soc Rev 47(2):309–321. https://doi.org/10.1039/C7CS00627F
Forbes R, Makhija V, Veyrinas K, Stolow A, Lee JWL, Burt M, Brouard M, Vallance C, Wilkinson I, Lausten R et al (2017) Time-resolved multi-mass ion imaging: femtosecond UV-VUV pump-probe spectroscopy with the PImMS camera. J Chem Phys 147(1):013911. https://doi.org/10.1063/1.4978923
Gemenetzidis E, Shavorskaya O, Xu Y-Z, Trigiante G (2013) Topical 4-thiothymidine is a viable photosensitiser for the photodynamic therapy of skin malignancies. J Dermatol Treat 24(3):209–214
Ghafur O, Crane SW, Ryszka M, Bockova J, Rebelo A, Saalbach L, De Camillis S, Greenwood JB, Eden S, Townsend D (2018) Ultraviolet relaxation dynamics in uracil: time-resolved photoion yield studies using a laser-based thermal desorption source. J Chem Phys 149(3):034301
Gobbo JP, Borin AC (2014) 2-Thiouracil deactivation pathways and triplet states population. Comput Theoret Chem 1040:195–201
Harada Y, Suzuki T, Ichimura T, Xu Y-Z (2007) Triplet formation of 4-thiothymidine and its photosensitization to oxygen studied by time-resolved thermal lensing technique. J Phys Chem B 111(19):5518–5524
Hudock HR, Levine BG, Thompson AL, Satzger H, Townsend D, Gador N, Ullrich S, Stolow A, Martinez TJ (2007) Ab initio molecular dynamics and time-resolved photoelectron spectroscopy of electronically excited uracil and thymine. Chem A Eur J 111(34):8500–8508
Jiang J, Zhang T-S, Xue J-D, Zheng X, Cui G, Fang W-H (2015) Short-time dynamics of 2-thiouracil in the light absorbing S2 ($\pi\pi*$) state. J Chem Phys 143(17):11B605_601
Jin Y-C, Zheng X-M (2017) UV absorption and resonance Raman spectra of 2, 4-Dithiouracil. Acta Phys -Chim Sin 33(10):1989–1997
Karran P, Attard N (2008) Thiopurines in current medical practice: molecular mechanisms and contributions to therapy-related cancer. Nat Rev Cancer 8(1):24
Kosma K, Schröter C, Samoylova E, Hertel IV, Schultz T (2009) Excited-state dynamics of cytosine tautomers. J Am Chem Soc 131(46):16939–16943. https://doi.org/10.1021/ja907355a
Kotsina N, Townsend D (2021) Improved insights in time-resolved photoelectron imaging. Phys Chem Chem Phys 23(18):10736–10755. https://doi.org/10.1039/D1CP00933H
Koyama D, Milner MJ, Orr-Ewing AJ (2017) Evidence for a double well in the first triplet excited state of 2-thiouracil. J Phys Chem B 121(39):9274–9280

Lever F, Mayer D, Picconi D, Metje J, Alisauskas S, Calegari F, Düsterer S, Ehlert C, Feifel R, Niebuhr M et al (2020) Ultrafast dynamics of 2-thiouracil investigated by time-resolved auger spectroscopy. J Phys B Atomic Mol Phys 54(1):014002. https://doi.org/10.1088/1361-6455/abc9cb

Lever F, Mayer D, Metje J, Alisauskas S, Calegari F, Düsterer S, Feifel R, Niebuhr M, Manschwetus B, Kuhlmann M et al (2021) Core-level spectroscopy of 2-Thiouracil at the sulfur L1- and L2,3-edges utilizing a SASE free-electron laser. Molecules 26

Mabbs R, Grumbling ER, Pichugin K, Sanov A (2009) Photoelectron imaging: an experimental window into electronic structure. Chem Soc Rev 38(8):2169–2177. https://doi.org/10.1039/B815748K

Mai S, Marquetand P, Richter M, González-Vázquez J, González L (2013) Singlet and triplet excited-state dynamics study of the keto and enol tautomers of cytosine. ChemPhysChem 14(13):2920–2931

Mai S, Marquetand P, González L (2015) A static picture of the relaxation and intersystem crossing mechanisms of photoexcited 2-thiouracil. Chem A Eur J 119(36):9524–9533

Mai S, Pollum M, Martínez-Fernández L, Dunn N, Marquetand P, Corral I, Crespo-Hernández CE, González L (2016a) The origin of efficient triplet state population in sulfur-substituted nucleobases. Nat Commun 7:13077

Mai S, Marquetand P, González L (2016b) Intersystem crossing pathways in the noncanonical nucleobase 2-thiouracil: a time-dependent picture. J Phys Chem Lett 7(11):1978–1983

Mai S, Ashwood B, Marquetand P, Crespo-Hernández CE, González L (2017) Solvatochromic effects on the absorption spectrum of 2-Thiocytosine. J Phys Chem B 121(20):5187–5196. https://doi.org/10.1021/acs.jpcb.7b02715

Mai S, Mohamadzade A, Marquetand P, González L, Ullrich S (2018) Simulated and experimental time-resolved photoelectron spectra of the intersystem crossing dynamics in 2-Thiouracil. Molecules 23(11):2836

Manae MA, Hazra A (2019) Triplet decay dynamics in sulfur substituted thymine: how position of substitution matters. Chem A Eur J 123(51):10862–10867

Martínez-Fernández L, Granucci G, Pollum M, Crespo-Hernández CE, Persico M, Corral I (2017) Decoding the molecular basis for the population mechanism of the triplet phototoxic precursors in UVA light-activated pyrimidine anticancer drugs. Chem–Eur J 23(11):2619–2627

Martinez-Fernandez L, Fahleson T, Norman P, Santoro F, Coriani S, Improta R (2017) Optical absorption and magnetic circular dichroism spectra of thiouracils: a quantum mechanical study in solution. Photochem Photobiol Sci 16(9):1415–1423

Massey A, Xu Y-Z, Karran P (2001) Photoactivation of DNA thiobases as a potential novel therapeutic option. Curr Biol 11(14):1142–1146

Mayer D, Picconi D, Robinson MS, Gühr M (2022a) Experimental and theoretical gas-phase absorption spectra of thionated uracils. Chem Phys 558:111500. https://doi.org/10.1016/j.chemphys.2022.111500

Mayer D, Lever F, Picconi D, Metje J, Alisauskas S, Calegari F, Düsterer S, Ehlert C, Feifel R, Niebuhr M et al (2022b) Following excited-state chemical shifts in molecular ultrafast x-ray photoelectron spectroscopy. Nat Commun 13(1):198. https://doi.org/10.1038/s41467-021-27908-y

Middleton CT, de La Harpe K, Su C, Law YK, Crespo-Hernández CE, Kohler B (2009) DNA excited-state dynamics: from single bases to the double helix. Annu Rev Phys Chem 60:217–239

Mohamadzade A (2022) Time-resolved photoelectron spectroscopy of uracil derivatives: femtosecond-nanosecond excited state dynamics. University of Georgia

Mohamadzade A, Ullrich S (2020) Internal conversion and intersystem crossing dynamics of uracil upon double thionation: A time-resolved photoelectron spectroscopy study in the gas phase. Phys Chem Chem Phys 22(27):15608–15615

Mohamadzade A, Bai S, Barbatti M, Ullrich S (2018) Intersystem crossing dynamics in singly substituted thiouracil studied by time-resolved photoelectron spectroscopy: micro-environmental effects due to sulfur position. Chem Phys 515:572–579

Mohamadzade A, Nenov A, Garavelli M, Conti I, Ullrich S (2023) Double thionated pyrimidine nucleobases: molecular tools with tunable photoproperties. J Am Chem Soc 145(22):11945–11958. https://doi.org/10.1021/jacs.2c12061

Nachtigallová D, Aquino AJ, Szymczak JJ, Barbatti M, Hobza P, Lischka H (2011) Nonadiabatic dynamics of uracil: population split among different decay mechanisms. Chem A Eur J 115(21):5247–5255

Nakayama A, Harabuchi Y, Yamazaki S, Taketsugu T (2013) Photophysics of cytosine tautomers: new insights into the nonradiative decay mechanisms from MS-CASPT2 potential energy calculations and excited-state molecular dynamics simulations. Phys Chem Chem Phys 15(29):12322–12339

Nenov A, Conti I, Borrego-Varillas R, Cerullo G, Garavelli M (2018) Linear absorption spectra of solvated thiouracils resolved at the hybrid RASPT2/MM level. Chem Phys 515:643–653

Nir E, Müller M, Grace LI, de Vries MS (2002) REMPI spectroscopy of cytosine. Chem Phys Lett 355(1–2):59–64

Ortiz-Rodríguez LA, Crespo-Hernández CE (2020) Thionated organic compounds as emerging heavy-atom-free photodynamic therapy agents. Chem Sci 11(41):11113–11123

Pollum M, Jockusch S, Crespo-Hernández CE (2014a) 2, 4-Dithiothymine as a potent UVA chemotherapeutic agent. J Am Chem Soc 136(52):17930–17933

Pollum M, Martínez-Fernández L, Crespo-Hernández CE (2014b) Photochemistry of nucleic acid bases and their thio-and aza-analogues in solution. In: Photoinduced phenomena in nucleic acids I. Springer, pp 245–327

Pollum M, Jockusch S, Crespo-Hernández CE (2015) Increase in the photoreactivity of uracil derivatives by doubling thionation. Phys Chem Chem Phys 17(41):27851–27861

Pridgeon S, Heer R, Taylor G, Newell D, O'toole K, Robinson M, Xu Y, Karran P, Boddy A (2011) Thiothymidine combined with UVA as a potential novel therapy for bladder cancer. Br J Cancer 104(12):1869

Richter M, Mai S, Marquetand P, González L (2014) Ultrafast intersystem crossing dynamics in uracil unravelled by ab initio molecular dynamics. Phys Chem Chem Phys 16(44):24423–24436

Robinson MS, Niebuhr M, Gühr M (2023) Ultrafast photo-ion probing of the relaxation dynamics in 2-Thiouracil. Molecules 28

Ruckenbauer M, Mai S, Marquetand P, González L (2016a) Photoelectron spectra of 2-thiouracil, 4-thiouracil, and 2, 4-dithiouracil. J Chem Phys 144(7):074303

Ruckenbauer M, Mai S, Marquetand P, González L (2016b) Revealing deactivation pathways hidden in time-resolved photoelectron spectra. Sci Rep 6(1):35522

Saenger W (1984) Principles of nucleic acid structure. https://doi.org/10.1007/978-1-4612-5190-3

Sanchez-Rodriguez JA, Mohamadzade A, Mai S, Ashwood B, Pollum M, Marquetand P, Gonzalez L, Crespo-Hernandez CE, Ullrich S (2017) 2-Thiouracil intersystem crossing Photodynamics studied by wavelength-dependent photoelectron and transient absorption spectroscopies. Phys Chem Chem Phys 19:19756

Schmitt M, Lochbrunner S, Shaffer JP, Larsen JJ, Zgierski MZ, Stolow A (2001) Electronic continua in time-resolved photoelectron spectroscopy. II. Corresponding ionization correlations. J Chem Phys 114(3):1206–1213. https://doi.org/10.1063/1.1331637

Schnappinger T, Jadoun D, Gudem M, Kowalewski M (2022) Time-resolved X-ray and XUV based spectroscopic methods for nonadiabatic processes in photochemistry. Chem Commun 58(92):12763–12781. https://doi.org/10.1039/D2CC04875B

Schuurman MS, Blanchet V (2022) Time-resolved photoelectron spectroscopy: the continuing evolution of a mature technique. Phys Chem Chem Phys 24(34):20012–20024. https://doi.org/10.1039/D1CP05885A

Solling TI (2021) Nonstatistical photoinduced processes in gaseous organic molecules. ACS Omega 6(44):29325–29344. https://doi.org/10.1021/acsomega.1c04035

Stavros VG, Verlet JRR (2016) Gas-phase femtosecond particle spectroscopy: a bottom-up approach to nucleotide dynamics. Annu Rev Phys Chem 67(1):211–232. https://doi.org/10.1146/annurev-physchem-040215-112428

Stolow A, Underwood JG (2008) Time-resolved photoelectron spectroscopy of nonadiabatic dynamics in polyatomic molecules. In: Advances in chemical physics, pp 497–584

Stolow A, Bragg AE, Neumark DM (2004) Femtosecond time-resolved photoelectron spectroscopy. Chem Rev 104(4):1719–1758. https://doi.org/10.1021/cr020683w

Suits AG (2018) Invited review article: photofragment imaging. Rev Sci Instrum 89(11):111101. https://doi.org/10.1063/1.5045325

Suzuki T (2006) Femtosecond time-resolved photoelectron imaging. Annu Rev Phys Chem 57(1):555–592. https://doi.org/10.1146/annurev.physchem.57.032905.104601

Teles-Ferreira DC, Conti I, Borrego-Varillas R, Nenov A, Van Stokkum IH, Ganzer L, Manzoni C, de Paula AM, Cerullo G, Garavelli M (2020) A unified experimental/theoretical description of the ultrafast photophysics of single and double thionated uracils. Chem–Eur J 26(1):336–343

Teles-Ferreira DC, van Stokkum IHM, Conti I, Ganzer L, Manzoni C, Garavelli M, Cerullo G, Nenov A, Borrego-Varillas R, de Paula AM (2022) Coherent vibrational modes promote the ultrafast internal conversion and intersystem crossing in thiobases. Phys Chem Chem Phys 24(36):21750–21758. https://doi.org/10.1039/D2CP02073D

Trigiante G, Xu Y-Z (2015) 4-thiothymidine and its analogues as UVA-activated photosensitizers

Ullrich S, Schultz T, Zgierski MZ, Stolow A (2004) Electronic relaxation dynamics in DNA and RNA bases studied by time-resolved photoelectron spectroscopy. Phys Chem Chem Phys 6(10):2796–2801

Ullrich S, Qu Y, Mohamadzade A, Shrestha S (2022) The effect of methylation on the triplet-state dynamics of 2-Thiouracil: time-resolved photoelectron spectroscopy of 2-Thiothymine. Chem A Eur J 126(44):8211–8217. https://doi.org/10.1021/acs.jpca.2c06051

Xie B-B, Wang Q, Guo W-W, Cui G (2017) The excited-state decay mechanism of 2, 4-dithiothymine in the gas phase, microsolvated surroundings, and aqueous solution. Phys Chem Chem Phys 19(11):7689–7698

Xu J, Blaga CI, Agostini P, DiMauro LF (2016) Time-resolved molecular imaging. J Phys B Atomic Mol Phys 49(11):112001. https://doi.org/10.1088/0953-4075/49/11/112001

Yang J, Zhu X, Nunes F, Yu JK, Parrish RM, Wolf TJA, Centurion M, Gühr M, Li R, Liu Y et al (2020) Simultaneous observation of nuclear and electronic dynamics by ultrafast electron diffraction. Science 368(6493):885–889. https://doi.org/10.1126/science.abb2235

Young L, Ueda K, Gühr M, Bucksbaum PH, Simon M, Mukamel S, Rohringer N, Prince KC, Masciovecchio C, Meyer M et al (2018) Roadmap of ultrafast x-ray atomic and molecular physics. J Phys B Atomic Mol Phys 51(3):032003. https://doi.org/10.1088/1361-6455/aa9735

Yu H, Sanchez-Rodriguez JA, Pollum M, Crespo-Hernández CE, Mai S, Marquetand P, González L, Ullrich S (2016) Internal conversion and intersystem crossing pathways in UV excited, isolated uracils and their implications in prebiotic chemistry. Phys Chem Chem Phys 18(30):20168–20176

Zou X, Dai X, Liu K, Zhao H, Song D, Su H (2014) Photophysical and photochemical properties of 4-thiouracil: time-resolved IR spectroscopy and DFT studies. J Phys Chem B 118(22):5864–5872

Chapter 4
Nucleobases as Molecular Fossils of Prebiotic Photoselection

Mattanjah S. de Vries and Nathan Svadlenak

Abstract The molecular inventory on a prebiotic earth, some four billion years ago, would likely have included a large variety of heterocyclic compounds. Many chemical parameters may have affected the eventual selection of the canonical bases as the building blocks for life as we know it today. One of the aspects may have been photochemistry since prior to life the absence of an ozone layer in the atmosphere would have enabled more energetic photochemistry than today. This chapter discusses elements of nucleobase photochemistry that can contribute to an explanation of a possible role of photochemical selection in the origin of life.

Keywords Nucleobase photochemistry · Gas phase spectroscopy · Conical intersections · Internal conversion · Multiphoton ionization · Excited-state dynamics · Excited-state lifetimes

4.1 Introduction: Prebiotic Photochemistry

The photochemistry of nucleobases is of interest for multiple reasons. The nucleobases are the main chromophores in DNA and RNA. The absorption of UV light by DNA is one of the mechanisms leading to photolesions and skin cancer. On an early earth, some four billion years ago, the interaction between UV radiation and nucleobases may have been especially important. It is likely that the planet started initially with a reducing atmosphere; free oxygen would only have become a component of the atmosphere in the great oxygenation event, following the evolution of photosynthetic organisms that produce oxygen (Bekker et al. 2004). This could not have occurred before life existed. Without free oxygen there would not have been an

M. S. de Vries (✉) · N. Svadlenak
Department of Chemistry and Biochemistry, University of California Santa Barbara, Santa Barbara, CA, USA
e-mail: devries@chem.ucsb.edu

S. Matsika, A. H. Marcus (eds.), *Nucleic Acid Photophysics and Photochemistry*, Nucleic Acids and Molecular Biology 36,
https://doi.org/10.1007/978-3-031-68807-2_4

ozone layer which today shields us from UV radiation at wavelengths shorter than about 280 nanometers. Therefore, prebiotic chemistry took place under conditions in which photochemistry would have been possible much further to the blue of the spectrum than today (Sagan 1973; Cockell 1998; Cockell 2000; Cnossen et al. 2007).

Cnossen et al. estimated the solar spectrum that would have reached the planet's surface, shown in Fig. 4.1 compared to the present (Cnossen et al. 2007). Models estimate that UV damage rates were three orders of magnitude greater in the Archean oceans of the early Earth than today (Cockell 2000). Therefore, for nearly one billion years preceding the great oxygenation event, life coped with intense UV surface irradiation. The consequences of this intense radiation for organic chemical evolution prior to the emergence of life are uncertain.

A primordial soup would like have contained not just the canonical bases but rather a wide variety of heterocyclic compounds and their isomers, analogs, and derivatives (Rios and Tor 2013). From this mixture, eventually the canonical bases emerged as the building blocks of life as we know it today (Cafferty et al. 2018). The prevalence of organic molecules is determined not just by their rates of synthesis, but also by how quickly they are degraded. Simpler chemical building blocks must persist long enough in their environments to be joined into the oligomers and complex molecular assemblies found in living organisms. For example, molecules synthesized near deep-sea vents must avoid the hydrolysis reactions that are promoted by superheated water, while molecules in sun-exposed environments must avoid photolysis. UV light is likely to have been a major, if not dominant, source of energy for primordial prebiotic reactions (Ranjan and Sasselov 2016; 2017; Xu et al. 2017; Green et al. 2021; Saladino et al. 2015; Becker et al. 2016; Becker et al.

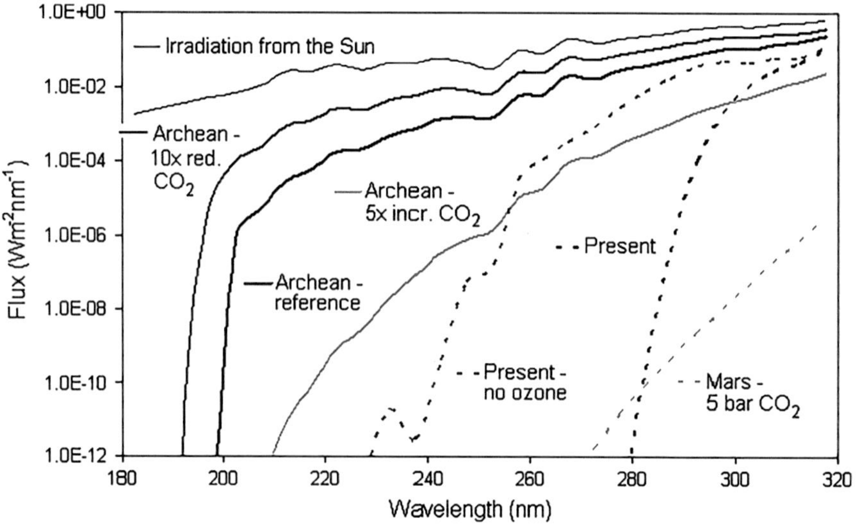

Fig. 4.1 Irradiation at the top of the atmosphere of the Archean Earth, and the surface irradiance for several sets of conditions (Cnossen et al. 2007)

2019; Teichert et al. 2019; Yadav et al. 2020; Bertram et al. 2022). However, in addition to a role in synthesis, solar radiation could also cause photochemical degradation and in both steps, light can provide selectivity (Xu et al. 2017; Powner et al. 2009; Xu et al. 2019; Roberts et al. 2018; Xu et al. 2020).

One form of this photoselectivity is suggested by the fact that the nucleobases that are involved in replication possess a form of high intrinsic stability against otherwise harmful UV photo-damage while many alternative nucleobases lack this property. UV protection comes about when the electronic excitation, which results from UV absorption, is converted to heat by internal conversion (IC) and safely dissipated to the environment at rates too fast for other more harmful reactive pathways to occur. This property would have been highly advantageous for the first self-replicating molecules in prebiotic times before modern enzymatic repair and before the formation of the ozone layer (Ribas et al. 2010). This intrinsic damage resistance is literally built into the canonical nucleobases and is exquisitely sensitive to molecular structure. Remarkably, the canonical RNA and DNA nucleobase monomers, that now make up life as we know it, have nearly ideal UV resistant properties for organic molecules of their size and complexity in the precise spectral region of intense UV irradiation that characterized the first half of Earth's existence.

The canonical nucleobases generally decay out of the excited state (ES) back to the ground state (GS) in the order of picoseconds. This safe deexcitation is orders of magnitude faster than in most other heterocyclic compounds, which would likely also have been present on an early earth and which could have been candidates for alternate genetic lexicons (Benner et al. 2016; Georgiadis et al. 2015; Yang et al. 2006). This distinction suggests that the natural bases were selected in part due to their high intrinsic photostability and that their photochemistry is a vital property of the set of prebiotic molecules that could give rise to life on earth or elsewhere in the universe. This hypothesis implies that the molecular photo-properties of the nucleobases, which have been preserved unchanged from prebiotic times, may be a relic of some of the earliest stages of prebiotic chemistry. *Thus, this photochemistry provides us with* ***molecular fossils*** *of 4-billion-year-old chemistry.*

As summarized in Fig. 4.2, the response times to UV irradiation vary somewhat between solution and gas phase; however, the trends remain the same. In both phases, the canonical bases are among the most photostable by virtue of their ultrashort excited-state lifetime. It is therefore very advantageous to study the bases in the gas phase where we can obtain by far the most detailed understanding of photochemical mechanisms that govern these chromophores. This approach allows for the most fundamental modeling with high-level theoretical computations. A number of reviews cover theoretical computations (Gustavsson et al. 2010; Barbatti et al. 2010; Cuellar-Zuquin et al. 2023; Improta et al. 2016; Barone and Puzzarini 2023; Szabla 2021) and gas phase spectroscopy with neutrals (Boldissar and de Vries 2018; Beckstead et al. 2016) and ions as well as solution phase spectroscopy (Prokhorenko et al. 2016; Middleton et al. 2009; Hoeh et al. 2022). This chapter summarizes findings in gas phase experiments with neutral molecules. Such experiments have become possible by the development of laser-desorption jet-cooling techniques that pulse vaporize fragile molecules intact and entrain them in a

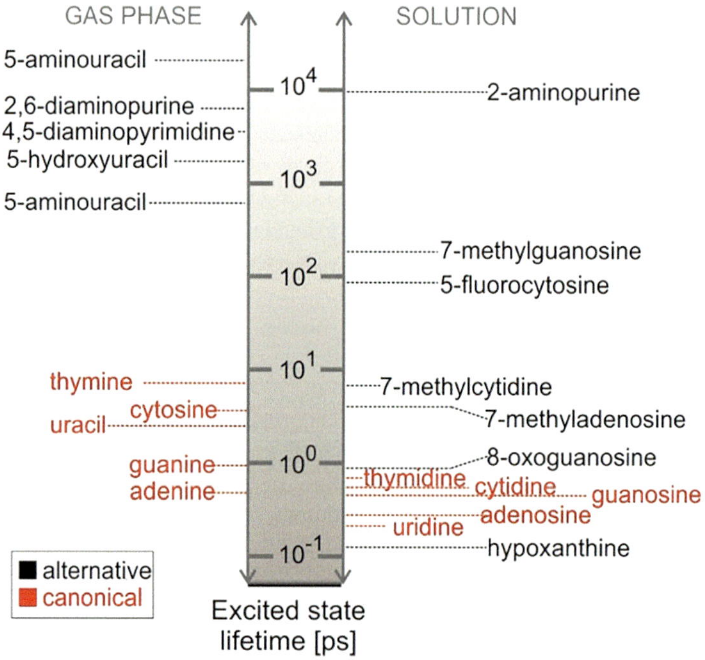

Fig. 4.2 Excited-state lifetimes of nucleobases

supersonic beam (Arrowsmith et al. 1988; Meijer et al. 1990). Spectroscopy is typically based on resonance-enhanced multiphoton ionization spectroscopy combined with pump–probe spectroscopy for lifetime measurements and IR–UV hole burning for isomeric selection and structure elucidation.

4.2 Photochemical Pathways

Photostability should not be viewed as a binary property because rates of photodegradation vary continuously and over a wide dynamic range. We know, for example, that the response to UV absorption depends on wavelength and that pyrimidines have a finite quantum yield for populating a triplet state (Ligare et al. 2015; Siouri et al. 2017; Lobsiger et al. 2013). The ultrashort lifetimes of the canonical nucleobases reduce, but do not eliminate, the probability of photochemical destruction (Peeters et al. 2003). Similarly, longer lifetimes do not guarantee photodamage but enhance its probability. Greater photostability (or slower photodecomposition) provides an advantage that could favor one kind of organic chemistry over another, leading eventually to its selection for the synthesis of self-replicating polymers. Examples of non-canonical bases such as hypoxanthine are also known that have ultrashort excited state (ES) lifetimes comparable to the natural bases, but which are not used in the genetic code today (but may have been constituents of protonucleic acids). In this sense, UV hardiness may be a necessary, but not a sufficient condition for selection of the genetic code.

In a number of cases, a single excited-state lifetime, measured at a single excitation wavelength, has been reported; however, this is an incomplete description. ES lifetimes depend on the absorption wavelength, especially when barriers exist for certain relaxation channels. When measured well above such barriers, the excited-state lifetime may be short, while it can be significantly longer upon excitation close to or below the barrier. Thus, it is possible for compounds to be well protected against UV photochemical damage except for a limited range of wavelengths in the UV spectrum. Furthermore, these heterocyclic compounds typically have a rich tautomeric variety and different tautomers can have different UV responses.

Following absorption of a photon, different photochemical processes are generally possible. Figure 4.3 shows schematically a number of those and indicates typical orders of magnitude of the associated lifetimes. In addition to fluorescence, which occurs typically in the nanosecond timescale, a variety of non-radiative processes could be possible. Conical intersections can mediate internal conversion, either to the electronic ground state or to another excited state. The latter often have lifetimes of the order of hundreds of picoseconds to nanoseconds. Non-radiative transitions can include population of dark states, such as an nπ* state or a triplet state (by intersystem crossing, ISC), both of which can occur in nucleobases. Charge transfer can occur in hydrogen bonded structures such as base pairs or nucleosides and can be followed by non-radiative return to the ground state by reverse charge transfer. Structures with base stacking can exhibit exciplex and exciton formation, followed by charge recombination (Crespo-Hernandez et al. 2005).

The outcome of the competition between all possible processes depends on the quant yields (QY) which are inversely proportional to the excited-state lifetimes. Photochemical pathways depend on the potential energy landscape. Excitation

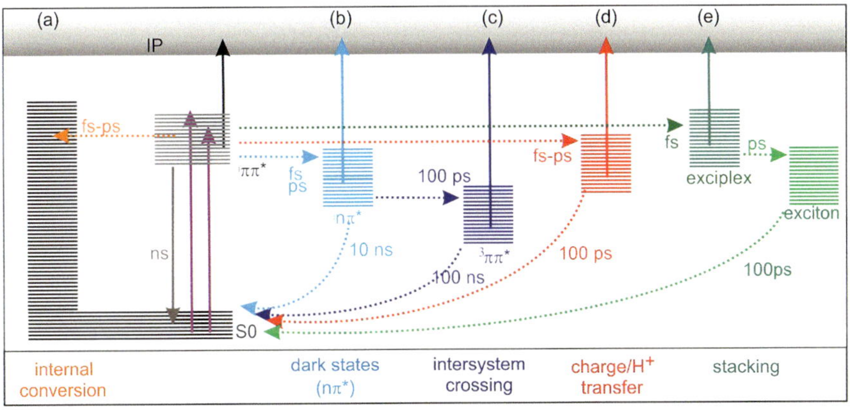

Fig. 4.3 Schematic Jablonski diagram to illustrate major photochemical pathways with typical orders of magnitude of associated lifetimes

positions the molecule on an excited-state potential in the Franck–Condon region. From that point the wavepacket will evolve on the potential energy surfaces. The outcome is dictated by the shapes of those potential energy surfaces, which may include features such as barriers and conical intersections between multiple surfaces and along different internuclear coordinates. Figure 4.4 schematically illustrates the idea with just one electronically excited state, S_1, in addition to the ground state, S_0, as a function of just two internuclear coordinates, q_1 and q_2. The blue arrow represents absorption, green represents fluorescence, which has a typical timescale of the order of a nanosecond. The other two arrows indicate two possible non-radiative pathways, with red indicating internal conversion to the ground state via a conical intersection. Depending on the details of the potential, this process can take place within on the order of a picosecond. Orange indicates other possible photochemical reactions, such as intersystem crossing, bond breaking, photodissociation, charge transfer, or tautomerization, which may involve other electronic states, proceed along other coordinates, and involve multiple timescales.

Multiple processes can compete with the fastest timescale providing the highest quantum yield. When internal conversion is significantly faster than other pathways it can outcompete other, more reactive processes and thus prevent radiative damage. Internal conversion does not remove any energy from the molecule but it converts the electronic excitation to internal excitation in the ground state, essentially

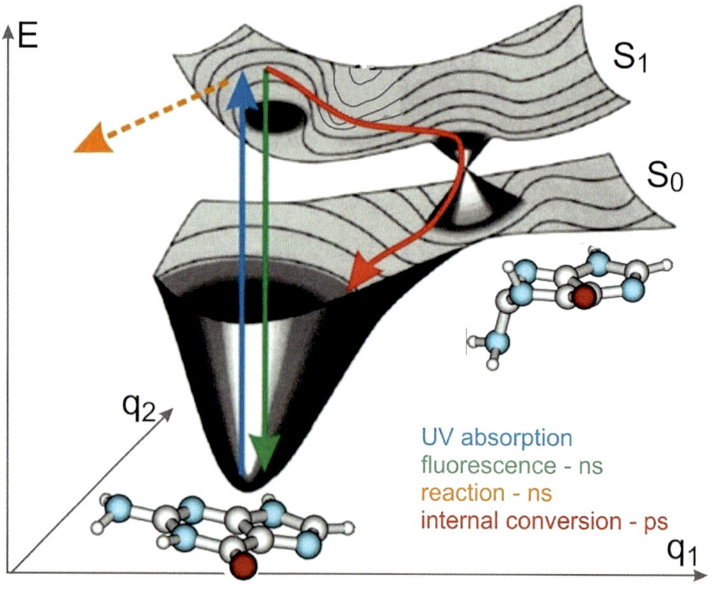

Fig. 4.4 Schematic potential energy surfaces as a function of two internuclear coordinates, q_1 and q_2. Insets represent structures of keto-guanine at ground state equilibrium and at the conical intersection

converting radiative energy to heat. When the hot ground state molecule can dissipate the heat to its environment, i.e. the solvent, this provides a safe pathway to diffuse the electronic excitation without radiative damage.

4.3 Structure Dependence

Whether such a "self-healing" pathway is available depends on the intricate detail of the excited-state potentials, particularly the presence of a barrierless path to conical intersections leading to the electronic ground state. As illustrated in Fig. 4.4, conical intersections occur at internuclear coordinate positions that are very different from the equilibrium structure. Generally, the molecular structure needs to deform significantly to accommodate a radiationless trajectory. As an example, Fig. 4.4 shows the structures of guanine at its ground state equilibrium and at the S_1/S_0 conical intersection (Marian 2007). This is a major reason for the exquisite dependence of the photochemistry on molecular structure.

Figure 4.5 illustrates this structure dependence with three of the canonical nucleobases and hypoxanthine in the left column and some of their derivatives in the right column (De Vries 2014, 2015; Kleinermanns et al. 2013; Nachtigallova et al. 2010; Gengeliczki et al. 2010). When probed at energies close to the onset of absorption, all compounds in the left column have excited-state lifetimes of the order of picoseconds or less, while those in the right column have lifetimes of the order of nanoseconds. This trend exists both in the gas phase and in solution (Middleton et al. 2009; Crespo-Hernandez et al. 2004, 2005; Hare et al. 2007). The red and blue shaded areas on the molecular frame indicate where the geometry change takes place that leads to a major conical intersection. It should be stressed that the excited-state potential energy landscape is a multidimensional function of many internuclear coordinates leading to many possible pathways, possibly involving a number of conical intersections and seams (Kistler and Matsika 2008; Matsika

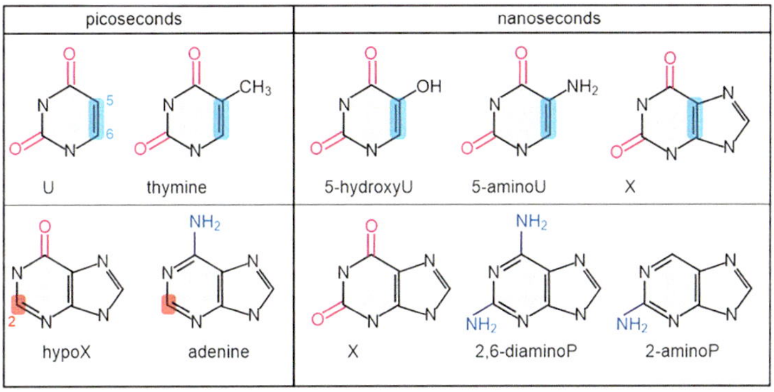

Fig. 4.5 Structural motifs affecting conical intersections. U = uracil, X = xanthine, P = purine

2005). Nevertheless, it is a useful simplification to highlight two structural motifs that appear to dominate the dynamics of these compounds. For the pyrimidines, the area shaded in blue indicates the C5=C6 bond that twists and stretches. For the purines, the most critical geometry change is puckering or pyramidalization at C2, indicated in red. Therefore, substitutions at C5 and C2 can often cause variations in excited-state lifetimes by orders of magnitude. The effect is due to modifications of the potential energy surfaces, which, in addition to modifying the conical intersections (CIs) themselves, can create barriers or otherwise change the trajectories toward them.

An example is the series of C5 derivatives of uracil (U), shown in the top row of Fig. 4.5. Both U and 5-methylU (thymine) have excited-state lifetimes of the order of ps while 5-hydroxyU and 5-aminoU have lifetimes of the order of 1 and 12 ns, respectively (Nachtigallova et al. 2010; Epifanovsky et al. 2008). Immobilizing the C5=C6 bond by a five membered ring, as in xanthine, leads to an excited-state lifetime of nanoseconds. Leutwyler and coworkers have demonstrated similar trends for cytosine (C), where immobilizing the C5=C6 bond by a five membered ring leads to an orders of magnitude longer excited-state lifetime (Trachsel et al. 2017).

The bottom row compares a series of purines in which those with an unsubstituted C2—adenine and hypoxanthine—have ultrashort lifetimes. Those with a substituent at C2 have much longer excited-state lifetimes. This pattern is strikingly exemplified in the difference between the isomers adenine (6-aminopurine) and 2-aminopurine. The former does not fluoresce and the latter does, while its excited-state lifetime is further affected by hydrogen bonding in solution (Lobsiger et al. 2014a). Therefore, 2-aminopurine is used as an alternative base that serves as a fluorescent marker (Tor et al. 2007).

Cytosine, with an amino substituent at C4, has two dominant pathways; in addition to "twisting" of the C5=C6 bond there is a CI associated with a C4–NH_2 out of plane "sofa" configuration (Kistler and Matsika 2007). The 5-methyl-2-pyrimidone analog has similar conical intersections but is fluorescent because the absorption maximum is at lower energies, preparing the molecule in the excite state below a barrier and with less vibrational energy (Trachsel et al. 2017; Kistler and Matsika 2007; Blaser et al. 2016; Lobsiger et al. 2013, 2014b, 2015; Frey et al. 2014; Lobsiger and Leutwyler 2012; Trachsel et al. 2016).

In spite of the trends described so far, the precise effect of structure modification is impossible to predict without high level computations. For example, guanine (G) has an amino group at C2, like 2-aminopurine, but some of its tautomers are short lived, while others are long-lived, as discussed below.

4.4 Purines

Figure 4.6 shows nine derivatives of purine, formed by all possible combinations of hydroxy and amino substitutions in C2 and C6. Two of these structures constitute the canonical bases adenine (6-aminopurine) and guanine (2-amino-6-oxopurine).

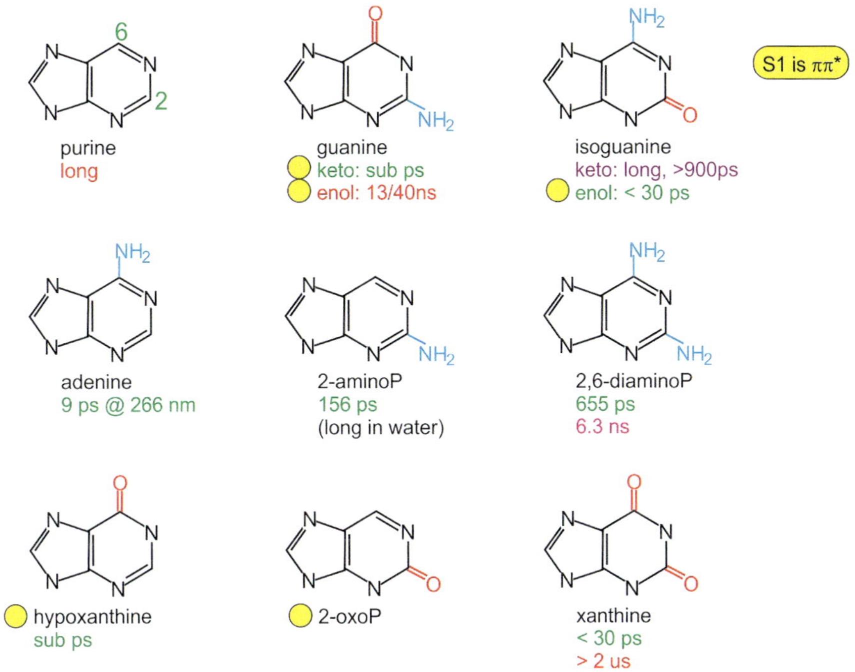

Fig. 4.6 Selected lifetimes for nine amino and oxo derivatives of purine at the origin of absorption. See also Table 4.1. Yellow dots mark the cases for which S1 is ππ* as opposed to nπ*

Purine itself is long lived as it populates a triplet state, which it reaches in two steps: a transition from $S_2(^1\pi\pi^*)$ to $S_1(^1n\pi^*)$ via a CI in about 100 fs followed by a near 100% QY for ISC to a $^3\pi\pi^*$ state. Therefore, the chromophore by itself is not responsible for the photostability of the purine nucleobases. This is similar for the pyrimidine nucleobases as Arpa et al. have pointed out that "The pyrimidine chromophore is not responsible for the photostability of the nucleobases." (Arpa et al. 2020) For purine, substitutions at the C_2 or C_6 position by oxo- or amino-groups modify the potential energy landscape, to produce different dynamics for each of the other eight structures, including producing photostability for selected ones. For pyrimidine, substitutions at the C_2 or C_4 position can create ultrafast pathways through CIs to the ground state. For purine-based structures, three major motifs have emerged, describing how the excited-state dynamics is affected by oxo- and amino-derivatives.

1. The first motif is the order of ππ* and nπ* states which is modified by the stabilizing and destabilizing effects of the added groups (Mburu and Matsika 2008). The ππ* and nπ* state are the bright and dark states, respectively. The nπ* state is lower in energy in the Frank-Condon region or its minimum is below that of the ππ* state, in all cases, except guanine, hypoxanthine, 2-oxopurine, and enol isoguanine. In those cases, there may be conical intersections between the two

Table 4.1 Excited-state lifetimes of purine derivatives as measured at the origin of lowest transition

Compound	R2	R6	Tautomer	Lifetime (ps)	S1	References
Purine	H	H		Long	nπ*	Schneider et al. (2009)
Guanine	NH_2	O	keto	<1	ππ*	Barbatti et al. (2011), Yamazaki et al. (2008)
			enol	13,000/40,000	ππ*	Siouri et al. (2017)
Isoguanine	O	NH_2	keto	>900	nπ*	Gate et al. (2019)
			enol	<30	nπ*	
Adenine	H	NH_2		9	nπ*	Luhrs et al. (2001)
2-AP	NH_2	H		156	nπ*	Lobsiger et al. (2014a)
2,6-dAP	NH_2	NH_2		655/6300	nπ*	Gate et al. (2021)
HypoX	H	O		<1	ππ*	Chen and Kohler (2012)
2-oxoP	O	H			ππ*	Martinez-Fernandez et al. (2019)
Xanthine	O	O		<30/>2 μs	nπ*	Gate et al. (2021)

that allow the nπ* state to be populated and to serve as a dark state (Perun et al. 2005; Picconi et al. 2013). For example, the nπ* state leads to intersystem crossing to a $^3ππ^*$ state in purine and 2-oxopurine, and possibly xanthine, 2,6-diaminouprine, and 2-aminopurine. This transition is El-Sayed allowed and large spin–orbit coupling can also contribute to this strength of this ISC pathway. However, even when the nπ* state is not accessible, there may or may not be a barrierless path to the ground state, as in keto vs. enol guanine (Marian 2007). Conversely, even when the $^1nπ^*$ state can be reached, it may subsequently lead to the ground state, as in adenine.

2. C2 puckering can characterize a CI geometry (Martinez-Fernandez et al. 2019; Broo 1998). The short lifetime of two structures without a C2 substituent, adenine and hypoxanthine has been attributed to a CI with this geometry. An example of this effect is the lifetime difference between adenine and 2-aminopurine. For 2-aminopurine, it has been proposed that the out of-plane rotation of the C2-amino group takes longer than the C2–H bond rotation in adenine, resulting in the longer observed lifetime. However, this trend does not hold for purine which has no C2 substituent but still has a long lifetime. It should also be noted that both ketoG and enol isoG with an amino- and oxo-substituent in C2, respectively, have fast excited-state lifetimes.
3. C6 puckering can also characterize conical intersection geometries (Serrano-Andres et al. 2006). This geometry provides an additional deactivation channel in adenine, which becomes significant only at higher excitation energies (Plasser et al. 2014). C6 substitution has also been proposed to play a role in reducing the barrier in the ππ* state. This effect is illustrated with the observation that adenine and hypoxanthine decay fast relative to purine. However, the observations that 2,6-dAP is slower than 2AP and that enol-G and keto-isoG are slow complicate this explanation.

Other substitutions are possible as well, such as sulfur substitution. Sulfur has been considered in prebiotic scenarios since the presence of sulfur must be assumed for the formation of the RNA polymer backbone. Generally sulfur increases intersystem crossing rates, as is indeed the case for thioguanine (Siouri et al. 2017). Siouri et al. reported data on two thiol tautomers, which decay to a dark state, possibly a triplet state, with rates depending on tautomer form and on excitation wavelength, with the fastest rate on the order of 10^{10} s^{-1}.

4.5 Analogs

Nucleobase analogs represent another class of molecules that represent possible alternative bases or protonucleic bases in prebiotic scenarios. These analog structures are of interest as building blocks of primitive informational polymers that preceded RNA/DNA on account of their versatile hydrogen bonding possibilities. Analogs with different arrangements of nitrogens on the aromatic ring include *pyrazines*, *triazoles*, and *triazines* (Zhang et al. 2016; Schuster et al. 2021).

Figure 4.7 shows these structures as analogs of cytosine; similar sets of structures exist for the other bases. With these structures and their derivatives, the position of nitrogens on the ring provides compounds that are isosteric with the pyrimidines for the base pairing part of the structure. Interestingly, most of these compounds are transparent in parts of the UV region where their canonical analogs absorb strongly, which could have been an advantage under prebiotic radiative conditions (Kolb et al. 1994).

Zhou et al. reported excited-state lifetimes for 5-azacytosine in solution of 1.5 ps and 15 ps, representing a 40-fold increase over cytosine (Zhou et al. 2017). The authors propose that the CI associated with the 5–6 internuclear coordinate is affected by the N5 substitution since the N5–C6 bond is shorter than the C5–C6 bond in cytosine. Zhang et al. found for melamine in solution that the excited-state

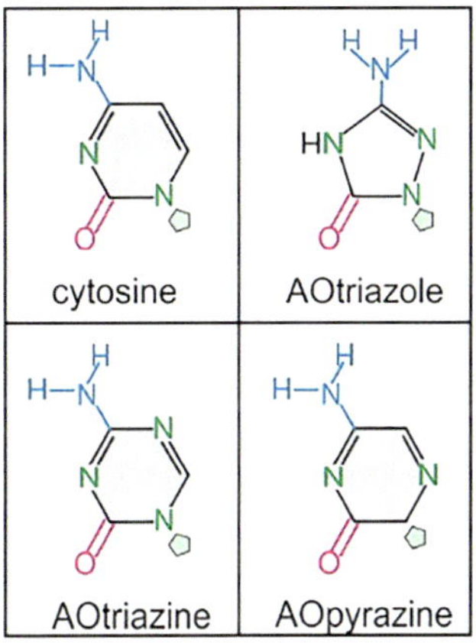

Fig. 4.7 Cytosine analogs

population decays by internal conversion with a lifetime of 13 ps. The excited-state lifetime of the lysine derivative is slightly longer (18 ps), but the dominant deactivation pathway is otherwise the same as for melamine. In both cases, a minor part of the excited-state population is trapped in a long-lived triplet state (Zhang et al. 2016).

Other potential components of a hypothetical predecessor of RNA (Cafferty et al. 2013, 2018; Cafferty and Hud 2015; Fialho et al. 2017) include triaminopyrimidine, studied by Brister et al. (Schuster et al. 2021; Brister et al. 2016), 2-aminooxazole (Todd et al. 2021; Szabla et al. 2013a, b), trioxopyrimidine, also known as barbituric acid, and triaminotriazine, also known as melamine (Schuster et al. 2021). Time-resolved experiments in solution have shown that the S1 excited-state lifetime of melamine in aqueous solution at neutral pH is 13 ps, or more than an order of magnitude longer than the canonical nucleobases (Improta et al. 2016; Zhang et al. 2016).

4.6 Nucleosides

According to a number of proposed synthetic scenarios, it is possible that nucleosides were synthesized directly without nucleobases as intermediate steps (Xu et al. 2017; Stairs et al. 2017). The phosphorylation of a directly synthesized nucleoside is straightforward, sidestepping the difficult problem of building a nucleotide from a free nucleobase (Xu et al. 2017; Powner et al. 2009; Stairs et al. 2017). Therefore, the photochemical prebiotic selection might have occurred at the level of nucleosides, rather than the bases.

The addition of the sugar moiety to the nucleobase chromophore can affect the excited-state dynamics in a number of interesting ways. For example, calculations by Tuna, Domcke, and Sobolewski describe a charge transfer state in two separate conformers of adenosine which intersects with the ground state and involves a proton transfer between the sugar and the base (Tuna et al. 2014; Zheng et al. 2017). The resulting conical intersections are about 0.6 eV lower in energy than those corresponding to ring puckering in the base, possibly leading to faster deexcitation and begging the question how the dynamics depends on decreasing excitation energy. It has also been speculated that following intramolecular vibrational redistribution, the system can be trapped in the excited state due to the increased number of vibrational modes in the larger compound. Calculations by Pepino et al. predict ultrafast photoinduced deactivations in water-solvated pyrimidine nucleosides (Pepino et al. 2017).

Nucleosides in solution exhibit sub-picosecond relaxation after excitation near 260 nm, typically a factor 2 shorter than those obtained for the bases, with the exception of G (Bertram et al. 2022; Canuel et al. 2005; Kang et al. 2002; Satzger et al. 2006; Peon and Zewail 2001; Pecourt et al. 2000, 2001; Zhang et al. 2014). De Camillis et al. found similar lifetimes in the gas phase for A, C, and T nucleosides and a longer time for guanosine, at 267 nm (De Camillis et al. 2015). However, these data are for a single wavelength of 267 nm, likely well above any possible barrier for internal conversion. Furthermore, the spectra were obtained for fragment ions rather than at the parent mass and there was no jet-cooling in the experiment and no option to observe longer time scale processes. There was no isomer selection, which may be crucially important as seen in the nucleobases (Abo-Riziq et al. 2007; Nir et al. 2000a, 2004), where large differences in excited-state behavior between tautomers occur. Furthermore, nucleosides can exist in at least two conformations, *anti* and *syn* (Abo-Riziq et al. 2007; Nir et al. 2000a, 2004; Asami et al. 2012). Another example, which is featured in prebiotic scenarios is inosine, which according to Kim et al. exhibits surprisingly rapid and accurate nonenzymatic RNA copying (Kim et al. 2018) while hypoxanthine is notably short lived in the excited state.

4.7 Base Pairing

Interstrand hydrogen bonding can also affect the excited-state dynamics. For example, Zeleny et al. found computationally that in the GC base pair the out of plane motions of the guanine NH_2 are restricted, affecting the associated CI (Zeleny et al. 2012). On the other hand, the base pair forms a charge transfer state along the N_1-H coordinate, which provides another fast decay trajectory (Nir et al. 2002a; Sobolewski and Domcke 2004). When base pairs form by hydrogen bonding, charge transfer states can couple via conical intersections. This process is at the heart of the model Sobolewski and Domcke used to describe findings on GC base pairing, implying a short-lived excited state for the Watson–Crick (WC) structure and long-lived excited states for non-WC structures (Sobolewski et al. 2005; Nir et al. 2000b). Other base pairs, such as GG and CC, exhibit similar strong structure dependence (Abo-Riziq et al. 2005; Plützer et al. 2003; Nir et al. 2002b, 2003). In addition, transient absorption experiments in aqueous solution also provide evidence for a similar charge transfer decay channel (Rottger et al. 2015). The role of a dark $n\pi^*$ charge transfer state in the safe deactivation of GC but not in AT was also discussed by Szkaradek et al. (2020).

Figure 4.8, based in part on work by Benner and coworkers, shows examples of base pairs formed with an alternative genetic lexicon (Georgiadis et al. 2015; Schuster et al. 2021; von Krosigk and Benner 2004; Leal et al. 2015; Benner 2004). It would, in principle, be possible to generate functional forms of RNA and DNA with different nucleobases. In (b)–(d), the electron pair donor and acceptor have been switched to form nucleobase pairs with the same triple hydrogen bonds and the same geometry as the guanine–cytosine (GC) base pair in panel (a) but with different purine and pyrimidine derivatives. Similarly, panel (f) provides the double hydrogen bonded alternative pair for A-U in panel (e). Panels (g) and (h) show examples with nucleobase analogs. Panel (g) shows the Z-P pair, composed of the two additional letters in the Hachimoji set (Krul et al. 2022). This recent eight character, genetic alphabet demonstrated by Hoshika et al. (Hoshika et al. 2019) is one of a growing number of examples of alternative nucleobases. In addition to the

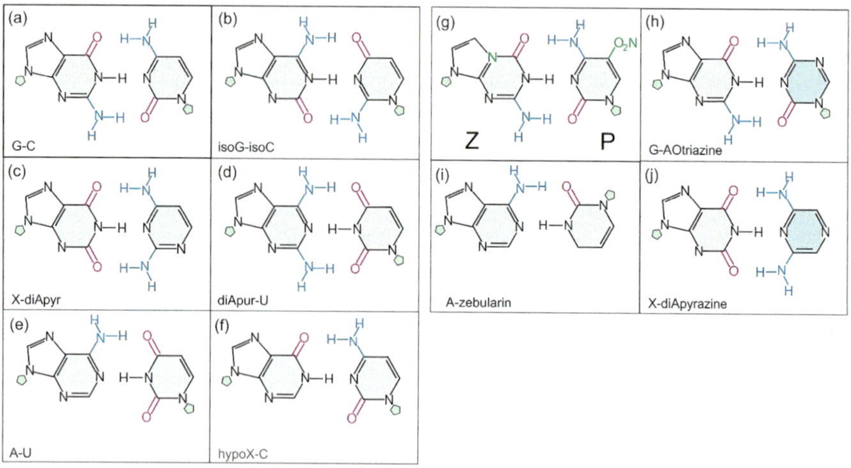

Fig. 4.8 Alternative base pair combinations

familiar C, G, A, and U genetic building blocks, this set proposes S, B, Z, and P. Here S and B are isoguanine and isocytosine, respectively. Work on isocytosine (B) suggests that the enol tautomer is significantly more stable than the keto tautomer (Berenbeim et al. 2017). It is intriguing that isocytosine behaves more like guanine, with which it is partially isosteric, than like its isomer cytosine, even though the guanine dynamics are not affected by the C5=C6 deformation available to both pyrimidine isomers. The UV-excited enol tautomers of isoguanine (S) are relatively short lived and therefore photostable. In contrast, the biologically more relevant keto forms are trapped in dark $n\pi^*$ states which are sufficiently long lived to participate in destructive photochemistry (Gate et al. 2019).

Generally, to rule out an alternative base pair, it may be enough that one of the alternative bases of a potential alternative pair is not UV-hardy, or even if both are individually, it may be that the required base pair structure is not.

4.8 Outlook

While in the field of nucleobase studies, logically the tendency is to focus on canonical bases, a critical need exists in the field of exobiology for studying plausible alternative bases and precursors. Photochemical viability of nucleobases is one of a number of parameters to be considered in prebiotic modeling, and the molecular response to UV radiation is an aspect of early chemistry that merits further study and better understanding. The sensitivity of excited-state lifetimes of nucleobases to molecular structure is one example of how UV radiation directs molecular inventories. Understanding the principles that govern whether a particular structure is UV damage resistant or not can support detailed modeling of prebiotic chemistry.

Insights into this functionality will help provide insight into the emergence of the biochemistry we know today.

Understanding the role of photochemistry may also constrain hypotheses for prebiotic conditions as well as for alternative self-replicating informational polymers that could be on exoplanets elsewhere in the universe. Knowledge of photochemical constraints is furthermore important to a wide variety of origin of life paradigms due to the generality of the interaction between prebiotic compounds and UV radiation.

References

Abo-Riziq A, Grace L, Nir E, Kabelac M, Hobza P, de Vries MS (2005) Photochemical selectivity in guanine-cytosine base-pair structures. Proc Natl Acad Sci USA 102(1):20–23

Abo-Riziq A, Crews BO, Compagnon I, Oomens J, Meijer G, Von Helden G, Kabelac M, Hobza P, de Vries MS (2007) The mid-IR spectra of 9-ethyl guanine, guanosine, and 2-Deoxyguanosine. J Phys Chem A 111(31):7529–7536

Arpa EM, Brister MM, Hoehn SJ, Crespo-Hernandez CE, Corral I (2020) On the origin of the photostability of DNA and RNA monomers: excited state relaxation mechanism of the pyrimidine chromophore. J Phys Chem Lett:5156–5161

Arrowsmith P, de Vries MS, Hunziker HE, Wendt HR (1988) Pulsed laser desorption near a jet orifice: concentration profiles of entrained perylene vapor. Applied Physics B 46:165–173

Asami H, Urashima S, Tsukamoto M, Motoda A, Hayakawa Y, Saigusa H (2012) Controlling glycosyl bond conformation of guanine nucleosides: stabilization of the anti conformer in 5 '-O-Ethylguanosine. J Phys Chem Lett 3(5):571–575

Barbatti M, Aquino AJA, Szymczak JJ, Nachtigallova D, Hobza P, Lischka H (2010) Relaxation mechanisms of UV-photoexcited DNA and RNA nucleobases. Proc Natl Acad Sci USA 107(50):21453–21458

Barbatti M, Szymczak JJ, Aquino AJA, Nachtigallova D, Lischka H (2011) The decay mechanism of photoexcited guanine – a nonadiabatic dynamics study. J Chem Phys 134:014304

Barone V, Puzzarini C (2023) Gas-phase computational spectroscopy: the challenge of the molecular bricks of life. Annu Rev Phys Chem 74:29–52

Becker S, Thoma I, Deutsch A, Gehrke T, Mayer P, Zipse H, Carell T (2016) A high-yielding, strictly regioselective prebiotic purine nucleoside formation pathway. Science 352(6287):833–836

Becker S, Feldmann J, Wiedemann S, Okamura H, Schneider C, Iwan K, Crisp A, Rossa M, Amatov T, Carell T (2019) Unified prebiotically plausible synthesis of pyrimidine and purine RNA ribonucleotides. Science 366:76–82

Beckstead AA, Zhang YY, de Vries MS, Kohler B (2016) Life in the light: nucleic acid photoproperties as a legacy of chemical evolution. Phys Chem Chem Phys 18(35):24228–24238

Bekker A, Holland HD, Wang PL, Rumble D, Stein HJ, Hannah JL, Coetzee LL, Beukes NJ (2004) Dating the rise of atmospheric oxygen. Nature 427(6970):117–120

Benner SA (2004) Redesigning genetics. Science 306(5296):625–626

Benner SA, Karalkar NB, Hoshika S, Laos R, Shaw RW, Matsuura M, Fajardo D, Moussatche P (2016) Alternative Watson-crick synthetic genetic systems. Cold Spring Harb Perspect Med 6(9):1–26

Berenbeim JA, Boldissar S, Siouri FM, Gate G, Haggmark MR, Aboulache B, Cohen T, De Vries MS (2017) Excited state dynamics of isocytosine; a hybrid case of canonical nucleobase photodynamics. J Phys Chem Lett 8:5184–5189

Bertram L, Roberts SJ, Powner MW, Szabla R (2022) Photochemistry of 2-thiooxazole: a plausible prebiotic precursor to RNA nucleotides. Phys Chem Chem Phys

Blaser S, Trachsel MA, Lobsiger S, Wiedmer T, Frey HM, Leutwyler S (2016) Gas-phase cytosine and cytosine-N-1-derivatives have 0.1-1 ns lifetimes near the S-1 state minimum. J Phys Chem Lett 7(5):752–757

Boldissar S, de Vries MS (2018) How nature covers its bases. Phys Chem Chem Phys 20:9701–9716

Brister MM, Pollum M, Crespo-Hernandez CE (2016) Photochemical etiology of promising ancestors of the RNA nucleobases. Phys Chem Chem Phys 18(30):20097–20103

Broo A (1998) A theoretical investigation of the physical reason for the very different luminescence properties of the two isomers adenine and 2-Aminopurine. JPhys Chem A 102:526–531

Cafferty BJ, Hud NV (2015) Was a pyrimidine-pyrimidine base pair the ancestor of Watson-Crick base pairs? Insights from a systematic approach to the origin of RNA. Israel J Chem 55(8):891–905

Cafferty BJ, Gallego I, Chen MC, Farley KI, Eritja R, Hud NV (2013) Efficient self-assembly in water of long noncovalent polymers by nucleobase analogues. J Am Chem Soc 135(7):2447–2450

Cafferty BJ, Fialho DM, Hud NV (2018) Searching for possible ancestors of RNA: the self-assembly hypothesis for the origin of proto-RNA. In: Prebiotic Chemistry and Chemical Evolution of Nucleic Acids, pp 143–174

Canuel C, Mons M, Piuzzi F, Tardivel B, Dimicoli I, Elhanine M (2005) Excited states dynamics of DNA and RNA bases: characterization of a stepwise deactivation pathway in the gas phase. J Chem Phys 122(7):074316–074317

Chen JQ, Kohler B (2012) Ultrafast nonradiative decay by hypoxanthine and several methylxanthines in aqueous and acetonitrile solution. Phys Chem Chem Phys 14(30):10677–10682

Cnossen I, Sanz-Forcada J, Favata F, Witasse O, Zegers T, Arnold NF (2007) Habitat of early life: solar X-ray and UV radiation at earth's surface 4-3.5 billion years ago. J Geophys Res-Planets 112(E2):1–10

Cockell CS (1998) Ultraviolet radiation, evolution and the pi-electron system. Biol J Linn Soc 63(3):449–457

Cockell CS (2000) Ultraviolet radiation and the photobiology of earth's early oceans. Orig Life Evol Biosph 30(5):467–499

Crespo-Hernandez CE, Cohen B, Hare PM, Kohler B (2004) Ultrafast excited-state dynamics in nucleic acids. Chem Rev 104(4):1977–2019

Crespo-Hernandez CE, Cohen B, Kohler B (2005) Base stacking controls excited-state dynamics in A-T DNA. Nature 436(7054):1141–1144

Cuellar-Zuquin J, Pepino AJ, Fdez Galvan I, Rivalta I, Aquilante F, Garavelli M, Lindh R, Segarra-Marti J (2023) Characterizing conical intersections in DNA/RNA Nucleobases with Multiconfigurational wave functions of varying active space size. J Chem Theory Comput 19(22):8258–8272

De Camillis S, Miles J, Alexander G, Ghafur O, Williams ID, Townsend D, Greenwood JB (2015) Ultrafast non-radiative decay of gas-phase nucleosides. Phys Chem Chem Phys 17:23643–23650

De Vries MS (2014) UV-excitation from an experimental perspective I: frequency resolved. In: Barbatti M, Borin A, Ullrich S (eds) Photoinduced phenomena in nucleic acids. Springer International Publishing, Cham

de Vries MS (2015) Gas-phase IR spectroscopy of Nucleobases. In: Rijs AM, Oomens J (eds) Gas-phase IR spectroscopy for the structural characterization of biomolecules. Springer, pp 271–298

Epifanovsky E, Kowalski K, Fan PD, Valiev M, Matsika S, Krylov AI (2008) On the electronically excited states of uracil. J Phys Chem A 112(40):9983–9992

Fialho D, Clarke K, Moore M, Schuster G, Krishnamurthy R, Hud N (2017) Glycosylation of a model proto-RNA nucleobase with non-ribose sugars: implications for the prebiotic synthesis of nucleosides. Org Biomol Chem 16:1263–1271

Frey JA, Ottiger P, Leutwyler S (2014) Watson-crick and sugar-edge base pairing of cytosine in the gas phase: UV and infrared spectra of cytosine center dot 2-Pyridone. J Phys Chem B 118(3):682–691

Gate G, Szabla R, Haggmark MR, Sponer J, Sobolewski AL, de Vries MS (2019) Photodynamics of alternative DNA base isoguanine. Phys Chem Chem Phys:13474–13485

Gate G, Williams A, Haggmark MR, Svadlenak N, Hill G, de Vries MS (2021) Nucleobases as molecular fossils of prebiotic photochemistry: excited-state dynamics of C2 and C6 substituted purines. In: Saija F, Cassone G (eds) Prebiotic photochemistry: from Urey–Miller-like experiments to recent findings. the Royal Society of Chemistry, pp 124–147

Gengeliczki Z, Callahan MP, Svadlenak N, Pongor CI, Sztaray B, Meerts L, Nachtigallova D, Hobza P, Barbatti M, Lischka H, de Vries MS (2010) Effect of substituents on the excited-state dynamics of the modified DNA bases 2,4-diaminopyrimidine and 2,6-diaminopurine. Phys Chem Chem Phys 12(20):5375–5388

Georgiadis MM, Singh I, Kellett WF, Hoshika S, Benner SA, Richards NGJ (2015) Structural basis for a six nucleotide genetic alphabet. J Am Chem Soc 137(21):6947–6955

Green NJ, Xu J, Sutherland JD (2021) Illuminating Life's origins: UV photochemistry in abiotic synthesis of biomolecules. J Am Chem Soc

Gustavsson T, Improta R, Markovitsi D (2010) DNA/RNA: building blocks of life under UV irradiation. J Phys Chem Lett 1(13):2025–2030

Hare PM, Crespo-Hernandez CE, Kohler B (2007) Internal conversion to the electronic ground state occurs via two distinct pathways for pyrimidine bases in aqueous solution. Proc Natl Acad Sci USA 104(2):435–440

Hoeh SJ, Caldero-Rodríguez NE, Crespo-Hernández CE (2022) Photochemistry of RNA, RNA monomers, and plausible prebiotic precursors. In: Improta R, Douki T (eds) DNA photodamage: from light absorption to cellular responses and skin cancer. the Royal Society of Chemistry

Hoshika S, Leal NA, Kim MJ, Kim MS, Karalkar NB, Kim HJ, Bates AM, Watkins NE, SantaLucia HA, Meyer AJ, DasGupta S, Piccirilli JA, Ellington AD, SantaLucia J, Georgiadis MM, Benner SA (2019) Hachimoji DNA and RNA: a genetic system with eight building blocks. Science 363(6429):884-+

Improta R, Santoro F, Blancafort L (2016) Quantum mechanical studies on the photophysics and the photochemistry of nucleic acids and nucleobases. Chem Rev 116(6):3540–3593

Kang H, Lee KT, Jung B, Ko YJ, Kim SK (2002) Intrinsic lifetimes of the excited state of DNA and RNA bases. J Am Chem Soc 124(44):12958–12959

Kim SC, O'Flaherty DK, Zhou LJ, Lelyveld VS, Szostak JW (2018) Inosine, but none of the 8-oxo-purines, is a plausible component of a primordial version of RNA. Proc Natl Acad Sci USA 115(52):13318–13323

Kistler KA, Matsika S (2007) Radiationless decay mechanism of cytosine: an ab initio study with comparisons to the fluorescent analogue 5-methyl-2-pyrimidinone. J Phys Chem A 111(14):2650–2661

Kistler KA, Matsika S (2008) Three-state conical intersections in cytosine and pyrimidinone bases. J Chem Phys 128(21):215102–215114

Kleinermanns K, Nachtigallova D, de Vries MS (2013) Excited state dynamics of DNA bases. Int Rev Phys Chem 32(2):308–342

Kolb VM, Dworkin JP, Miller SL (1994) Alternative bases in the RNA world - the prebiotic synthesis of urazole and its ribosides. J Mol Evol 38(6):549–557

Krul SE, Costa GJ, Hoehn SJ, Valverde D, Oliveira LMF, Borin AC, Crespo-Hernandez CE (2022) Resolving ultrafast photoinitiated dynamics of the Hachimoji 5-aza-7-deazaguanine nucleobase: impact of synthetically expanding the genetic alphabet(1). Photochem Photobiol 99(2):693–705

Leal NA, Kim HJ, Hoshika S, Kim MJ, Carrigan MA, Benner SA (2015) Transcription, reverse transcription, and analysis of RNA containing artificial genetic components. ACS Synth Biol 4(4):407–413

Ligare M, Siouri F, Bludsky O, Nachtigallova D, de Vries MS (2015) Characterizing the dark state in thymine and uracil by double resonant spectroscopy and quantum computation. Phys Chem Chem Phys 17:24336–24340

Lobsiger S, Leutwyler S (2012) The adiabatic ionization energy and triplet T-1 energy of jet-cooled Keto-amino cytosine. J Phys Chem Lett 3(23):3576–3580

Lobsiger S, T'rachsel MA, Frey HM, Leutwyler S (2013) Excited-state structure and dynamics of Keto-amino cytosine: the $1\pi\pi^*$ state is nonplanar and its radiation less decay is not ultrafast. J Phys Chem B 117(20):6106–6115

Lobsiger S, Blaser S, Sinha RK, Frey HM, Leutwyler S (2014a) Switching on the fluorescence of 2-aminopurine by site-selective microhydration. Nat Chem 6(11):989–993

Lobsiger S, Trachsel MA, Den T, Leutwyler S (2014b) Excited-state structure, vibrations, and nonradiative relaxation of jet-cooled 5-Fluorocytosine. J Phys Chem B 118(11):2973–2984

Lobsiger S, Etinski M, Blaser S, Frey HM, Marian C, Leutwyler S (2015) Intersystem crossing rates of S-1 state keto-amino cytosine at low excess energy. J Chem Phys 143(23):23430101–23430112

Luhrs DC, Viallon J, Fischer I (2001) Excited state spectroscopy and dynamics of isolated adenine and 9-methyladenine. Phys Chem Chem Phys 3:1827–1831

Marian CM (2007) The guanine tautomer puzzle: quantum chemical investigation of ground and excited states. J Phys Chem A 111(8):1545–1553

Martinez-Fernandez L, Arslancan S, Ivashchenko D, Crespo-Hernandez CE, Corral I (2019) Tracking the origin of photostability in purine nucleobases: the photophysics of 2-oxopurine. Phys Chem Chem Phys 21:13467–13473

Matsika S (2005) Three-state conical intersections in nucleic acid bases. J Phys Chem A 109(33):7538–7545

Mburu E, Matsika S (2008) An ab initio study of substituent effects on the excited states of purine derivatives. J Phys Chem A 112(48):12485–12491

Meijer G, de Vries MS, Hunziker HE, Wendt HR (1990) Laser desorption jet-cooling of organic-molecules: cooling characteristics and detection sensitivity. Appl Phys B-Photophysics Laser Chem 51(6):395–403

Middleton CT, de La Harpe K, Su C, Law YK, Crespo-Hernandez CE, Kohler B (2009) DNA excited-state dynamics: from single bases to the double helix. Annu Rev Phys Chem 60:217–239

Nachtigallova D, Lischka H, Szymczak JJ, Barbatti M, Hobza P, Gengeliczki Z, Pino G, Callahan MP, de Vries MS (2010) The effect of C5 substitution on the photochemistry of uracil. Phys Chem Chem Phys 12(19):4924–4933

Nir E, Imhof P, Kleinermanns K, de Vries MS (2000a) REMPI spectroscopy of laser desorbed guanosines. J Am Chem Soc 122(33):8091–8092

Nir E, Kleinermanns K, de Vries MS (2000b) Pairing of isolated nucleic-acid bases in the absence of the DNA backbone. Nature 408(6815):949–951

Nir E, Janzen C, Imhof P, Kleinermanns K, de Vries MS (2002a) Pairing of the nucleobase guanine studied by IR-UV double-resonance spectroscopy and ab initio calculations. Phys Chem Chem Phys 4(5):740–750

Nir E, Janzen C, Imhof P, Kleinermanns K, de Vries MS (2002b) Pairing of the nucleobases guanine and cytosine in the gas phase studied by IR-UV double-resonance spectroscopy and ab initio calculations. Phys Chem Chem Phys 4(5):732–739

Nir E, Hünig I, Kleinermanns K, de Vries MS (2003) The nucleobase cytosine and the cytosine dimer investigated by double resonance laser spectroscopy and ab initio calculations. Phys Chem Chem Phys 5(21):4780–4785

Nir E, Hünig I, Kleinermanns K, de Vries MS (2004) Conformers of guanosines and their vibrations in the electronic ground and excited states, as revealed by double-resonance spectroscopy and ab initio calculations. ChemPhysChem 5(1):131–137

Pecourt JML, Peon J, Kohler B (2000) Ultrafast internal conversion of electronically excited RNA and DNA nucleosides in water. J Am Chem Soc 122(38):9348–9349

Pecourt JML, Peon J, Kohler B (2001) DNA excited-state dynamics: ultrafast internal conversion and vibrational cooling in a series of nucleosides. J Am Chem Soc 123(42):10370–10378

Peeters Z, Botta O, Charnley SB, Ruiterkamp R, Ehrenfreund P (2003) The astrobiology of nucleobases. Astrophys J 593(2):L129–L132

Peon J, Zewail AH (2001) DNA/RNA nucleotides and nucleosides: direct measurement of excited-state lifetimes by femtosecond fluorescence up-conversion. Chem Phys Lett 348(3–4):255–262

Pepino AJ, Segarra-Marti J, Nenov A, Improta R, Garavelli M (2017) Resolving ultrafast photoinduced deactivations in water-solvated pyrimidine nucleosides. J Phys Chem Lett 8(8):1777–1783

Perun S, Sobolewski AL, Domcke W (2005) Ab initio studies on the radiationless decay mechanisms of the lowest excited singlet states of 9H-adenine. J Am Chem Soc 127(17):6257–6265

Picconi D, Ferrer FJA, Improta R, Lami A, Santoro F (2013) Quantum-classical effective-modes dynamics of the $\pi\pi^*$ -> $n\pi^*$ decay in 9H-adenine. A quadratic vibronic coupling model. Faraday Discuss 163:223–242

Plasser F, Crespo-Otero R, Pederzoli M, Pittner J, Lischka H, Barbatti M (2014) Surface hopping dynamics with correlated single-reference methods: 9H-adenine as a case study. J Chem Theory Comput 10(4):1395–1405

Plützer C, Hünig I, Kleinermanns K, Nir E, de Vries MS (2003) Pairing of isolated nucleobases: double resonance laser spectroscopy of adenine-thymine. ChemPhysChem 4(8):838–842

Powner MW, Gerland B, Sutherland JD (2009) Synthesis of activated pyrimidine ribonucleotides in prebiotically plausible conditions. Nature 459(7244):239–242

Prokhorenko VI, Picchiotti A, Pola M, Dijkstra AG, Miller RJ (2016) New insights into the photophysics of DNA nucleobases. J Phys Chem Lett 7(22):4445–4450

Ranjan S, Sasselov DD (2016) Influence of the UV environment on the synthesis of prebiotic molecules. Astrobiology 16(1):68–88

Ranjan S, Sasselov DD (2017) Constraints on the early terrestrial surface UV environment relevant to prebiotic chemistry. Astrobiology 17(3):169–204

Ribas I, de Mello GFP, Ferreira LD, Hebrard E, Selsis F, Catalan S, Garces A, do Nascimento JD, de Medeiros JR (2010) Evolution of the solar activity over time and effects on planetary atmospheres. II. kappa(1) ceti, an analog of the sun when life arose on earth. Astrophys J 714(1):384–395

Rios AC, Tor Y (2013) On the origin of the canonical nucleobases: an assessment of selection pressures across chemical and early biological evolution. Israel J Chem 53(6–7):469–483

Roberts SJ, Szabla R, Todd ZR, Stairs S, Bucar DK, Sponer J, Sasselov DD, Powner MW (2018) Selective prebiotic conversion of pyrimidine and purine anhydronucleosides into Watson-crick base-pairing arabino-furanosyl nucleosides in water. Nat Commun 9:1–10

Rottger K, Marroux HJB, Grubb MP, Coulter PM, Bohnke H, Henderson AS, Galan MC, Temps F, Orr-Ewing AJ, Roberts GM (2015) Ultraviolet absorption induces hydrogen-atom transfer in G center dot C Watson-crick DNA Base pairs in solution. Angew Chem-Int Ed 54(49):14719–14722

Sagan C (1973) Ultraviolet selection pressure on earliest organisms. J Theor Biol 39(1):195–200

Saladino R, Carota E, Botta G, Kapralov M, Timoshenko GN, Rozanov AY, Krasavin E, Di Mauro E (2015) Meteorite-catalyzed syntheses of nucleosides and of other prebiotic compounds from formamide under proton irradiation. Proc Natl Acad Sci USA 112(21):E2746–E2755

Satzger H, Townsend D, Zgierski MZ, Patchkovskii S, Ullrich S, Stolow A (2006) Primary processes underlying the photostability of isolated DNA bases: adenine. Proc Natl Acad Sci USA 103(27):10196–10201

Schneider M, Hain T, Fischer I (2009) Resonance-enhanced multiphoton ionisation of purine. ChemPhysChem 10(4):634–636

Schuster GB, Cafferty BJ, Karunakaran SC, Hud NV (2021) Water-soluble supramolecular polymers of paired and stacked heterocycles: assembly, structure, properties, and a possible path to pre-RNA. J Am Chem Soc 143:9279–9296

Serrano-Andres L, Merchan M, Borin AC (2006) Adenine and 2-aminopurine: paradigms of modern theoretical photochemistry. Proc Natl Acad Sci USA 103(23):8691–8696

Siouri FM, Boldissar S, Berenbeim JA, de Vries MS (2017) Excited state dynamics of 6-Thioguanine. J Phys Chem A 121(28):5257–5266

Sobolewski AL, Domcke W (2004) Ab initio studies on the photophysics of the guanine–cytosine base pair. Phys Chem Chem Phys 6:2763–2771

Sobolewski AL, Domcke W, Hättig C (2005) Tautomeric selectivity of the excited-state lifetime of guanine/cytosine base pairs: the role of electron-driven proton-transfer processes. PNAS 102:17903–17906

Stairs S, Nikmal A, Bucar DK, Zheng SL, Szostak JW, Powner MW (2017) Divergent prebiotic synthesis of pyrimidine and 8-oxo-purine ribonucleotides. Nat Commun 8:15270

Szabla R (2021) Rethinking UV-induced prebiotic selection of biomolecules. In: Saija F, Cassone G (eds) Prebiotic photochemistry: from Urey–Miller-like experiments to recent findings. the Royal Society of Chemistry, p 79

Szabla R, Sponer JE, Sponer J, Gora RW (2013a) Theoretical studies of the mechanism of 2-aminooxazole formation under prebiotically plausible conditions. Phys Chem Chem Phys 15(20):7812–7818

Szabla R, Tuna D, Gora RW, Sponer J, Sobolewski AL, Domcke W (2013b) Photochemistry of 2-Aminooxazole, a hypothetical prebiotic precursor of RNA nucleotides. J Phys Chem Lett 4(16):2785–2788

Szkaradek KE, Stadlbauer P, Šponer J, Góra RW, Szabla R (2020) UV-induced hydrogen transfer in DNA base pairs promoted by dark $n\pi^*$ states. Chem Commun 56:201–204

Teichert JS, Kruse FM, Trapp O (2019) Direct prebiotic pathway to DNA nucleosides. Angew Chem-Int Ed 58(29):9944–9947

Todd ZR, Szostak JW, Sasselov DD (2021) Shielding from UV photodamage: implications for surficial origins of life chemistry on the early earth. ACS Earth Space Chem 5(2):239–246

Tor Y, Del Valle S, Jaramillo D, Srivatsan SG, Rios A, Weizman H (2007) Designing new isomorphic fluorescent nucleobase analogues: the thieno[3,2-d]pyrimidine core. Tetrahedron 63(17):3608–3614

Trachsel MA, Wiedmer T, Blaser S, Frey HM, Li QS, Ruiz-Barragan S, Blancafort L, Leutwyler S (2016) The excited-state structure, vibrations, lifetimes, and nonradiative dynamics of jet-cooled 1-methylcytosine. J Chem Phys 145(13):13430701–13430715

Trachsel MA, Lobsiger S, Schar T, Blancafort L, Leutwyler S (2017) Planarizing cytosine: the S-1 state structure, vibrations, and nonradiative dynamics of jet-cooled 5,6-trimethylenecytosine. J Chem Phys 146(24):24430801–24430813

Tuna D, Sobolewski AL, Domcke W (2014) Mechanisms of ultrafast excited-state deactivation in adenosine. J Phys Chem A 118(1):122–127

von Krosigk U, Benner SA (2004) Expanding the genetic alphabet: pyrazine nucleosides that support a donor-donor-acceptor hydrogen-bonding pattern. Helv Chim Acta 87(6):1299–1324

Xu J, Tsanakopoulou M, Magnani CJ, Szabla R, Sponer JE, Sponer J, Gora RW, Sutherland JD (2017) A prebiotically plausible synthesis of pyrimidine beta-ribonucleosides and their phosphate derivatives involving photoanomerization. Nat Chem 9(4):303–309

Xu JF, Green NJ, Gibard C, Krishnamurthy R, Sutherland JD (2019) Prebiotic phosphorylation of 2-thiouridine provides either nucleotides or DNA building blocks via photoreduction. Nat Chem 11(5):457–462

Xu J, Chmela V, Green NJ, Russell DA, Janicki MJ, Gora RW, Szabla R, Bond AD, Sutherland JD (2020) Selective prebiotic formation of RNA pyrimidine and DNA purine nucleosides. Nature 582(7810):60–66

Yadav M, Kumar R, Krishnamurthy R (2020) Chemistry of abiotic nucleotide synthesis. Chem Rev 120(11):4766–4805

Yamazaki S, Domcke W, Sobolewski AL (2008) Nonradiative decay mechanisms of the biologically relevant tautomer of guanine. J Phys Chem A 112(47):11965–11968

Yang ZY, Hutter D, Sheng PP, Sismour AM, Benner SA (2006) Artificially expanded genetic information system: a new base pair with an alternative hydrogen bonding pattern. Nucleic Acids Res 34(21):6095–6101

Zeleny T, Ruckenbauer M, Aquino AJA, Muller T, Lankas F, Drsata T, Hase WL, Nachtigallova D, Lischka H (2012) Strikingly different effects of hydrogen bonding on the photodynamics of individual nucleobases in DNA: comparison of guanine and cytosine. J Am Chem Soc 134(33):13662–13669

Zhang YY, Improta R, Kohler B (2014) Mode-specific vibrational relaxation of photoexcited guanosine 5 '-monophosphate and its acid form: a femtosecond broadband mid-IR transient absorption and theoretical study. Phys Chem Chem Phys 16(4):1487–1499

Zhang YY, Beckstead AA, Hu YS, Piao XJ, Bong D, Kohler B (2016) Excited-state dynamics of melamine and its lysine derivative investigated by femtosecond transient absorption spectroscopy. Molecules 21(12):1645–1656

Zheng D, Zhang M, Zhao G (2017) Combined TDDFT and AIM insights into photoinduced excited state intramolecular proton transfer (ESIPT) mechanism in hydroxyl- and amino-anthraquinone solution. Sci Rep 7(1):13766

Zhou Z, Zhou X, Wang X, Jiang B, Li Y, Chen J, Xu J (2017) Ultrafast excited-state dynamics of cytosine Aza-derivative and analogues. J Phys Chem A 121(14):2780–2789

Chapter 5
Excited-State Dynamics in Silver-DNA Assemblies

Supriya Ghosh and Bern Kohler

Abstract The already rich supramolecular chemistry of DNA becomes even richer when silver(I) ions are added to nucleobase monomers or DNA strands. Silver(I) ions coordinate preferably to nitrogen atoms on the nucleobases, giving rise to supramolecular nanostructures through the formation of nucleobase coordination polymers and silver-ion-mediated base pairs (silver base pairs for short). These structures form by self-assembly in nearly activationless reactions. Silver-DNA assemblies have promising photoproperties, but realizing their potential requires understanding how their atomistic structures and the covalent and non-covalent interactions among their subunits shape their photophysics. In this chapter, we provide an overview of the diverse supramolecular structures that form when silver(I) ions are combined with DNA nucleobases and DNA strands. We then describe our initial studies of some representative silver-DNA assemblies using ultrafast spectroscopy. Improved structural understanding of silver-DNA assemblies together with the experimental and computational approaches used to study excited states in regular DNA systems during the past 25 years creates a promising research landscape for seeking first-principles understanding of photoproperty emergence in silver-DNA assemblies, which are of great interest for DNA nanotechnology.

Keywords Metal ion-mediated base pairs · DNA nanowires and nanotechnology · Photoluminescence · Coordination polymers · Self-assembled nanostructures

S. Ghosh · B. Kohler (✉)
Department of Chemistry and Biochemistry, The Ohio State University, Columbus, OH, USA
e-mail: kohler@chemistry.ohio-state.edu

S. Matsika, A. H. Marcus (eds.), *Nucleic Acid Photophysics and Photochemistry*, Nucleic Acids and Molecular Biology 36,
https://doi.org/10.1007/978-3-031-68807-2_5

5.1 Introduction

Silver ions are in a class of metal ions that form directional and moderately strong coordinate bonds to nucleobases (Zhou et al. 2017). This allows them to direct structure formation in ways that are not possible for metal ions that interact only with the charged phosphate groups of DNA's sugar-phosphate backbone through non-directional, Coulombic interactions. The latter metal ions include alkali metal ions like Na^+ and K^+ that are the ubiquitous counterions for regular DNA strands. The ability of silver ions to form mononuclear and polynuclear coordination complexes with nucleobases leads to a wide range of supramolecular structures when silver ions are combined with nucleobase monomers or DNA strands in aqueous solution. For simplicity, we will refer to these structures collectively as "silver-DNA assemblies," even when they are made from components that lack the sugar-phosphate backbone of true DNA strands. Representative structures to be discussed are summarized in Fig. 5.1. Hereafter, we will use the term metal-free DNA (or silver-free DNA) as shorthand to refer to conventional DNA structures without coordinating metal (silver) ions.

The synthesis of silver-DNA assemblies could not be much simpler. Like other metal–organic assemblies, they form spontaneously by self-assembly without catalysts (Vohlídal and Hissler 2019). A typical synthesis is performed by adding a small volume of a concentrated aqueous silver salt solution to an aqueous solution of the DNA system of interest, stirring briefly, and possibly followed by thermal annealing of the solution. Metal–DNA assemblies can be classified into structures formed by mixing DNA strands (Fig. 5.1a–d) or monomeric nucleobases (Fig. 5.1e–g) with a silver salt like $AgNO_3$. Atomistic structures of the former can be found as PDB entries in the Protein Data Bank (wwPDB consortium 2019), while crystal structures of the latter are located in the Cambridge Structural Database using a Cambridge Crystallographic Data Centre (CCDC) reference number (Groom et al. 2016).

Metal–DNA assemblies like the silver-DNA assemblies that are the focus of this chapter are of great interest for DNA nanotechnology and bioimaging on account of their novel photophysical and other properties (Copp et al. 2015; Eidelshtein et al. 2016; Müller 2019; Sun et al. 2022), which are almost endlessly tunable. Silver-DNA assemblies made from DNA strands can acquire metal-like character (Fig. 5.1a–d), while metal-organic solids made from Ag^+ ions and nucleobases (Fig. 5.1e–g) can have DNA-like attributes. Some metal–nucleobase crystals form nucleobase-incorporated metal-organic frameworks (MOFs) (Zhang et al. 2014; Thomas-Gipson et al. 2015; Lopez and Liu 2017; Anderson et al. 2019; Mistry et al. 2021; Gupta et al. 2022). Like MOFs made from non-nucleobase linkers, some of these crystals have high porosity with the added feature that their DNA functional groups allow them to recognize and interact with biomolecules. Adenine-based MOFs built with cobalt ions (An et al. 2010) or Ni^{2+} ions (Stylianou et al. 2011) have excellent CO_2 uptake characteristics.

Silver ions interact not just with nucleobases but also with themselves (argentophilic interactions) and these combined interactions give rise to spectacular new

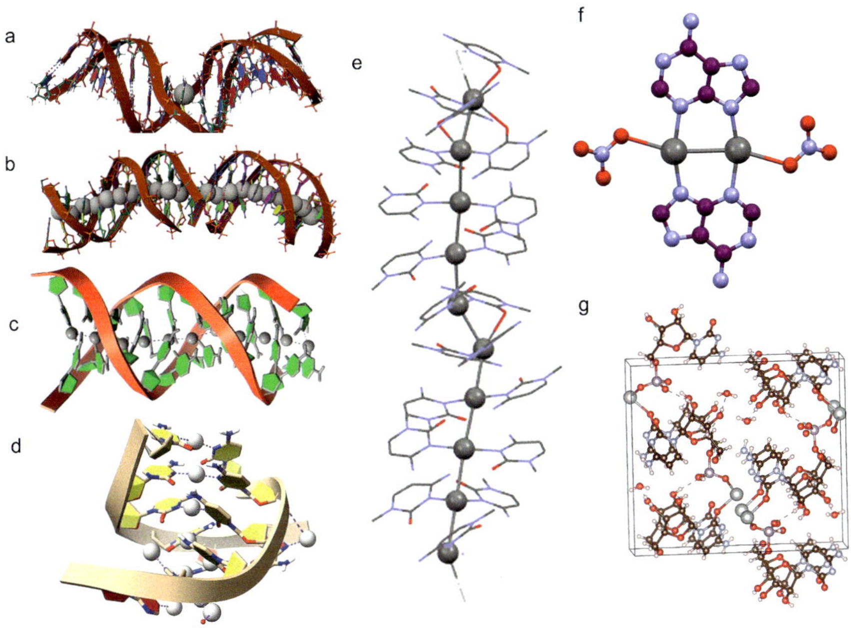

Fig. 5.1 Atomistic structures of typical silver-DNA assemblies ordered from more DNA-like to less DNA-like structures. All structures were determined by X-ray diffraction except for ones in panels a (a nuclear magnetic resonance (NMR) structure) and c (molecular dynamics (MD) simulation). Shown are (**a**) a B-form DNA duplex, PDB 2RVP (Dairaku et al. 2016), (**b**) a silver-DNA nanorod, PDB 7XKM (Atsugi et al. 2022), (**c**) a parallel-stranded $(Ag)_{10}dC_{20}$ duplex (Chen et al. 2017), (**d**) a DNA-encapsulated 8-atom silver nanocluster, PDB 6NIZ (Huard et al. 2019), (**e**) a left-handed helical complex made of Ag^+ ions and N1-methylcytosine, CCDC reference number 1847831 (Linares et al. 2019), (**f**) a dinuclear silver complex with protonated adenines and nitrate ions, CCDC reference number 772592 (Mishra et al. 2010), and (**g**) a silver–cytidine metal–organic framework, CCDC reference number (Mistry et al. 2021)

structures like DNA nanowires (Fig. 5.1b, c, e) and nanoclusters (Fig. 5.1d). DNA nanowires form when every base pair in a duplex captures a metal ion, creating an atomically thin wire or one-dimensional array of ions (Kondo et al. 2017; Swasey et al. 2018; Vecchioni et al. 2019; Atsugi et al. 2022). Metal-DNA assemblies are also the basis for novel silver ion sensors (Shi et al. 2015; Kang et al. 2016; Xie et al. 2023).

Silver-DNA assemblies often exhibit intriguing luminescence. Particularly noteworthy are the DNA-encapsulated silver clusters (Fig. 5.1d). DNA oligonucleotides act as three-dimensional ligands that organize small numbers of silver ions into compact clusters that emit brightly at visible wavelengths after chemical reduction (Petty et al. 2004; Copp et al. 2015; Gwinn et al. 2015; Gonzàlez-Rosell et al. 2021). Unfortunately, predicting and optimizing desirable photophysical properties in these and other silver-DNA assemblies is extremely challenging. On the one hand, the complex supramolecular chemistry of metal–DNA assemblies and their flexible and diverse packing interactions create a rich materials space. On the other hand, promiscuous and spontaneous self-assembly reactions make structural characterization difficult. The problem is particularly acute for solution-phase samples for which the underlying structures may differ significantly from known crystal structures.

Combining coordinating metal ions like Ag^+ ions and DNA can affect excited states of the latter in three main ways. First, bound metal ions can give rise to excitations with metal character such as metal-to-ligand charge transfer (MLCT) and ligand-to-metal charge transfer (LMCT) excited states. Second, a bound metal ion can enhance intersystem crossing (ISC) by introducing substantial spin–orbit coupling through the heavy atom effect (Sasikumar et al. 2020). The resulting triplet excited states can initiate desirable or undesirable photochemical reactions or be useful for creating luminescent materials. Third, metal ions can bring nucleobases together, changing their mutual electronic coupling. This allows metal ions to create new excited states by spatially organizing the bases in motifs that give rise to new excited states through excitonic and charge transfer (CT) interactions. These metal ion-induced structural changes can alter absorption and emission spectra as well as excited-state lifetimes and decay mechanisms.

In this chapter, we review our work using steady-state and time-resolved spectroscopy to investigate the excited-state dynamics of some representative silver-DNA assemblies. The supramolecular systems that we have studied are formed by combining a simple silver(I) salt like $AgNO_3$ with either nucleobase monomers (Snyder et al. 2019) or defined-sequence DNA oligonucleotides (Kohl et al. 2020; Martínez-Fernández et al. 2024). Our aim is to understand the nature of the excited states and their deactivation pathways to advance first-principles understanding of structure–photoproperty relationships. When possible, we have also sought insight from high-level quantum chemical modeling.

Even without coordinating metal ions, the supramolecular chemistry of DNA can be customized in myriad ways by changing the base sequence, incorporating modified nucleobases or other molecules, or altering the temperature or ionic strength. DNA strands can associate into duplex, triplex, and tetraplex structures through hydrogen bonding (base pairing) between nucleobases. The strands that associate through hydrogen bonding can adopt parallel or antiparallel orientations. The strands do not even need to be separate molecules and intramolecular folding can bring two (base pairs), three (base triples), or four (base quartets) nucleobases together to create hairpins and higher-order structures. Importantly, DNA structures are polymorphic and dynamic. For example, DNAs with the same base sequence can form multiple duplex (double-stranded) structures (e.g., A-, B-, and Z-DNA), depending on conditions such as temperature, ionic strength, and the level of hydration. Silver-DNA assemblies share this dynamism and expand the space of possible structures. As discussed below, one of the most interesting aspects is the ability of coordinating metal ions to assemble monomeric nucleobases into hierarchical assemblies. This contrasts with the behavior of metal-free DNA where nucleobases must generally be built into strands before they can assemble by hydrogen bonding, hydrophobic, and other forces.

The first papers to describe metal ions that bind directly and strongly to the DNA nucleobases appeared in the 1950s. In 1952, Katz reported large changes in the viscosity of calf thymus DNA solutions with added Hg^{2+} ions (Katz 1952). To induce similar changes with alkali metal ions required much higher concentrations. Ten years later, Yamane and Davidson reported that Ag^{+} ions bind strongly to several natural DNAs (Yamane and Davidson 1962). Complexes formed between Ag^{+} ions and monomer purine nucleobases were studied subsequently (Gillen et al. 1964). These and other early studies showed that adding certain transition metal ions leads to significant changes in the UV-vis absorption spectra (Thomas 1954; Yamane and Davidson 1962; Daune et al. 1966; Tu and Reinosa 1966; Weser and Dönnicke 1970) The release of protons upon metal ion binding to strands (Yamane and Davidson 1961, 1962) and single nucleobases (Gillen et al. 1964) was also noted. Together, these observations provide evidence that metal ions bind directly to the nucleobases and not to phosphate ions.

Work from the 1950s and 1960s on metal-DNA assemblies formed from DNA strands was limited to the higher molecular weight DNA samples available at the time such as natural calf thymus DNA. The advent of automated phosphoramidite DNA synthesizers allowed later investigators to study how coordinating metal ions interact with defined-sequence oligonucleotides. This led to a renaissance of interest in metallo-base pairs in the first years of this century (Atwell et al. 2001; Clever et al. 2007). The better-defined assemblies also made it possible to obtain the first crystal structures of metal ions bound to DNA strands containing non-canonical (Atwell et al. 2001; Johannsen et al. 2010) and canonical bases (Kondo et al. 2015, 2017; Atsugi et al. 2022).

The remainder of this chapter is organized as follows. We first discuss the building blocks of silver-DNA assemblies: Silver-ion-mediated base pairs and coordination polymers. Not surprisingly, the richest structural information comes from X-ray diffraction analyses, so we highlight some crystal structures obtained for representative silver-DNA assemblies formed when metal ions react with nucleobase monomers or strands. Because spectroscopic measurements using steady-state and time-resolved spectroscopy are usually carried out on liquid samples, we discuss the challenges of structure determination in solution and the evidence that structural motifs similar to those seen in crystals are present in solution. Next, we summarize our studies of silver-DNA assemblies made from purine nucleobase monomers or all-cytosine-containing DNA strands, discussing what excited-state dynamics reveal about the nexus between structural motifs and photoproperties. We end with conclusions and some suggestions for future study directions.

5.2 Silver-DNA Structures

5.2.1 Building Blocks of Silver-DNA Assemblies

Silver Coordination Polymers Silver ions bind preferably to nitrogen atoms of the nucleobases, although they can sometimes coordinate to oxygen atoms. Most nucleobases have multiple nitrogen atoms with available electron lone pairs, allowing them to form coordinate bonds simultaneously to two or more metal ions and act as di- or multitopic ligands. When the nucleobases act as bridging ligands between metal ions, polynuclear complexes known as coordination polymers can form (Amo-Ochoa and Zamora 2014; Lopez and Liu 2017). In coordination polymers, metal ions serve as nodes that assemble nucleobases into one-, two-, or three-dimensional networks of ligands. Coordination polymers that extend indefinitely in one or more dimensions are commonly seen in crystals obtained by reacting nucleobase monomers with a silver salt like $AgNO_3$.

Silver Base Pairs Complexes in which two nucleobases are arranged in an approximately linear coordination geometry around a Ag^+ ion are especially important as these can closely mimic native base pairs both in size and hydrogen-bonding possibilities. Silver-ion-mediated base pairs (hereafter, we will refer simply to *silver base pairs*) can replace or co-exist alongside natural base pairs in DNA and DNA-like structures. Like natural base pairs, they can assemble through π–π stacking interactions. They can orient nucleobase functional groups into the major or minor grooves, enabling the specific recognition of other molecules. Notably, silver base pairs can be so similar to natural ones that they can be processed by enzymes such as polymerases (Park et al. 2010; Urata et al. 2010; Kaul et al. 2011; Funai et al. 2012, 2020; Flamme et al. 2020) and ligases (Vecchioni et al. 2019).

Example silver base pairs are shown in Fig. 5.2. Additional details are provided in recent reviews (Müller 2019; Naskar et al. 2020; Sun et al. 2022). In most silver base pairs, a nitrogen atom on each base is symmetrically ligated to a central Ag^+ ion. To achieve this N-Ag^+-N bonding pattern, a Ag^+ ion can displace a proton in an N–H · · · N hydrogen bond. A significant number of silver base pairs were discovered by applying this strategy to known base pairs. In 1966, Jensen and Davidson proposed that a Ag^+ ion could take the place of the proton in the central N–H · · · N bond of a G·C base pair (Fig. 5.2h). This far-sighted proposal was finally validated a half century later by X-ray crystallography: Compare Fig. 5.10 in (Jensen and Davidson 1966) with Fig. 5.1d in (Kondo et al. 2017).

The strategy of replacing an N–H · · · N interaction by N-Ag^+-N is nicely illustrated for non-canonical base pairs by Polonius and Müller (2007). These authors demonstrated Ag^+ ion-mediated duplex formation between d(D_{19}A) and d(T_{20}), where D stands for 1-deaza-2′-deoxyadenosine (Polonius and Müller 2007). Because D lacks the N1 atom of A, Watson–Crick base pairing is not possible, and D must base pair with T using its Hoogsteen face (Fig. 5.2g). Polonius and Müller

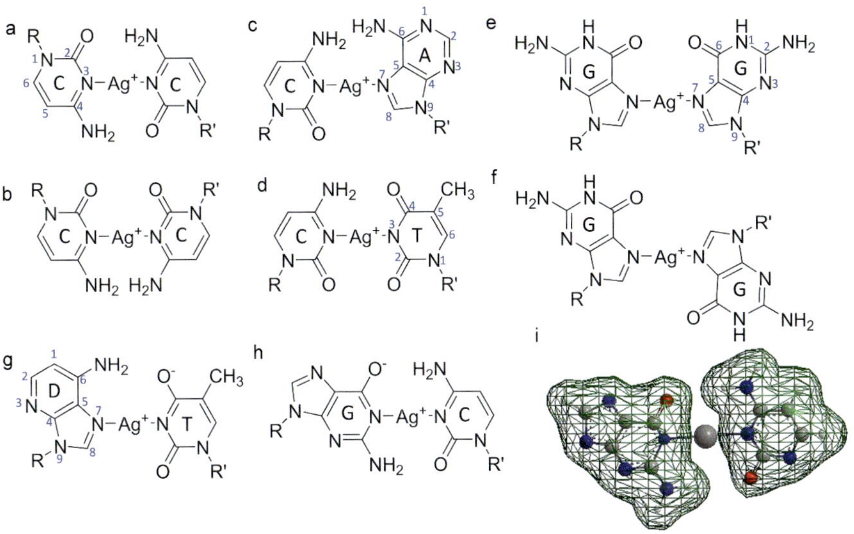

Fig. 5.2 Silver base pairs seen in crystal structures include (**a**) transoid and (**b**) cisoid C-Ag$^+$-C; (**c**) C-Ag$^+$-A; (**d**) C-Ag$^+$-T(-H$^+$)$^-$; (**e**) cisoid and (**f**) transoid G-Ag$^+$-G; (**g**) 1-deazaadenine-Ag$^+$-T(-H$^+$)$^-$; and (**h**) G(-H$^+$)-Ag$^+$-C, which is also drawn with a charge density surface map (**i**). Atom ring numbering is illustrated for various nucleobases

showed that adding one Ag$^+$ ion per D·T base pair allows D-Ag$^+$-T base pairs to form in which the N3 proton of T is displaced by a Ag$^+$ ion. Nearly identical circular dichroism (CD) spectra were recorded with and without added Ag$^+$ ions, indicating that the same secondary structure is formed. The D-Ag$^+$-T base pairs are much more stable than the D·T base pairs, and the thermal denaturation temperature of the duplex increased by an impressive 36 °C (Polonius and Müller 2007).

In some cases, it is unnecessary to displace a proton because a Ag$^+$ ion can join bare nitrogen atoms on different bases. Silver homo-base pairs form in this way as when two cytosines are connected through their N3 atoms (Fig. 5.2a, b) or two guanines are joined by N7-Ag bonds (Fig. 5.2e, f). The former base pair is particularly important for fluorescent silver nanoclusters (Sect. 5.2.2.1) and appears often in recent crystal structures. For most of the base pairs in Fig. 5.2, one of the bases can be flipped about its glycosidic bond. This is illustrated for the C-Ag$^+$-C base pair, which can be transoid (Fig. 5.2a) or cisoid (Fig. 5.2b), depending on whether the amino groups of C are on the opposite or same sides, respectively, of a line through the N3 atoms. Because of conformational preferences, the cisoid geometry is favored in antiparallel-stranded duplexes, while transoid base pairs are usually observed in parallel-stranded DNA duplexes.

Additional silver base pairs are analogs of ones that form at acidic or alkaline pH. The high pK_a of cytosine (~4) makes this nucleobase a preferred proton acceptor in acidic conditions, and protonated cytosine (CH$^+$) is present in triplex DNA and tetra-stranded i-motif DNA. In one of the two most common base triplets in triplex DNA, CH^+ hydrogen bonds to the Hoogsteen face of a GC base pair. Ihara and co-workers (Ihara et al. 2009) showed that a Ag^+ ion can displace the proton in CH^+, allowing triplex structures to form at neutral pH.

The larger size of the bound Ag^+ ion compared to the bound proton has significant structural consequences. In Fig. 5.3, a C-Ag^+-C base pair is overlayed with a C•CH^+ base pair to compare their geometries. Both were extracted from crystal structures and oriented to achieve maximal overlap of the bases shown on the right in Fig. 5.3a. The N3–N3 distance in the C•CH^+ base pair increases from 2.78 Å to 4.29 Å when the proton is replaced by the larger Ag^+ ion. This makes it impossible to close the two hydrogen bonds on either side of a line through the N3 atoms. Calculations and X-ray structures confirm that the C-Ag^+-C base pair opens and can close only one of the two possible H bonds in the C-Ag^+-C base pair (Megger et al. 2011). The loss of one or more in-plane hydrogen bonds is compensated, on the one hand, by the greater strength of the N-Ag-N bonds, and, on the other hand, by hydrogen bonds that connect bases in adjacent base pairs as will be discussed below.

In i-motif DNA (Gehring et al. 1993; Abou Assi et al. 2018), CH^+ pairs with a second cytosine to form a hemiprotonated CH^+-C base pair (Fig. 5.4c). The possibility that silver ions can fold i-motif-forming DNA strands into similar tetraplex structures was considered but rejected in 2010 (Megger and Müller 2010). However, Waller and co-workers reported a few years later that i-motif-like structures are formed by an oligonucleotide with a different base sequence in the presence of Ag^+ ions at neutral pH (Day et al. 2013). This possibility continues to be invoked often (Goncharova 2014; Xu et al. 2015; Shi et al. 2015; Kang et al. 2016; Wang et al. 2019), even though it has been questioned (Naskar et al. 2020; Kohl et al. 2020; Martínez-Fernández et al. 2024) and a crystal structure of an i-motif-like structure in which Ag^+ ions take the place of protons has never been obtained. It is worth noting that crystal structures of i-motif structures have overwhelmingly planar hemiprotonated base pairs in contrast to the strongly propeller-twisted C-Ag^+-C base pairs that are ubiquitous in parallel-stranded silver-DNAs as we describe next.

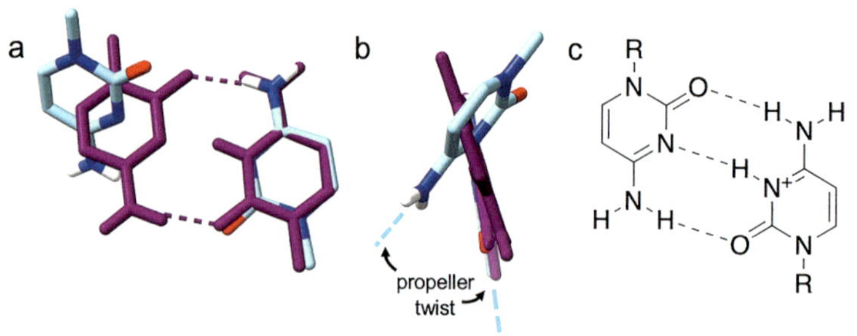

Fig. 5.3 (a) Top-down and (b) side views of an overlay of a C•CH^+ (purple) and C-Ag^+-C (colored by atoms, the Ag^+ ion is not shown) base pairs extracted from crystal structures. The C•CH^+ base pair contains residues C2 and C14 in PDB 1ELN. The C-Ag^+-C base pair is made from C5 residues on chains A and B in PDB 6NIZ. The propeller twist angle (negative in the drawing) is illustrated. (c) Structure of the hemiprotonated base pair C•CH^+

Twisted Base Pairs and Inter-Base-Pair Hydrogen Bonds A recurring motif in crystal structures that contain runs of consecutive silver base pairs is a nearly straight "wire" of silver atoms with short Ag–Ag distances. Silver nanowire formation resulting from argentophilic interactions has been seen in crystal structures of DNA duplexes with added Ag^+ ions (Kondo et al. 2017; Atsugi et al. 2022). Ag–Ag distances of less than 3.44 Å are generally considered to indicate argentophilic interactions (Schmidbaur and Schier 2015). This bonding interaction is surprising given the closed shell character of the ions and their nominal Coulombic repulsion.

C-Ag^+-C base pairs in crystal structures often have a high propeller twist angle. This angle is illustrated in Fig. 5.3b. Details about this and other standard nucleic acid conformational parameters can be found in the literature (Dickerson 1989; Olson et al. 2001). The need for each canonical Watson–Crick base pair to form two or three hydrogen bonds prevents paired bases in silver-free DNA from deviating too much from co-planarity. Some propeller twisting ($\leq 15°$) is observed in silver-free B- and (especially) A-form duplexes, where it promotes base stacking (Levitt 1978). The in-plane hydrogen bonds can still form through modest out-of-plane deformations of the hydrogen donor and acceptor groups. However, in crystal structures of silver-DNA assemblies, it is common to observe much larger propeller twist angles.

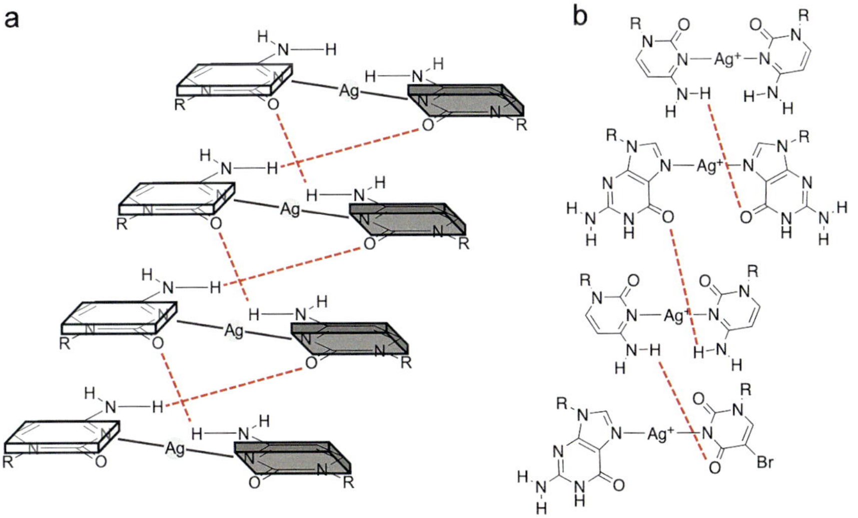

Fig. 5.4 (**a**) Schematic structure of a parallel-stranded $(Ag)_n dC_{2n}$ duplex with high and negative propeller twist angles, showing inter-base-pair hydrogen bonds (red dashed lines). The visible face of each base is colored black or white, according to the convention in Lavery et al. (1992). The 5′→3′ strand direction for both strands runs from the bottom to the top of the page. (**b**) Excerpt showing 4 base pairs from the silver-ion mediated duplex PDB 7XKM (Atsugi et al. 2022). The strands in this duplex are antiparallel with the strand on the left running in the 5′→3′ direction from top to bottom

High propeller twist angles are especially common in runs of consecutive C-Ag^+-C base pairs as can be seen in the structures shown in Fig. 5.1c–e. This geometry maintains the strong N–Ag coordinate bonds that are the source of their high thermodynamic stability. At the same time, the in-plane hydrogen bonds that can no longer form because of the size of the Ag^+ ion are replaced by cross-strand hydrogen bonds between a base on one strand and a second base on the opposite strand that is one base pair away. First identified in computer simulations (Espinosa Leal et al. 2015), these were termed interplanar hydrogen bonds by Espinosa Leal and co-workers. We prefer to call them "inter-base-pair hydrogen bonds" as this terminology, which has been used by others (Atsugi et al. 2022), more accurately describes bonds that form between (adjacent) base pairs.

Inter-base-pair hydrogen bonds have precedent in the "cross-strand diagonal hydrogen bonds" first observed in silver-free DNAs with consecutive A·T base pairs (Nelson et al. 1987). The latter generally form bifurcated, three-center hydrogen bonds (Nelson et al. 1987; Mohan and Yathindra 1991; Shibata and Zielinski 1994) as their smaller propeller twist angles allow the in-plane hydrogen bonds in A·T base pairs to be maintained. Inter-base-pair hydrogen bonding is only possible when there is significant propeller twisting in consecutive metal base pairs, allowing the donor and acceptor groups to reach each other. Propeller-twisted base pairs facilitate inter-base-pair hydrogen bonding while allowing the bases on each strand to maintain good π–π stacking interactions. Inter-base-pair hydrogen bonding is also reinforced by the favorable argentophilic interactions that are responsible for DNA nanowires (Atsugi et al. 2022).

To date, inter-base-pair hydrogen bonds in consecutive C-Ag^+-C base pairs have only been detected in silver-DNA strand assemblies when the base pairs are transoid (Huard et al. 2019), a geometry observed in parallel-stranded duplexes. In transoid C-Ag^+-C base pairs, propeller twist causes both amino groups to point in the direction of the next base pair on the 3′ side, while the two carbonyls point in the opposite direction toward the 5′ base pair (Fig. 5.4a). However, inter-base-pair hydrogen bonding is not exclusive to parallel-stranded structures and has been observed in antiparallel duplexes (Atsugi et al. 2022) (Fig. 5.4b). In the structure shown in Fig. 5.4b, cisoid C-Ag^+-C base pairs with high propeller twist are adjacent to other types of silver base pairs, allowing inter-base-pair hydrogen bonds to form, albeit with a smaller number of H bonds per nucleobase compared to the structure in Fig. 5.4a. The special thermal stability of parallel-stranded C-Ag^+-C structures (Swasey et al. 2015; Liu et al. 2017) is likely due to their higher density of hydrogen bonds. We discuss the unusual photophysics seen in these structures in Sect. 5.3.3. There is disagreement about whether similar motifs exist in parallel-stranded Ag-dG_n duplexes. Inter-base-pair hydrogen bonds were observed in calculations on a base tetramer (Chen et al. 2018) but were not seen in ab initio molecular dynamics simulations of a longer duplex of G-Ag^+-G base pairs (Swasey et al. 2018).

5.2.2 *Solid-State Structures*

5.2.2.1 Silver–DNA Oligonucleotide Assemblies

B-form Duplexes Supramolecular assemblies formed by Ag^+ ions and DNA strands (Fig. 5.1a–d) generally retain the DNA-like structural traits (e.g., helicity and the presence of stacked base pairs) of the latter. How much their structures depart from metal-free DNA structures depends on how many metal base pairs are present. When the fraction of metal base pairs is small, only modest changes are seen. For example, a duplex containing a single C-Ag^+-C base pair (Fig. 5.1a) retains B-form conformation, according to a solution NMR study (Dairaku et al. 2016). This single cisoid silver base pair is propeller twisted by −18.3° to reduce the steric clash between the amino groups (Dairaku et al. 2016). Although the secondary structure is hardly altered by this single silver base pair, the melting temperature increases by 8 °C (Ono et al. 2008). The enhanced thermal stability of silver-DNA assemblies is of great interest for DNA nanotechnology (Swasey and Gwinn 2016; Swasey et al. 2018; Bethur et al. 2024).

DNA-encapsulated Silver Clusters Combining DNA oligonucleotides of suitable base sequence with Ag^+ ions is the first step in the synthesis of DNA-stabilized few-atom metal clusters (Petty et al. 2004; New et al. 2016; Xie et al. 2020; Gonzàlez-Rosell et al. 2021; Yang et al. 2023). The DNA strand folds into a binding pocket that stabilizes a small number of silver ions. Chemically reducing the silver ion–DNA complexes with $NaBH_4$(aq) forms partially oxidized silver atom clusters with typically 12 or fewer atoms. The final metal clusters are encapsulated and stabilized by the DNA strand, preventing them from combining with other clusters to form larger silver nanoparticles. The clusters are often called "nanoclusters" even though the smallest ones are <1 nm in diameter. Their sparse energy level structure gives them molecular photophysical properties that differ dramatically from those of bulk metals. A number of investigators have identified DNA sequences that lead to excellent fluorophores with emission suitable for bioimaging and biosensing applications (Copp and Gonzàlez-Rosell 2021; Gonzàlez-Rosell et al. 2021).

Optimizing the luminescence of DNA-encapsulated silver clusters has proceeded largely empirically. A common workflow is to combine silver salts with single-stranded DNA of diverse base sequence, purify the clusters after chemical reduction, and test. However, a growing number of crystal structures has begun to appear in the last 5 years (Huard et al. 2019; Cerretani et al. 2019, 2020). This augurs well for using expanding structural insights to understand silver nanocluster photoproperties. Many of the crystal structures seen to date feature runs of consecutive C-Ag^+-C base pairs in parallel-stranded arrangements, indicating that the structure illustrated in Fig. 5.4a is an important motif for synthesizing silver nanoclusters. Although considerable evidence suggests that Ag-dC_n and Ag-dG_n assemblies form parallel-stranded duplexes in solution (Swasey et al. 2015, 2018; Swasey and Gwinn 2016), crystal structures are not yet available for these systems.

It is interesting that reducing the Ag^+ ions to Ag^0, which is necessary to produce emissive structures, does not appear to significantly change the crystal structure. For example, reducing crystals grown from Ag^+ and N1-methylcytosine by cold hydrogen plasma does not change the crystal structure (Linares et al. 2019). Separately, Huard and co-workers noted that the crystal structures of their Ag nanoclusters before and after reduction of Ag^+ to Ag^0 are superposable (Huard et al. 2019).

5.2.2.2 Silver–Nucleobase Monomer Assemblies

Nucleobase-Derived MOFs Single crystals grown from nucleobase monomers and Ag^+ ions exhibit remarkably varied structures. Here, "nucleobase monomer" is understood to include nucleobases, nucleoside, and nucleotide forms of either the canonical bases or their derivatives. Discussion of some of the possible structures involving Ag^+ and other metal ions can be found in the literature (Munakata et al. 1998; Purohit and Verma 2006; Purohit et al. 2007; Verma et al. 2010; Amo-Ochoa and Zamora 2014; Thomas-Gipson et al. 2015; Mohapatra et al. 2017). Here, we highlight only a few structures with unusual characteristics. Crystals grown from cytidine 5′-monophosphate with $AgNO_3$ feature coordination polymers that extend throughout the crystals (Terrón et al. 2017b). The coordination polymers feature one-dimensional zigzag chains of Ag^+ ions in which each nucleotide coordinates through its N_3 and O_2 atoms to two Ag^+ ions that are nearest neighbors in the chain. This is notable because Ag^+ usually coordinates only to nitrogen atoms in crystal structures of assemblies made from DNA strands.

Nucleobase-derived MOFs with a great many structures are known. The five nitrogen atoms of adenine create many possible ligation sites. Ag^+ ion binding to a number of these sites has been documented in coordination polymers observed in crystal structures. Crystals grown from protonated 9-methyladenine and $AgNO_3$ at low pH contain 1D-coordination polymers with N1–N7 linkages (Gagnon and Beauchamp 1975). Dimeric complexes of $AdeH^+$ (7H tautomer) are seen in crystals grown from Ade and $AgNO_3$ at low pH. The two adenine cations are joined by N-Ag^+-N bonds joining N3 and N9 atoms (Fig. 5.1f, (Mishra et al. 2010)). The dimers are arranged in π-stacked layers in the crystal with a perpendicular distance between planes of 3.27 Å, indicating that π–π stacking is an important structural element in this and in many other nucleobase-MOFs (Stylianou et al. 2011). We discuss below whether this is likely to occur in solution structures and then show spectroscopic evidence in Sect. 5.3.2 that this is the case.

Crystals with DNA-like Helices Remarkably, DNA-like double helices have been observed in crystals grown from Ag^+ ions and several cytosine derivatives, even though the latter components lack the sugar-phosphate backbone of DNA (Terrón et al. 2017a; Linares et al. 2019; Mistry et al. 2019a, 2021). Strictly speaking, these structures are also MOFs, but we discuss them separately here due to their unusual, DNA-like structures. A typical example is shown in Fig. 5.1e (Linares et al. 2019). In this and related structures, the silver base pairs "rotate" around a silver nanowire that serves as an internal backbone, forming a double helix (the solid samples typically contain equal numbers of left- and right-handed helices). Like the unsuccessful Pauling–Corey model of DNA (Pauling and Corey 1953) that preceded the correct structure published by Watson and Crick (Watson and Crick 1953), these unusual structures have their bases on the outside. Whereas the backbone of the Pauling–Corey structure directs the helicity, the silver pole or wire in the silver structures serves mostly to center the base pairs under one another through favorable argentophilic interactions. The twist angle is determined not by the conformational characteristics of the backbones but by the geometries necessary for inter-base-pair hydrogen bonding.

Mistry et al. observed that cytosine is unable to form helices in the solid state, and they attributed this to interference by N1–H hydrogen bonding (Mistry et al. 2019b). Two distinct structures were obtained from crystals grown from cytosine and $AgNO_3$. Both lack the short Ag–Ag distances that are a signature of the argentophilic interactions found in DNA nanowire structures. One crystal structure contains transoid C-Ag^+-C base pairs that form layered two-dimensional sheets through extensive hydrogen bonding. The two cytosines in these C-Ag^+-C base pairs make a modest propeller twist angle of −15.5°. The sheets are slipped with respect to each other and the observed minimum Ag–Ag distance of ~3.6 Å is greater than the perpendicular distance between stacked cytosines of 3.24 Å. In the second crystal structure, the cisoid C-Ag^+-C base pairs have an even larger propeller twist angle of −47.0°. In this crystal structure there is again good π–π stacking between cytosine planes with a perpendicular distance of 3.3 Å, but even greater minimum Ag–Ag distances >7 Å are observed.

Very short Ag–Ag distances ≤3 Å have been observed in the nanowire crystals made with cytidine or N1-substituted cytosine nucleobases (Terrón et al. 2017b; Mistry et al. 2019a, b, 2021; Linares et al. 2019). Although these distances are only a few percent greater than the distance between nearest silver atoms in metallic silver at ambient conditions, conductivity measurements made on crystals show that these supramolecular structures are insulators (Mistry et al. 2019a, 2021).

5.2.3 Solution Structures

Supramolecular Structures Form in Dilute Solution Adding coordinating metal ions makes it possible for nucleobases to self-assemble into supramolecular structures even in dilute solution. Base pairing and base stacking, the hallmark supramolecular motifs of metal-free DNA structures are generally not observed unless the nucleobases are covalently joined into strands. Some purine nucleobases are exceptions. Due to their greater hydrophobic surface area, these can aggregate at relatively low concentrations (~10 mM), and quadruplex structures can self-assemble from guanine derivatives in some cases (Gellert et al. 1962; Peters and Davis 2016).

Unlike silver–DNA structures, silver-free DNA strands are "hard to make" from their building blocks by spontaneous reactions. How to accomplish the abiotic synthesis of DNA is after all one of the defining questions of origin of life science, and the ability of metal ions to organize discrete nucleobases into DNA-like structures has been suggested as a possible answer (Bernal 1949; Ferris et al. 1996). We note that the difficulty of spontaneously synthesizing DNA has a valuable corollary: The lack of promiscuous self-assembly reactions gives enzymes exquisite and essentially exclusive control over DNA synthesis.

Hydrogels, Nanofibers, and Soft-Matter Structures Mixing Ag^+ ions and DNA strands or nucleobases in aqueous solution forms colloidal or other soft-matter structures. Hydrogels form when $AgNO_3$ is added to alkaline solutions of adenine, cytosine, thymine, and uracil (Sharma et al. 2016). The silver-to-nucleobase ratio (β) was 1.25 in these experiments. Electron microscopy revealed dense networks of nanofibers that are 20–35 nm in diameter and several microns long. Although hydrogels were synthesized using a high nucleobase concentration (80 mM), the fibers were still detectable when the nucleobase concentration was lowered to 10 mM. In Sect. 5.3.2, we present our findings that nanofibers form in aqueous solutions of $AgNO_3$ that contain either adenine or 2-aminopurine (Snyder et al. 2019).

Nanofibers can also form when Ag^+ ions are added to DNA strands. The binding of Ag^+ ions to dC_n oligonucleotides allows them to form long fibers when one strand "slips" relative to another, allowing them to join out of register. The overhanging ends allow the duplex to be continued indefinitely by forming interlocked strands through multimerization (Fig. 5.5). Extended G-wire structures based on guanine quartets form by an analogous mechanism (Marsh and Henderson 1994; Marsh et al. 1995; Krishnan-Ghosh et al. 2004; Bose et al. 2018). Vecchioni et al. (2019) inferred that nanofibers form from dC_{30} strands and added Ag^+ ions based on the low electrophoretic mobilities of the silver-DNA assemblies. A stoichiometric excess of Ag^+ ions appears to promote nanofiber formation (Vecchioni et al. 2019; Bethur et al. 2024).
Hierarchical structure formation in silver-DNA assemblies is undoubtedly influenced by the charge of the coordinating ions. When metal ions bind to a DNA strand, they alter the local charge density and destabilize the repulsive interactions between backbone phosphate ions that contribute to the rigidity of the DNA double helix. When enough base pairs bind one or more metal ions, the reduction of linear charge density can be sufficient to induce DNA condensation or compaction (Jiang and Ran 2018). This can lead to amorphous soft-matter structures such as nanoparticles (Wei et al. 2007; Nishiyabu et al. 2009; Venkatesh et al. 2016; Al-Mahamad et al. 2017; Lopez and Liu 2019) and nanofibers (Sharma et al. 2016; Mistry et al. 2019a; Snyder et al. 2019). The latter may undergo crosslinking to give hydrogels or 'metallogels' (Dash et al. 2011; Peters and Davis 2016; Sharma et al. 2016; Xu et al. 2018).

The Unusual CD Spectra of Many Silver-DNA Assemblies CD spectra are highly sensitive to secondary structure, but it is difficult to predict structure given only the CD spectrum and base sequence. Moreover, the factors that cause a CD spectrum to look as it does can be obscure, a point that is well illustrated by silver-DNA assemblies. Strikingly, many silver–DNA oligonucleotide assemblies have nearly unisignate CD spectra. For example, a CD spectrum with twin negative peaks near 270 nm and 220 nm has been seen in silver(I) assemblies made with d(AGGTCGCCGCCC) (Petty et al. 2004), d(TAACCC)$_4$ (Day et al. 2013), d($C_4G_4C_4$) (Swasey and Gwinn 2016), 5′-(CCCTAA)$_3$CCC-3′ (Kang et al. 2016), and variable-length dC_n oligonucleotides (Swasey et al. 2018). Some of these sequences fold into i-motif structures at pH 5.5. The fact that the i-motif CD spectrum looks nothing like the approximately unisignate CD spectrum seen when Ag^+ ions are added suggests, however, that silver ion binding causes the DNA to adopt a very different secondary structure. Nearly unisignate and negative CD spectra have also been observed in duplexes with G-Ag^+-G base pairs (Swasey et al. 2015; Swasey and Gwinn 2016; Bethur et al. 2024).

Unisignate CD spectra were first observed in silver-DNA assemblies more than fifty years ago (Arya and Yang 1975; Walter and Luck 1977). Notably, the CD spectrum of Ag-poly(C) reported in 1975 (Arya and Yang 1975) closely resembles the CD spectra of the above-mentioned Ag–oligonucleotide assemblies. Two competing explanations emerged in the 1970s to explain the negative CD spectra observed in Ag–polynucleotide and Ag–genomic DNA assemblies. The first explanation is that the negative CD spectra arise from aggregation or condensation of the DNA due to metal ion (Ag^+ or Hg^{2+}) binding (Walter and Luck 1977). Unisignate CD spectra are characteristic of the so-called psi-type CD spectra, which are observed when DNA interacts with salt or polymers (Jordan et al. 1972). Here, psi, which is sometimes written as ψ, is an acronym for "polymer- and salt-induced." Adding polymer or salt promotes the aggregation and compaction of DNA strands by screening or neutralizing DNA's negative charges. These aggregated structures can exhibit mesoscale chirality on length scales comparable to the wavelength of light, giving rise to strong and negative CD signals (Kim et al. 1986; Livolant and Maestre 1988). It is known that Ag^+ ion binding can assemble dC_n (Vecchioni et al. 2019) and dG_n oligonucleotides (Bethur et al. 2024) into long fibers, possibly by producing slipped strand structures with overhangs like the ones illustrated in Fig. 5.5. Conceivably, these fibers, or higher-order aggregates generated from them, could yield psi-type CD spectra.

The second explanation is that the observed unisignate CD spectra arise from strong tilting and propeller twisting of the base pairs and not from aggregation

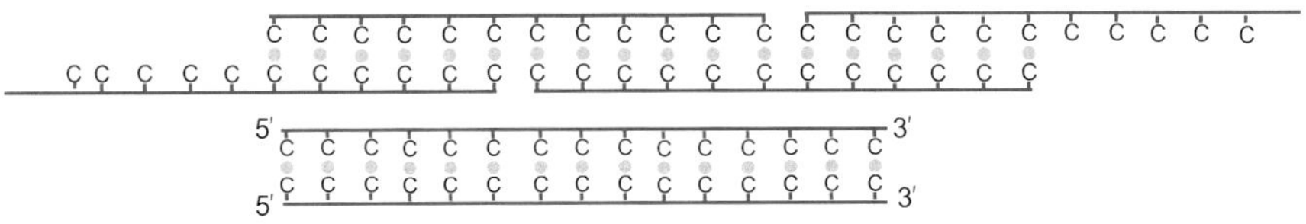

Fig. 5.5 Cartoon showing the formation of interlocked fibers by strand "slippage" in Ag-dC_n duplexes (top). The sketch at the bottom shows two strands in full registry

(Zavriev et al. 1979; Nordén et al. 1986). Base pair tilting and twisting cause the transition dipole moments of the $\pi\pi^*$ transitions of the nucleobases to no longer lie in a plane perpendicular to the helix axis and leads to large changes in the CD signal magnitudes (Yang and Samejima 1968; Studdert and Davis 1974a, b; Johnson et al. 1981). Today's crystal structures showing runs of consecutive silver base pairs with very high propeller twist (Sect. 5.2.1) would appear to be a strong argument in favor of this explanation. However, aggregation effects, which are easy to observe in a number of silver–DNA oligonucleotide assemblies (Vecchioni et al. 2019; Bethur et al. 2024), cannot be ruled out. More work is needed to understand how strong base twisting/tilting and aggregation, which may or may not occur together in a given silver-DNA assembly, contribute to the ubiquitous negative CD spectra seen recently in many silver–oligonucleotide assemblies with consecutive C-Ag^+-C or G-Ag^+-G base pairs.

Are Solution Structures the Same Ones Observed in Solids by X-ray Crystallography? This question has been asked about metal-free DNAs for decades, and the many techniques that are sensitive to the secondary structure of nucleic acids in solution can be profitably used to study silver-DNA assemblies. These techniques include hypo/hyperchromism measurements (thermal denaturation), viscosity, flow orientation, linear and circular dichroism, atomic force microscopy (AFM), native negative ion mode electrospray ionization-mass spectrometry (ESI-MS), and multi-dimensional NMR.

Determining the supramolecular structures that are present in a solution sample is a grand challenge. In some cases, silver-DNA assemblies can be isolated from a dispersion through flocculation or solvent evaporation. The solid isolates can then be studied by techniques such as X-ray diffraction, IR spectroscopy, atomic force microscopy (AFM), or dynamic light scattering (DLS). Some of these techniques do not provide atomically resolved structures but can still be useful for detecting nano- and mesoscale structural motifs such as nanofiber formation (see Sect. 5.3.2).

Even when crystals can be grown from solutions, there is no guarantee that the same structures are present in solution. Fortunately, a number of studies have provided evidence that solution-phase structures are in fact highly similar, if not identical, to ones seen in crystals (Kondo et al. 2017; Ono et al. 2019; Mistry et al. 2019a; Atsugi et al. 2022). Changes in nucleobase proton resonances measured by NMR during Ag^+ ion titration and thermal denaturation ("DNA melting") experiments provide evidence that the same silver base pairs seen in the crystal structure of a silver–DNA nanowire are present in solution (Kondo et al. 2017). NMR resonances broaden greatly when the Ag:nucleobase ratio is high due to the increasing molecular weight of the aggregates (Kondo et al. 2017). This may signal the formation of aggregated duplexes and longer wires (Atsugi et al. 2022). CD spectra and ESI-MS measurements on short T-rich oligonucleotides and Hg^{2+} ions are consistent with the presence of mercury–DNA assemblies in solution (Ono et al. 2019). Electrospray ionization-mass spectrometry (ESI-MS) results on solutions also show peaks that correspond to metal ion-bound oligonucleotides (O'Neill et al. 2009; Schultz and

Gwinn 2012; Swasey et al. 2015; Ono et al. 2019; Atsugi et al. 2022), providing insight into the types of structures present in solution.

CD, NMR, and ESI-MS were used to argue that the silver–DNA nanowire seen in the solid state (Fig. 5.1b) is also present in solution (Atsugi et al. 2022). As described in Sect. 5.3.2, our pump–probe measurements on solution-phase samples made by combining silver nitrate with purine nucleobases reveal photophysical signatures of base stacking, a motif that we conclude is present in the nanofibers seen in AFM measurements on evaporated material (Snyder et al. 2019). These examples illustrate how the supramolecular structures seen in crystals appear to be present in solution for experimentally relevant conditions.

5.3 Spectroscopic Studies of Excited-State Dynamics in Silver-DNA Assemblies

Our work to understand the photophysics of silver-DNA assemblies has sought to leverage the past quarter century of experimental and computational progress toward elucidating excited-state dynamics of nucleobases and DNA strands (Crespo-Hernández et al. 2004; Middleton et al. 2009; Gustavsson et al. 2010; Matsika and Krause 2011; Chen et al. 2015; Improta et al. 2016). Studies of excited-state dynamics of metal-free DNA strands using femtosecond pump–probe spectroscopy have drawn attention to the very different character of their excited-state deactivation pathways compared to their building blocks (Crespo-Hernández et al. 2005; Middleton et al. 2009). New excited-state deactivation pathways in strands include the formation of charge transfer (CT) states (Takaya et al. 2008; Martinez-Fernandez et al. 2016; Kufner et al. 2020) and photoinduced proton-coupled electron transfer (PCET) (Zhang et al. 2015, 2019). These effects arise whenever two or more nucleobases are positioned close enough to be electronically coupled. Excited-state dynamics are sensitive to base pairing (Röttger et al. 2015; Böhnke et al. 2019) and base stacking (Crespo-Hernández et al. 2005). The manifold ways that Ag^+ ions can spatially organize nucleobases through facile reactions suggest the possibility of interesting photophysics. After describing how spectroscopic techniques for studying metal-free DNA have been adapted to the silver-DNA systems, two case studies of excited-state dynamics in silver-DNA assemblies are presented.

5.3.1 *Methods for Studying Silver-DNA Assemblies by Optical Spectroscopy*

In our experiments, the silver-DNA assemblies are dispersed in a solvent. Steady-state and time-resolved spectroscopic measurements are easiest to perform on solution-phase samples in transmission. Recirculating samples through an optical flow cell reduces artifacts in femtosecond pump–probe experiments by continuously bringing fresh sample into the interaction region defined by the crossed and focused laser beams. In particular, this reduces problems due to sample photodegradation, sample re-excitation by successive laser pulses, and heat buildup.

Solid, polycrystalline samples are difficult to study due to light scattering. Signals from colloidal particles smaller than ~100 nm are dominated by absorption and light scattering can usually be neglected. Colloidal dispersions can thus be studied using the same instrumentation used to study molecular solutes. The most desirable samples for experiments are colloidally stable DNA–metal assemblies of known structure. This creates a challenge because the most detailed structural information comes from crystal structures, which may differ from the colloidal structures present in solution.

Silver-DNA samples are typically prepared by adding a few drops of a concentrated metal salt such as $AgNO_3$ to an aqueous solution of a DNA oligonucleotide or a nucleobase monomer. As is commonly done with silver-free DNA, samples are often thermally annealed to promote the formation of desired secondary structures. Copp and co-workers showed that thermal annealing can produce parallel-stranded Ag-dG$_n$ duplexes with minimal interference from G quadruplex structures (Bethur et al. 2024). Although nucleobase monomers could in principle form mononuclear silver complexes, there is no clear evidence for this yet, and many samples appear to form soluble coordination polymers and hierarchically assembled nanostructures.

Given the ease of forming coordination polymers with Ag^+ ions and nucleobases, it might be anticipated that samples would precipitate and not form stable colloids. The low nucleobase concentrations ($\leq$10 mM) used in ultrafast pump–probe experiments likely favor the formation of supramolecular structures over crystal growth. Nonetheless, it is important to look out for a visually turbid sample or one in which small, suspended particles can be seen by eye. Even before a sample appears turbid by eye, red tailing may be detected in absorption spectra due to light scattering by structures that are comparable in size to optical wavelengths. The presence of nanoscale structures or fibers can also be sensed through changes in viscosity (Loo et al. 2010).

When studying assemblies with Ag^+ ions it is important to avoid conditions that lead to the growth of Ag^0 nanoparticles. Laser irradiation readily reduces Ag^+ ions and neutral silver atoms have a strong tendency to aggregate. Silver cluster and nanoparticle growth can be detected by gray deposits at the solution–glass interface of the flow cell. Bound Ag^+ ions positioned inside silver base pairs are protected from the solvent and do not easily aggregate, so this problem is caused by excess, unbound Ag^+ ions. We found that silver nanoparticle growth can be prevented by working at stoichiometries that ensure that all Ag^+ ions are bound. Adding a chelating agent like ethylenediaminetetraacetic acid (EDTA) can also eliminate any unbound Ag^+ ions (Martínez-Fernández et al. 2024).

5.3.2 Silver-Adenine and Silver-2-Aminopurine Assemblies

We investigated silver–nucleobase assemblies prepared by adding concentrated aq. $AgNO_3$ to an aqueous solution of adenine (A) or its structural isomer 2-aminopurine (2AP) (Snyder et al. 2019). The samples were thermally annealed for 5 min at a temperature of 90 °C before cooling to room temperature. The mole ratio of Ag^+ to nucleobase was fixed at 2.0 for all measurements. Earlier, Nordén and co-workers showed that aqueous solutions of adenine with 2 moles of Ag^+ ions per mole of nucleobase produce supramolecular structures that can be oriented by flow. This indicates that the structures are longer than 50 nm in at least one dimension (Matsuoka et al. 1984). Inspired by these experiments, we investigated putative coordination polymers formed by Ag^+ ions and purine nucleobases by steady-state and time-resolved spectroscopy. A further interest in these systems springs from the well-characterized excited-state dynamics of aqueous solutions of the silver-free nucleobases A and 2AP. The remarkably different photophysics of these two bases has made them paradigmatic for how covalent modification can alter nucleobase nonradiative decay by orders of magnitude (Serrano-Andrés et al. 2006). The excited-state lifetime of 2AP is approximately 11.5 ns (Jones and Neely 2015), while the excited-state lifetime of the 9H tautomer of adenine is only 180 fs (Cohen et al. 2003).

We performed AFM measurements by evaporating a drop of Ag-purine solution on a mica substrate. The solution had the same concentration (1 mM) used in the femtosecond transient absorption experiments described below but was diluted by a factor of 10 with ultrapure water immediately before deposition. This was done to reduce the number of assemblies in the imaging area. The AFM measurements revealed slender, micrometer-long nanofibers (Fig. 5.6). Although drying produces high concentrations that could favor aggregation, we believe that the structures observed by AFM are actually present in solution based on the previously mentioned in situ evidence obtained by Nordén and co-workers of >50 nm nanostructures in their 79μM solutions that were considerably more dilute than ours. The fibers in Fig. 5.6 are nominally 30 nm in diameter, although the true width may be obscured by the width of the probe tip. The nanofibers from Ag-2AP are visibly thinner than ones from Ag-Ade (Fig. 5.6), indicating that the latter structures may form large aggregates more easily.

The precise structures are uncertain, but it seems likely that extended coordination polymers are present. Ade and 2AP each has five N atoms that a Ag^+ ion could bind to. Crystal structures of silver–adenine coordination polymers have demonstrated Ag^+ ion binding to N1, N3, N7, and/or N9 atoms, see Table 1 in (Amo-Ochoa and Zamora 2014). Nordén and co-workers concluded from their linear dichroism and other measurements that Ag-Ade forms linear polymers with N1–N1′, N9–N9′, and N1–N9′ connectivity (Matsuoka et al. 1985). However, the 1:1 Ag^+:adenine stoichiometry of their preferred structures is at odds with our silver ion-electrode measurements, showing that all Ag^+ ions are bound to adenine in our $\beta = 2.0$ conditions. It is conceivable that the higher Ag^+ ion concentrations used in our experiments cause 1D-chains like the ones in Matsuoka et al. (1985) to be further cross-linked into the nanofibers that we observe.

Spectroscopic observables provide clear evidence that nanostructures are present. The absorption spectra of the Ag-Ade and Ag-2AP samples are red-shifted compared to silver-free solutions of the nucleobases. This is illustrated for Ag-Ade in Fig. 5.7. The red shift can be partially explained by deprotonation of the nucleobase. The pK_a for loss of the N9 proton of neutral adenine is 9.87 (Izatt et al. 1971), but Ag^+ binding increases the positive charge of the nucleobase, increasing its acidity and causing deprotonation at pH values below the pK_a (Daune et al. 1966; Eichhorn et al. 1967). In our samples, adding $AgNO_3$ causes the pH to decrease to approximately 3.0 due to proton release from the nucleobases. The absorption spectrum of deprotonated adenine is red-shifted compared to the neutral parent molecule (Roberts et al. 2014). As the red shift in the Ag-Ade sample is considerably larger than for $Ade(-H^+)^-$, there may be an additional red shift due to excitonic coupling among proximal Ade molecules.

Ultrafast Excited-State Dynamics of Ag-Ade and Ag-2AP Femtosecond transient absorption measurements reveal that the Ag-Ade and Ag-2AP samples differ remarkably in their excited-state dynamics compared to the silver-free nucleobases (Figs. 5.8 and 5.9). Multiexponential decay kinetics are seen for both samples. In general terms, Ag^+ ion coordination causes the average lifetimes of Ade and 2AP to increase and decrease, respectively, by a factor of roughly one hundred. For Ag-Ade,

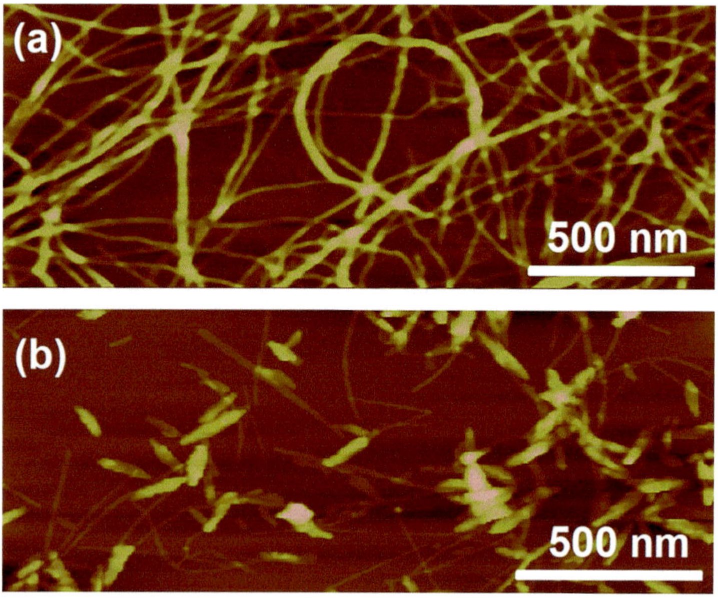

Fig. 5.6 AFM images obtained from solutions of (**a**) Ag-Ade ($\beta = 2$) and (**b**) Ag-2AP ($\beta = 2$) by placing a drop on a mica substrate and allowing the solvent to evaporate. Reprinted with permission from Snyder et al. (2019). Copyright 2019 American Chemical Society

lifetimes of 5.0 ± 0.1 ps, 150 ± 10 ps, and >3 ns are observed. These lifetimes are quite similar to ones of 1.3 ± 0.1 ps, 140 ± 10 ps, and >3 ns observed for the homopolymer poly(A) in the absence of silver under otherwise identical conditions. Rather similar lifetimes of 12.0 ± 0.5 ps, 180 ± 10 ps, and >3 ns were observed for Ag-2AP, even though the transient spectra differ greatly in appearance. As expected from the ns lifetime of 2AP in aqueous solution, we observed a dominant decay component of >3 ns (the long-time limit of our transient absorption spectrometer) for a silver-free solution of 2AP. The approximately 100× shorter lifetime seen for Ag-2AP compared to 2AP is consistent with the 98% drop in emission seen for the former sample. It is known that π-stacking with nucleobases in DNA strands quenches 2AP fluorescence (Jones and Neely 2015).

The dominant lifetimes of between 100 and 200 ps seen in both Ag-Ade and Ag-2AP are assigned to excimers in π-stacked structures formed by the nucleobases. Characteristic lifetimes of 100 to 200 ps are observed in femtosecond transient absorption measurements on aqueous solutions of single-stranded, silver-free dA_n oligonucleotides with excitation at 267 nm (Su et al. 2012) The observed excimer lifetimes are nearly three orders of magnitude longer than the 180 fs excited-state lifetime of the 9H tautomer of adenine (Cohen et al. 2003).

Excimers arise from excitations in stacks of two or more bases because they are quenched when base stacking is disrupted by raising the temperature or adding an organic co-solvent (Su et al. 2012). The excimer states have CT character that results from transferring an electron from one nucleobase to its π-stacked neighbor. Recent experiments have demonstrated that these states form in hundreds of femtoseconds (Petropoulos et al. 2023). Excimer formation extends the sub-picosecond lifetime of adenine but reduces the multi-nanosecond lifetime of 2AP. Their long decay times and ultrafast formation times have been explained by back electron transfer that takes place in the Marcus inverted region (Chen et al. 2015).

We propose that the Ag–purine nanofibers feature extensive π–π stacking interactions between the nucleobases, mirroring the structures seen in many coordination polymer crystals. Ag^+ ion binding perturbs the electronic structure of the bases

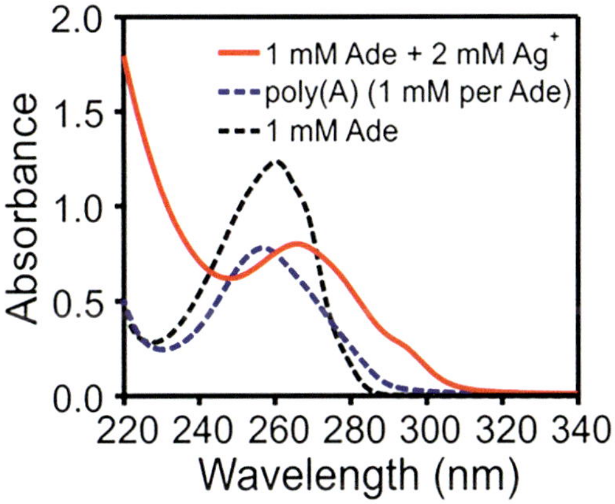

Fig. 5.7 Absorption spectrum of adenine before (black dashed curve) and after adding 2 equivalents of Ag^+ ions (red curve). For comparison, the absorption spectrum of poly(A) with the same nucleotide concentration is shown (blue dashed curve). Adapted with permission from Snyder et al. (2019). Copyright 2019 American Chemical Society

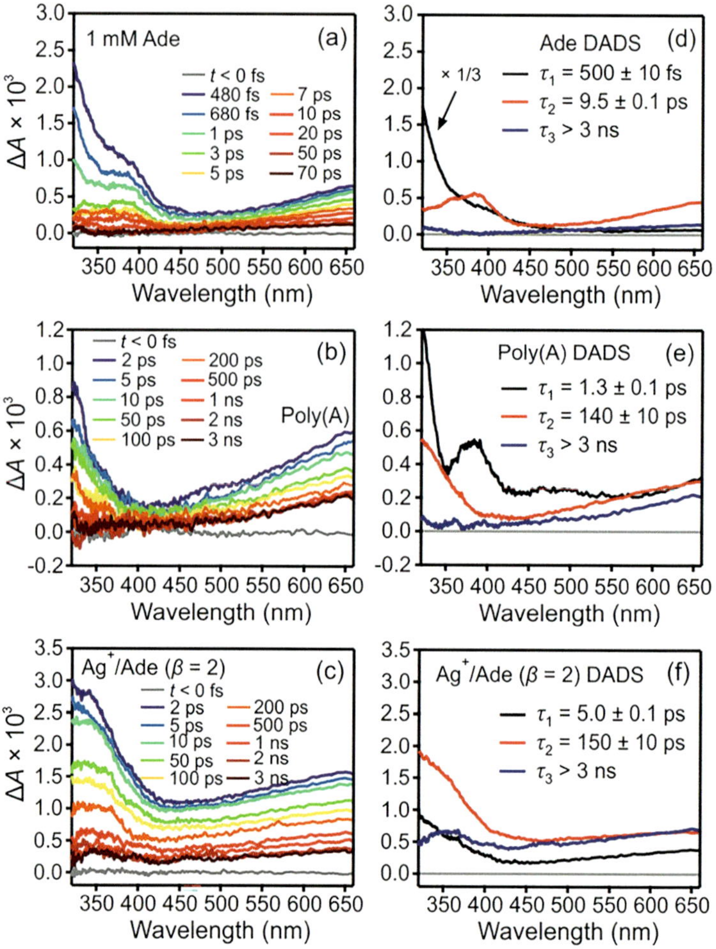

Fig. 5.8 Transient absorption spectra of (**a**) Ade in H_2O at pH 7, (**b**) poly(A) in H_2O at pH 7, and (**c**) Ag+/Ade ($\beta = 2$) in H_2O. The nucleobase concentration in all solutions was 1 mM and the pH was approximately 3. Excitation was at 265 nm. The decay-associated difference spectra (DADS) with the exponential lifetimes indicated are shown in panels (**d**–**f**). Reprinted with permission from Snyder et al. (2019). Copyright 2019 American Chemical Society

relatively modestly as seen from the minor changes to the steady-state absorption spectrum. Instead, the most important effect is the structure-forming ability of Ag^+ ions as confirmed by the AFM measurements. As in silver-free strands of nucleobases, π–π stacking is critical for excimer formation, and we propose that this is a key structural motif in the Ag-Ade and Ag-2AP nanofibers. Crystal structures provide ample evidence of how Ag^+ ions may form coordination polymers that assemble by π–π interactions from 2AP derivatives (Mohapatra and Verma 2016) and from adenine (Mishra et al. 2010).

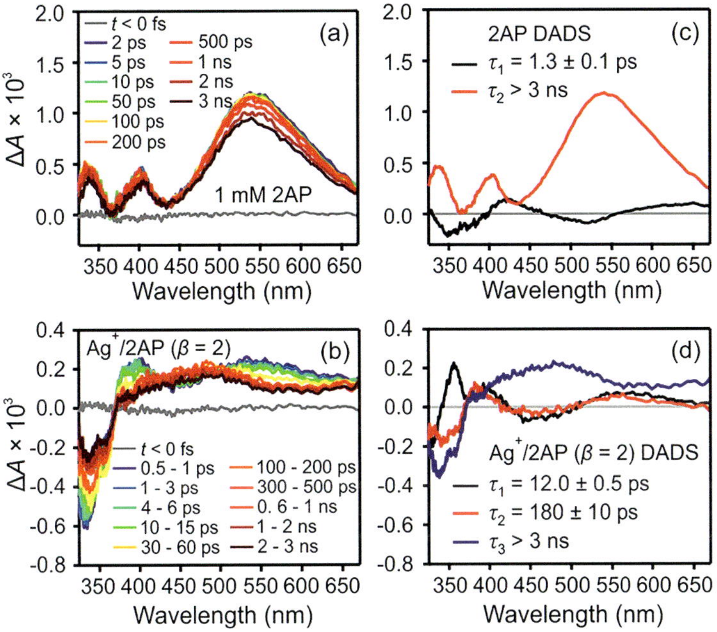

Fig. 5.9 Transient absorption spectra of 1 mM 2AP in H_2O measured at the indicated pump–probe delay times (**a**) without and (**b**) with 2 mM $AgNO_3$ ($\beta = 2$). The pH for these measurements was approximately 3, and excitation was at 265 nm. The DADS obtained from global fits using multi-exponential functions are shown in panels c and d. The lifetimes are shown in each graph. Reprinted with permission from Snyder et al. (2019). Copyright 2019 American Chemical Society

The fastest component of 5.0 ± 0.1 ps observed in Ag-Ade accounts for 22% of the total decay and is 10-fold longer than the lifetime of 0.5 ps observed for the Ade monomer. The absence of a sub-picosecond decay for Ag-Ade suggests that there are no free (i.e., uncomplexed) Ade molecules in the Ag-Ade sample. We propose that all Ade molecules are present in Ag-Ade assemblies (nanofibers). Free 2AP molecules are also present at a level of <2% in the Ag-2AP sample based on the observation that >98% of the emission from an aqueous solution of 2AP is quenched when $\beta = 2$. If the Ag-2AP assemblies have detectable levels of emission, then the percentage of free nucleobase would be even lower.

Based on these considerations, we conclude that Ade and 2AP are present overwhelmingly as subunits in self-assembled nanofibers that experience strong π–π stacking interactions. This has interesting implications for the nonradiative decay mechanism. In purine homopolymers like poly(A), many bases are not π stacked with other nucleobases because random coil sections coexist in equilibrium with runs of stacked bases (Applequist and Damle 1966; Su et al. 2012). The sub-picosecond component seen in experiments on these systems is attributed to unstacked bases that have excited states that can decay by monomer-like mechanisms. Longer lifetimes are observed when bases are stacked, but ultrafast deactivation may nevertheless still take place in regions containing stacked bases (Chen et al. 2013). The slower 5.0 ± 0.1 ps lifetime compared to adenine could come about because steric hindrance impedes access to the conical intersection that connects the excited state to the electronic ground state.

In summary, these measurements, which are the first to use ultrafast laser spectroscopy to study metal–DNA assemblies, show that strand-like excitations are formed in the self-assembled nanostructures created by Ag^+ ion binding. The results strongly suggest that the nanofibers are built of π-stacked coordination polymers. At the present time, we do not have evidence for hydrogen bonding in these systems, although it is likely present. Nevertheless, the most important factor governing excited-state deactivation appears to be the presence of π-stacked layers of the purine nucleobases. Previous work on silver-free DNA has shown that nucleobase excimer lifetimes depend on the chemical identity of the two stacked bases and are quite insensitive to the detailed stacking geometry (Chen and Kohler 2014). The common lifetimes seen in DNA strands and in stacked Ag–purine coordination polymers support this paradigm.

5.3.3 Silver-dC_n Assemblies

Silver-dC_{20} assemblies were studied by UV-vis femtosecond transient absorption spectroscopy (Kohl et al. 2020) and femtosecond time-resolved IR (TRIR) measurements (Martínez-Fernández et al. 2024). Unlike the Ag–purine assemblies (Sect. 5.3.2), the supramolecular structures formed by adding Ag^+ ions to cytosine homo-oligonucleotides are better characterized. Megger and Müller first proposed that Ag-dC_n assemblies form parallel-stranded duplexes (Megger and Müller 2010). This conclusion was confirmed later by Gwinn and co-workers in clever Förster resonance energy transfer (FRET) measurements of dye-labeled dC_n strands made in the presence of Ag^+ ions (Swasey et al. 2018). There is no crystal structure yet of a Ag-dC_n oligonucleotide, but the crystal structure of a fluorescent Ag cluster has a trio of transoid C-Ag^+-C base pairs (Fig. 5.1d, (Huard et al. 2019)). The structure contains strongly propeller-twisted base pairs like ones seen in an ab initio molecular dynamics (MD) simulation of Ag^+ ion-mediated duplexes with dC_{11} and dC_{20} oligonucleotides (Fig. 5.1c, (Chen et al. 2017)).

We are confident that the Ag-dC_{20} sample ($\beta = 0.5$) in our experiments contains parallel-stranded duplexes with inter-base-pair hydrogen bonding due to inclined C-Ag^+-C base pairs with a high degree of propeller twist (see Sect. 5.2.1 and Fig. 5.4a). Knowing the atomistic structures made it possible to interpret our experimental results with high-level quantum chemical calculations carried out by our collaborator, Lara Martínez-Fernández. The results provide a rare example of a directly detected triplet excited state in DNA and the first observation of a cytosine-based triplet state (Martínez-Fernández et al. 2024).

Ultrafast Excited-State Dynamics of Ag-dC_{20} Assemblies Experiments were performed on dC_{20} strands with 1 mole of added Ag^+ ions for every 2 moles of nucleotides ($\beta = 0.5$). The results were compared with experiments performed on a pH 8.5 solution of dC_{20} without Ag^+ ions. At this pH, the secondary structure of dC_{20} is single-stranded (ss-dC_{20}). When the pH is lowered to 4.5, stable i-motif structures form (i-dC_{20}) as indicated by the appearance of the characteristic i-motif CD spectrum. Transient absorption kinetic traces are compared for these samples in Fig. 5.10. The transient absorption signal at a probe wavelength of 350 nm shows an initial decay that turns into a slow rise, beginning approximately 100 ps after the pump pulse (Fig. 5.10a, gray circles). This signal is still rising by 3 ns, which is the longest delay time accessible with our spectrometer. The long-lived excited state was proposed to either be a long-lived CT excited state or a triplet excited state (Kohl et al. 2020). No long-lived signals were observed for the silver-free i-motif form of dC_{20}. The different dynamics seen for the Ag-dC_{20} sample supports the conclusion that Ag^+ ions do not fold an i-motif-like secondary structure.

To gain more insight into the nature of the long-lived signals, time-resolved IR (TRIR) measurements were conducted (Martínez-Fernández et al. 2024). These experiments were carried out in heavy water (D_2O) due to its better transparency in the mid-IR spectral region of interest (1400–1800 cm^{-1}). Under these conditions, all exchangeable protons are replaced by deuterons. This strongly affects the appearance of the ground- and excited-state vibrational spectra in the double-bond stretching region probed in our experiments, so calculations were conducted on deuterated substrates. It was also necessary to study more concentrated solutions (the final concentrations of $AgNO_3$ and nucleotides in dC_{20} were 2.7 mM and 5.4 mM, respectively) than was the case for the experiments in Kohl et al. (2020) because the cross sections of vibrational transitions are much weaker than electronic ones. Nevertheless, neither the increased concentration nor sample deuteration altered the secondary structure, and the same CD spectrum was observed in H_2O and D_2O solutions.

The TRIR spectra of Ag-dC_{20} do not decay to baseline and persist beyond 3 ns (Fig. 5.11). Kinetic traces reveal a slowly rising band at 1586 cm^{-1} band after photoexcitation of Ag-dC_{20} at 265 nm (Fig. 5.12a, red circles). The rise is slower in D_2O (1300 ps) than in H_2O solution (515 ps). However, the origin of this kinetic isotope effect ($\tau_{D_2O} / \tau_{H_2O}$) of 2.5 is not understood. The 1586 cm^{-1} marker band in Ag-dC_{20} vanishes in the TRIR spectrum of Ag-Cyt (Martínez-Fernández et al. 2024). The transient band at 1586 cm^{-1} is shifted from the known vibrational signature of the $^1n\pi^*$ state at 1574 cm^{-1} and also has a much longer lifetime than the characteristic $^1n\pi^*$ state lifetime of several tens of picoseconds (Keane et al. 2014). In fact, a signal from the $^1n\pi^*$ state is missing altogether in the Ag-dC_{20} TRIR data, likely because Ag^+ ion coordination destabilizes $^1n\pi^*$ excited states. Indeed, calculations indicate that the lowest $^1n\pi^*$ is associated with the carbonyl oxygen atom. However, its location as the seventh excited state (S_7) makes it too high in energy to affect the excited-state dynamics. Significantly, no rising signals or long-lived species were observed for a solution of 10.1 mM $AgNO_3$ and 20.2 mM cytosine ($\beta = 0.5$), which we refer to as the Ag–Cyt sample.

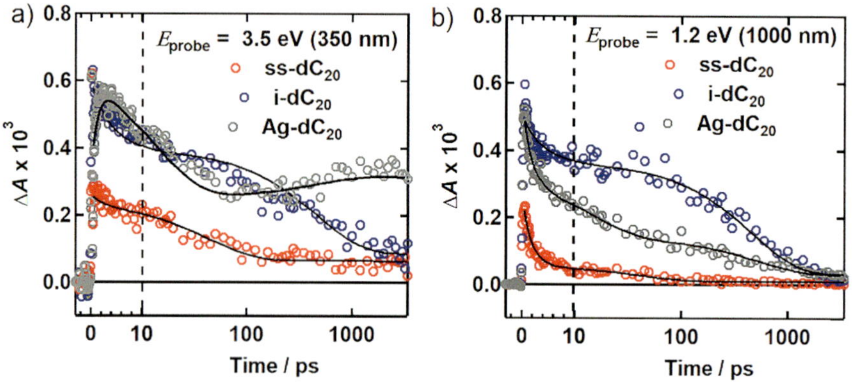

Fig. 5.10 Kinetic traces of the Ag-dC_{20} (gray), ss-dC_{20} (red), and i-dC_{20} (blue) samples in H_2O solution recorded after 265 nm (4.7 eV) excitation at probe pulse photon energies of (**a**) 3.5 eV (350 nm) and (**b**) 1.2 eV (1000 nm). All dC_{20} samples were 1.8 mM in nucleotides. Reproduced from Kohl et al. (2020), with the permission of AIP Publishing

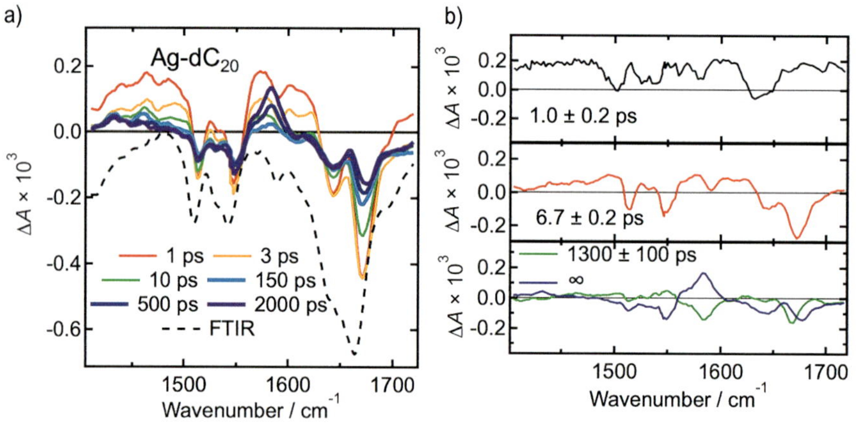

Fig. 5.11 (**a**) TRIR spectra for a D_2O solution of Ag-dC_{20} (5.4 mM in nucleotides) excited by a femtosecond pump pulse at 265 nm. (**b**) Decay-associated difference spectra (DADS) for the four exponential time constants (lifetimes indicated in each graph) used to globally fit the TRIR data in panel (**a**). Adapted with permission from Martínez-Fernández et al. (2024). Copyright 2024 American Chemical Society

To better understand excited-state deactivation in the Ag-dC_{20} assemblies, calculations were carried out on a model system consisting of two stacked C-Ag^+-C base pairs, arranged as in the parallel-stranded duplex. The optimized base pairs have a high degree of propeller twist consistent with the structures discussed earlier for assemblies with consecutive C-Ag^+-C base pairs and sketched in Fig. 5.4a. In the optimized structure an intermolecular hydrogen bond joins the amino group of one base in the lower base pair with the carbonyl oxygen of one base in the upper base pair.

The calculations provide a self-consistent description of the dynamics. In particular, the initial excited state populated by the 265 nm pump pulse can decay to a singlet state with significant CT character (S_{CT}), which does not involve CT between cytosines. Instead, 40% of the photoexcited electron is transferred from Ag^+ to C, according to the calculations. The CT state acts as a gateway to a triplet excimer state (T_{exc}), which is the proposed species responsible for the long-lived signals. The excimer state is spread over two cytosine bases that are stacked. Our assignment is supported by the good agreement between the calculated TRIR difference spectrum of the T_{exc} state and the experimental spectrum of the long-lived excited state (Fig. 5.12b). In particular, the calculations reproduce the red shifting of the strong ground state band near 1650 cm^{-1}, which has a major contribution from C=C stretching. Due to lengthening of the C5=C6 bond in the triplet excimer state, this mode red shifts to 1586 cm^{-1}. This results in a positive peak at this frequency and a negative peak near 1650 cm^{-1} in the triplet state difference spectrum (Fig. 5.12b, top). Similar features are seen in the calculated difference spectrum (Fig. 5.12b, bottom).

The observation of a long-lived triplet state in Ag-dC_{20} but not in Ag–Cyt rules out ISC due to a heavy atom effect. This is because our measurements made with a silver ion-sensitive electrode reveal that the added silver ions are fully bound in both samples, indicating that each has cytosines with bound silver ions. We propose that it is not the presence of a bound silver ion per se but rather the special secondary structure of the Ag-dC_{20} duplex that leads to its enhanced spin–orbit coupling compared to Ag-Cyt. Recent calculations predict ISC rates of $\leq 10^8$ s^{-1} for the DNA-encapsulated silver clusters Ag_3^+ and Ag_4^{2+} (Ramasanoff and Sokolov 2022), showing that a silver ion that is bound to a nucleobase does not always enhance

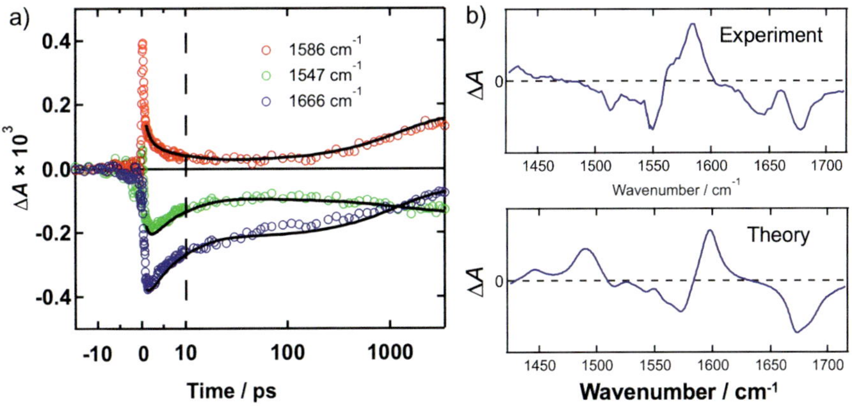

Fig. 5.12 (**a**) TRIR kinetic traces for a D_2O solution of Ag-dC_{20} (5.4 mM in nucleotides) excited by a femtosecond pump pulse at 265 nm. The pD (= $-\log[D_3O^+]$) of the solution was 6.2. The mid-IR probe frequencies are indicated in the graph. Circles show data markers, and solid curves are global fits to multiexponential functions. (**b**) TRIR difference spectrum for the triplet excimer state from experiment (top) and theory (bottom). Adapted with permission from Martínez-Fernández et al. (2024). Copyright 2024 American Chemical Society

ISC. These rates are orders of magnitude too low to explain the rapid ISC in our experiments. The precise reasons are unknown but may be due to differences in secondary structure. Inspection of some of the structures in the paper by Ramasanoff and Sokolov suggests that the calculations were performed on antiparallel-stranded DNAs, which likely differ in conformation compared to our Ag-dC_{20} duplex.

We propose that the propeller-twisted C-Ag^+-C base pairs that are a ubiquitous structural motif in silver assemblies with C-rich oligonucleotides play a key role in the formation of the triplet excimer state. This geometry is proposed to be absent in the Ag-Cyt assemblies in aqueous solution. These are proposed to have more planar C-Ag^+-C base pairs consistent with the known crystal structures of $AgNO_3$ and cytosine, which feature sheets of hydrogen-bonded C-Ag^+-C base pairs that lack close Ag–Ag distances (Mistry et al. 2019b).

It is noteworthy that excited states in the Ag-dC_{20} sample, which may contain many more than 20 consecutive C-Ag-C base pairs due to the possibility of forming nanofibers from interlocked strands, are well accounted for by models of excited states in a stack of just two C-Ag^+-C base pairs. Earlier, we showed that excited states in stacks of single-stranded dA_n oligonucleotides have dynamics that are nearly independent of the length of the strand when $n \geq 2$ (Su et al. 2012). Furthermore, excited-state dynamics in mini-duplexes containing just two GC base pairs and in oligonucleotides containing 20 GC base pairs are very similar (Zhang et al. 2016). This may reflect the importance and prevalence of CT excited states in DNA strands that are strongly localized on adjacent bases and base pairs.

Finally, we considered the photochemical possibilities of the triplet excited states. Excimer formation is accompanied by a strong decrease in the C6–C6′ distance, according to calculations. The close approach of these atoms could facilitate the formation of a cyclobutane pyrimidine dimer (CPD) photoproduct. However, the absorption spectrum of the Ag-dC_{20} sample changed minimally after laser experiments, ruling out significant levels of photodegradation. Earlier, Rahn and Landry (Rahn and Landry 1973) showed that CPD formation in UV-irradiated poly(C) is not enhanced by Ag^+ binding. Consistent with the experimental observations, calculations detailed in Martínez-Fernández et al. (Martínez-Fernández et al. 2024) found no evidence that excimer states of Ag-dC_{20} decay to CPD photoproducts. The possibility of photodamage is not completely ruled out as the calculated triplet energy of 1.6 eV for the excimer state of Ag-dC_{20} is high enough to sensitize singlet oxygen. Further experiments, including a measurement of the lifetime of the triplet excimer state as a function of dissolved oxygen concentration, would provide insight into whether the twisted base pairs in Ag-dC_{20} assemblies are capable of generating harmful reactive oxygen species.

5.4 Conclusions

In this chapter, we have reviewed the rich and growing structural information about the assemblies that form when Ag^+ ions are added to DNA oligonucleotides or monomeric nucleobases. Although some of these systems have been studied since the 1950s, reliable structural information is now finally coming into focus thanks to a growing number of solved crystal structures. We studied the excited-state dynamics of some representative silver-DNA assemblies for the first time using steady-state (UV-Vis, CD, photoluminescence, and FTIR) and time-resolved spectroscopy with UV-vis-NIR and mid-IR probing. Assemblies made from Ag^+ ions and the purine bases adenine and 2-aminopurine form nanofibers in solution, and the transient absorption signals from these systems closely resemble ones seen in strands containing stacked adenines. We propose that excimers form in both systems. This underscores the importance of π-stacking in silver-DNA assemblies made from monomeric nucleobases. The observation of excimer formation in these systems furthermore is evidence of the primacy of π–π interactions in shaping DNA excited-state dynamics. By comparing TRIR experiments with high-level quantum mechanical (QM) calculations, we were also able to detect novel triplet excimer states in Ag-dC_{20} duplexes. These assemblies have distinctive inter-base-pair hydrogen bonds that are a hallmark of silver base pairs and that enforce a high degree of propeller twisting.

Several directions appear ripe for future study. More efforts are needed to understand how assemblies form and whether they attain thermodynamic equilibrium or get stuck in kinetically trapped states. As in other areas of nanoscience, models are needed to describe the aggregation and spontaneous assembly of the supramolecular structures. It has been pointed out that very little is known about how coordination polymer nucleate and grow (James 2014). New methods like cryo-EM (Cheng 2015; Lyumkis 2019) that can determine the atomistic structures of soft-matter systems hold much promise. The fascinating field of DNA origami (Hong et al. 2017) can also be a source of new paradigms. DNA origami can organize pre-formed silver nanoparticles (Pal et al. 2010), or Ag^+ ions can create new structures that can then be chemically reduced, yielding chiral, metallized DNA nanostructures (Zhang et al. 2021).

Our experiments have been performed on as-prepared solutions without any attempt to isolate or purify the final structures. Researchers studying DNA-stabilized silver clusters have shown that the raw synthesis mixtures in their experiments contain distributions of structures, which can differ dramatically in their photophysical properties (Schultz and Gwinn 2012; Cerretani et al. 2017; Gonzàlez-Rosell et al. 2021). Distributions of structures are almost certainly present in many silver–DNA samples, but little is known about their detailed speciation and monodispersity. Better understanding of the nucleation and growth mechanisms that lead to colloidal metal-DNA assemblies will benefit efforts to understand and control heterogeneity. Approaches used by workers in the field of DNA-encapsulated silver clusters such as the use of high-performance liquid chromatography (HPLC) to isolate particular structures may help. Better understanding of the heterogeneity present in solutions of assemblies may offer ways to synthesize or isolate desired structures and provide access to more precise model systems for probing photophysics.

There are also challenging questions related to chirality. How does chirality emerge in frameworks with helical complexes? Does chirality emerge over mesoscale distances in self-assembled nanofibers as suggested by the mysterious unisignate CD spectra? Or do these effects arise from the highly twisted base pairs that are omnipresent in many silver-DNA duplexes? Ab initio calculations that explore these questions are needed. In addition, there is likely much that can be learned from recording anisotropic absorption spectra of oriented samples. This approach was first implemented a half century ago but would benefit from the more precise nanostructures that can be synthesized using today's custom-sequence oligonucleotides and advanced computational tools. All in all, the improving prospects for atomistic structural characterization combined with increasingly powerful spectroscopic approaches make it easy to predict future advances in fundamental understanding of the photoproperties of silver–DNA nanostructures.

Acknowledgments This work was performed with partial financial support by the US National Science Foundation (CHE-1800471). Structures in some figures were generated using the ChimeraX software (Pettersen et al. 2021) or the Mercury 4.0 program (Macrae et al. 2020).

References

Abou Assi H, Garavís M, González C, Damha MJ (2018) i-Motif DNA: structural features and significance to cell biology. Nucleic Acids Res 46:8038–8056. https://doi.org/10.1093/nar/gky735

Al-Mahamad LLG, El-Zubir O, Smith DG et al (2017) A coordination polymer for the site-specific integration of semiconducting sequences into DNA-based materials. Nat Commun 8:720. https://doi.org/10.1038/s41467-017-00852-6

Amo-Ochoa P, Zamora F (2014) Coordination polymers with nucleobases: from structural aspects to potential applications. Coord Chem Rev 276:34–58. https://doi.org/10.1016/j.ccr.2014.05.017

An J, Geib SJ, Rosi NL (2010) High and selective CO_2 uptake in a cobalt adeninate metal–organic framework exhibiting pyrimidine- and amino-decorated pores. J Am Chem Soc 132:38–39. https://doi.org/10.1021/ja909169x

Anderson SL, Boyd PG, Gładysiak A et al (2019) Nucleobase pairing and photodimerization in a biologically derived metal-organic framework nanoreactor. Nat Commun 10:1612. https://doi.org/10.1038/s41467-019-09486-2

Applequist J, Damle V (1966) Thermodynamics of one-stranded helix-coil equilibrium in polyadenylic acid. J Am Chem Soc 88:3895–3900

Arya SK, Yang JT (1975) Optical rotatory dispersion and circular dichroism of silver(I):Polyribonucleotide complexes. Biopolymers 14:1847–1861. https://doi.org/10.1002/bip.1975.360140907

Atsugi T, Ono A, Tasaka M et al (2022) A novel Ag^{I}-DNA rod comprising a one-dimensional array of 11 silver ions within a double helical structure. Angew Chem Int Ed 61:e202204798. https://doi.org/10.1002/anie.202204798

Atwell S, Meggers E, Spraggon G, Schultz PG (2001) Structure of a copper-mediated base pair in DNA. J Am Chem Soc 123:12364–12367. https://doi.org/10.1021/ja011822e

Bernal JD (1949) The physical basis of life. Proc Phys Soc Lond Sect A 62:537–558
Bethur E, Guha R, Zhao Z et al (2024) Formation and nanomechanical properties of silver-mediated guanine DNA duplexes in aqueous solution. ACS Nano 18:3002–3010. https://doi.org/10.1021/acsnano.3c08008
Böhnke H, Röttger K, Ingle RA et al (2019) Electronic relaxation dynamics of UV-photoexcited 2-aminopurine–thymine base pairs in Watson–Crick and Hoogsteen conformations. J Phys Chem B 123:2904–2914. https://doi.org/10.1021/acs.jpcb.9b02361
Bose K, Lech CJ, Heddi B, Phan AT (2018) High-resolution AFM structure of DNA G-wires in aqueous solution. Nat Commun 9:1959. https://doi.org/10.1038/s41467-018-04016-y
Cerretani C, Carro-Temboury MR, Krause S et al (2017) Temperature dependent excited state relaxation of a red emitting DNA-templated silver nanocluster. Chem Commun 53:12556–12559. https://doi.org/10.1039/c7cc06785b
Cerretani C, Kanazawa H, Vosch T, Kondo J (2019) Crystal structure of a NIR-emitting DNA-stabilized Ag_{16} nanocluster. Angew Chem Int Ed 58:17153–17157. https://doi.org/10.1002/anie.201906766
Cerretani C, Kondo J, Vosch T (2020) Removal of the A_{10} adenosine in a DNA-stabilized Ag_{16} nanocluster. RSC Adv 10:23854–23860. https://doi.org/10.1039/d0ra02672g
Chen J, Kohler B (2014) Base stacking in adenosine dimers revealed by femtosecond transient absorption spectroscopy. J Am Chem Soc 136:6362–6372. https://doi.org/10.1021/ja501342b
Chen J, Thazhathveetil AK, Lewis FD, Kohler B (2013) Ultrafast excited-state dynamics in hexaethyleneglycol-linked DNA homoduplexes made of AT base pairs. J Am Chem Soc 135:10290–10293. https://doi.org/10.1021/ja4049459
Chen J, Zhang Y, Kohler B (2015) Excited states in DNA strands investigated by ultrafast laser spectroscopy. Top Curr Chem 356:39–88. https://doi.org/10.1007/128_2014_570
Chen X, Karpenko A, Lopez-Acevedo O (2017) Silver-mediated double helix: structural parameters for a robust DNA building block. ACS Omega 2:7343–7348. https://doi.org/10.1021/acsomega.7b01089
Chen X, Makkonen E, Golze D, Lopez-Acevedo O (2018) Silver-stabilized guanine duplex: structural and optical properties. J Phys Chem Lett 9:4789–4794. https://doi.org/10.1021/acs.jpclett.8b01908
Cheng Y (2015) Single-particle cryo-EM at crystallographic resolution. Cell 161:450–457. https://doi.org/10.1016/j.cell.2015.03.049
Clever GH, Kaul C, Carell T (2007) DNA-metal base pairs. Angew Chem-Int Ed 46:6226–6236. https://doi.org/10.1002/anie.200701185
Cohen B, Hare PM, Kohler B (2003) Ultrafast excited-state dynamics of adenine and monomethylated adenines in solution: implications for the nonradiative decay mechanism. J Am Chem Soc 125:13594–13601
Copp SM, Gonzàlez-Rosell A (2021) Large-scale investigation of the effects of nucleobase sequence on fluorescence excitation and stokes shifts of DNA-stabilized silver clusters. Nanoscale 13:4602–4613. https://doi.org/10.1039/D0NR08300C
Copp SM, Schultz DE, Swasey S, Gwinn EG (2015) Atomically precise arrays of fluorescent silver clusters: a modular approach for metal cluster photonics on DNA nanostructures. ACS Nano 9:2303–2310. https://doi.org/10.1021/nn506322q
Crespo-Hernández CE, Cohen B, Hare PM, Kohler B (2004) Ultrafast excited-state dynamics in nucleic acids. Chem Rev 104:1977–2019. https://doi.org/10.1021/cr0206770
Crespo-Hernández CE, Cohen B, Kohler B (2005) Base stacking controls excited-state dynamics in A·T DNA. Nature 436:1141–1144. https://doi.org/10.1038/nature03933
Dairaku T, Furuita K, Sato H et al (2016) Structure determination of an Ag^{I}-mediated cytosine-cytosine base pair within DNA duplex in solution with $^{1}H/^{15}N/^{109}Ag$ NMR spectroscopy. Chem Eur J 22:13028–13031. https://doi.org/10.1002/chem.201603048
Dash J, Patil AJ, Das RN et al (2011) Supramolecular hydrogels derived from silver ion-mediated self-assembly of 5′-guanosine monophosphate. Soft Matter 7:8120–8126. https://doi.org/10.1039/C1SM05839H

Daune M, Dekker CA, Schachman HK (1966) Complexes of silver ion with natural and synthetic polynucleotides. Biopolymers 4:51–76. https://doi.org/10.1002/bip.1966.360040107
Day HA, Huguin C, Waller ZAE (2013) Silver cations fold i-motif at neutral pH. Chem Commun 49:7696–7698. https://doi.org/10.1039/C3CC43495H
Dickerson RE (1989) Definitions and nomenclature of nucleic acid structure components. Nucleic Acids Res 17:1797–1803. https://doi.org/10.1093/nar/17.5.1797
Eichhorn GL, Butzow JJ, Clark P, Tarien E (1967) Interaction of metal ions with polynucleotides and related compounds. X. Studies on the reaction of silver(I) with the nucleosides and polynucleotides, and the effect of silver(I) on the zinc(II) degradation of polynucleotides. Biopolymers 5:283–296. https://doi.org/10.1002/bip.1967.360050306
Eidelshtein G, Fardian-Melamed N, Gutkin V et al (2016) Synthesis and properties of novel silver-containing DNA molecules. Adv Mater 28:4839–4844. https://doi.org/10.1002/adma.201505049
Espinosa Leal LA, Karpenko A, Swasey S et al (2015) The role of hydrogen bonds in the stabilization of silver-mediated cytosine tetramers. J Phys Chem Lett 6:4061–4066. https://doi.org/10.1021/acs.jpclett.5b01864
Ferris JP, Hill AR, Liu R, Orgel LE (1996) Synthesis of long prebiotic oligomers on mineral surfaces. Nature 381:59–61. https://doi.org/10.1038/381059a0
Flamme M, Levi-Acobas F, Hensel S et al (2020) Enzymatic construction of artificial base pairs: the effect of metal shielding. Chembiochem 21:3398–3409. https://doi.org/10.1002/cbic.202000402
Funai T, Miyazaki Y, Aotani M et al (2012) Ag^{I} Ion mediated formation of a C–A mispair by DNA polymerases. Angew Chem Int Ed 51:6464–6466. https://doi.org/10.1002/anie.201109191
Funai T, Aotani M, Kiriu R et al (2020) Silver(I)-ion-mediated cytosine-containing base pairs: metal ion specificity for duplex stabilization and susceptibility toward DNA polymerases. Chembiochem 21:517–522. https://doi.org/10.1002/cbic.201900450
Gagnon C, Beauchamp AL (1975) Metal binding at N1 and N7 in a silver nitrate-9-methyladenine complex. Inorganica Chim Acta 14:L52–L52. https://doi.org/10.1016/s0020-1693(00)85709-6
Gehring K, Leroy J-L, Guéron M (1993) A tetrameric DNA structure with protonated cytosine-cytosine base pairs. Nature 363:561–565. https://doi.org/10.1038/363561a0
Gellert M, Lipsett MN, Davies DR (1962) Helix formation by guanylic acid. Proc Natl Acad Sci 48:2013–2018. https://doi.org/10.1073/pnas.48.12.2013
Gillen K, Jensen R, Davidson N (1964) Binding of silver ion by adenine and substituted adenines. J Am Chem Soc 86:2792–2796. https://doi.org/10.1021/ja01068a008
Goncharova I (2014) Ag (I)-mediated homo and hetero pairs of guanosine and cytidine: monitoring by circular dichroism spectroscopy. Spectrochim Acta A Mol Biomol Spectrosc 118:221–227
Gonzàlez-Rosell A, Cerretani C, Mastracco P et al (2021) Structure and luminescence of DNA-templated silver clusters. Nanoscale Adv 3:1230–1260. https://doi.org/10.1039/D0NA01005G
Groom CR, Bruno IJ, Lightfoot MP, Ward SC (2016) The Cambridge structural database. Acta Crystallogr Sect B Struct Sci Cryst Eng Mater 72:171–179. https://doi.org/10.1107/S2052520616003954
Gupta RK, Riaz M, Ashafaq M et al (2022) Adenine-incorporated metal–organic frameworks. Coord Chem Rev 464:214558. https://doi.org/10.1016/j.ccr.2022.214558
Gustavsson T, Improta R, Markovitsi D (2010) DNA/RNA: building blocks of life under UV irradiation. J Phys Chem Lett 1:2025–2030. https://doi.org/10.1021/jz1004973
Gwinn E, Schultz D, Copp SM, Swasey S (2015) DNA-protected silver clusters for nanophotonics. Nanomaterials 5:180–207. https://doi.org/10.3390/nano5010180
Hong F, Zhang F, Liu Y, Yan H (2017) DNA origami: scaffolds for creating higher order structures. Chem Rev 117:12584–12640. https://doi.org/10.1021/acs.chemrev.6b00825
Huard DJE, Demissie A, Kim D et al (2019) Atomic structure of a fluorescent Ag_8 cluster templated by a multistranded DNA scaffold. J Am Chem Soc 141:11465–11470. https://doi.org/10.1021/jacs.8b12203

Ihara T, Ishii T, Araki N et al (2009) Silver ion unusually stabilizes the structure of a parallel-motif DNA triplex. J Am Chem Soc 131:3826–3827. https://doi.org/10.1021/ja809702n

Improta R, Santoro F, Blancafort L (2016) Quantum mechanical studies on the photophysics and the photochemistry of nucleic acids and nucleobases. Chem Rev 116:3540–3593. https://doi.org/10.1021/acs.chemrev.5b00444

Izatt RM, Christensen JJ, Rytting JH (1971) Sites and thermodynamic quantities associated with proton and metal ion interaction with ribonucleic acid, deoxyribonucleic acid, and their constituent bases, nucleosides, and nucleotides. Chem Rev 71:439–482

James SL (2014) Tackling a difficult question: how do crystals of coordination polymers form? IUCrJ 1:263–264. https://doi.org/10.1107/S2052252514018624

Jensen RH, Davidson N (1966) Spectrophotometric potentiometric and density gradient ultracentrifugation studies of binding of silver ion by DNA. Biopolymers 4:17–32. https://doi.org/10.1002/bip.1966.360040104

Jiang W-Y, Ran S-Y (2018) Two-stage DNA compaction induced by silver ions suggests a cooperative binding mechanism. J Chem Phys 148:205102. https://doi.org/10.1063/1.5025348

Johannsen S, Megger N, Bohme D et al (2010) Solution structure of a DNA double helix with consecutive metal-mediated base pairs. Nat Chem 2:229–234. https://doi.org/10.1038/nchem.512

Johnson BB, Dahl KS, Tinoco I Jr et al (1981) Correlations between deoxyribonucleic acid structural parameters and calculated circular dichroism spectra. Biochemistry 20:73–78. https://doi.org/10.1021/bi00504a013

Jones AC, Neely RK (2015) 2-aminopurine as a fluorescent probe of DNA conformation and the DNA-enzyme interface. Q Rev Biophys 48:244–279. https://doi.org/10.1017/s0033583514000158

Jordan CF, Venable JH, Lerman LS (1972) Structure and circular-dichroism of DNA in concentrated polymer-solutions. Nat-New Biol 236:67–70. https://doi.org/10.1038/newbio236067a0

Kang BH, Gao ZF, Li N et al (2016) Thiazole orange as a fluorescent probe: label-free and selective detection of silver ions based on the structural change of i-motif DNA at neutral pH. Talanta 156–157:141–146. https://doi.org/10.1016/j.talanta.2016.05.006

Katz S (1952) The reversible reaction of sodium thymonucleate and mercuric chloride. J Am Chem Soc 74:2238–2245. https://doi.org/10.1021/ja01129a023

Kaul C, Müller M, Wagner M et al (2011) Reversible bond formation enables the replication and amplification of a crosslinking salen complex as an orthogonal base pair. Nat Chem 3:794–800. https://doi.org/10.1038/nchem.1117

Keane PM, Wojdyla M, Doorley GW et al (2014) Long-lived excited states in i-motif DNA studied by picosecond time-resolved IR spectroscopy. Chem Commun 50:2990–2992. https://doi.org/10.1039/c3cc46594b

Kim MH, Ulibarri L, Keller D et al (1986) The psi-type circular dichroism of large molecular aggregates. 3. calculations. J Chem Phys 84:2981–2989. https://doi.org/10.1063/1.450279

Kohl FR, Zhang Y, Charnay AP et al (2020) Ultrafast excited state dynamics of silver ion-mediated cytosine–cytosine base pairs in metallo-DNA. J Chem Phys 153:105104. https://doi.org/10.1063/5.0020463

Kondo J, Tada Y, Dairaku T et al (2015) High-resolution crystal structure of a silver(I)–RNA hybrid duplex containing Watson–Crick-like C-silver(I)-C metallo-base pairs. Angew Chem Int Ed 54:13323–13326. https://doi.org/10.1002/anie.201507894

Kondo J, Tada Y, Dairaku T et al (2017) A metallo-DNA nanowire with uninterrupted one-dimensional silver array. Nat Chem 9:956–960. https://doi.org/10.1038/nchem.2808

Krishnan-Ghosh Y, Liu D, Balasubramanian S (2004) Formation of an interlocked quadruplex dimer by d(GGGT). J Am Chem Soc 126:11009–11016. https://doi.org/10.1021/ja049259y

Kufner CL, Zinth W, Bucher DB (2020) UV-induced charge-transfer states in short guanosine-containing DNA oligonucleotides. Chembiochem 21:2306–2310. https://doi.org/10.1002/cbic.202000103

Lavery R, Zakrzewska K, Sun JS, Harvey SC (1992) A comprehensive classification of nucleic-acid structural families based on strand direction and base pairing. Nucleic Acids Res 20:5011–5016. https://doi.org/10.1093/nar/20.19.5011
Levitt M (1978) How many base-pairs per turn does DNA have in solution and in chromatin? Some theoretical calculations. Proc Natl Acad Sci 75:640–644. https://doi.org/10.1073/pnas.75.2.640
Linares F, Garcia-Fernandez E, Lopez-Garzon FJ et al (2019) Multifunctional behavior of molecules comprising stacked cytosine-Ag-I-cytosine base pairs; towards conducting and photoluminescence silver-DNA nanowires. Chem Sci 10:1126–1137. https://doi.org/10.1039/c8sc04036b
Liu HH, Shen FS, Haruehanroengra P et al (2017) A DNA structure containing Ag^{I}-mediated G:G and C:C base pairs. Angew Chem-Int Ed 56:9430–9434. https://doi.org/10.1002/anie.201704891
Livolant F, Maestre MF (1988) Circular dichroism microscopy of compact forms of DNA and chromatin in vivo and in vitro: cholesteric liquid-crystalline phases of DNA and single dinoflagellate nuclei. Biochemistry 27:3056–3068. https://doi.org/10.1021/bi00408a058
Loo K, Degtyareva N, Park J et al (2010) Ag^{+}-mediated assembly of 5'-guanosine monophosphate. J Phys Chem B 114:4320–4326. https://doi.org/10.1021/jp908085s
Lopez A, Liu JW (2017) Self-assembly of nucleobase, nucleoside and nucleotide coordination polymers: from synthesis to applications. Chemnanomat 3:670–684. https://doi.org/10.1002/cnma.201700154
Lopez A, Liu J (2019) Coordination nanoparticles formed by fluorescent 2-aminopurine and Au3+: stability and nanozyme activities. J Anal Test 3:219–227. https://doi.org/10.1007/s41664-019-00112-0
Lyumkis D (2019) Challenges and opportunities in cryo-EM single-particle analysis. J Biol Chem 294:5181–5197. https://doi.org/10.1074/jbc.REV118.005602
Macrae CF, Sovago I, Cottrell SJ et al (2020) Mercury 4.0: from visualization to analysis, design and prediction. J Appl Crystallogr 53:226–235. https://doi.org/10.1107/S1600576719014092
Marsh TC, Henderson E (1994) G-wires: self-assembly of a telomeric oligonucleotide, d(GGGGTTGGGG), into large superstructures. Biochemistry 33:10718–10724. https://doi.org/10.1021/bi00201a020
Marsh TC, Vesenka J, Henderson E (1995) A new DNA nanostructure, the G-wire, imaged by scanning probe microscopy. Nucleic Acids Res 23:696–700. https://doi.org/10.1093/nar/23.4.696
Martinez-Fernandez L, Zhang Y, de La Harpe K et al (2016) Photoinduced long-lived charge transfer excited states in AT-DNA strands. Phys Chem Chem Phys 18:21241–21245. https://doi.org/10.1039/C6CP04550B
Martínez-Fernández L, Kohl FR, Zhang Y et al (2024) Triplet excimer formation in a DNA duplex with silver ion-mediated base pairs. J Am Chem Soc 146:1914–1925. https://doi.org/10.1021/jacs.3c08793
Matsika S, Krause P (2011) Nonadiabatic events and conical intersections. Annu Rev Phys Chem 62:621–643. https://doi.org/10.1146/annurev-physchem-032210-103450
Matsuoka Y, Nordén B, Kurucsev T (1984) Formation of silver adenine long-chain aggregates in neutral aqueous solution. Study of flow linear dichroism. J Chem Soc-Chem Commun:1573–1574. https://doi.org/10.1039/c39840001573
Matsuoka Y, Nordén B, Kurucsev T (1985) Nucleic acid-metal interactions. III. Complexes of Ag(I) with adenine and 1-methyladenine from studies of UV and IR dichroic spectra. J Crystallogr Spectrosc Res 15:545–560. https://doi.org/10.1007/BF01164770
Megger DA, Müller J (2010) Silver(I)-mediated cytosine self-pairing is preferred over Hoogsteen-type base pairs with the artificial nucleobase 1,3-Dideaza-6-Nitropurine. Nucleosides Nucleotides Nucleic Acids 29:27–38. https://doi.org/10.1080/15257770903451579
Megger DA, Fonseca Guerra C, Bickelhaupt FM, Müller J (2011) Silver(I)-mediated Hoogsteen-type base pairs. J Inorg Biochem 105:1398–1404. https://doi.org/10.1016/j.jinorgbio.2011.07.005
Middleton CT, de La Harpe K, Su C et al (2009) DNA excited-state dynamics: from single bases to the double helix. Annu Rev Phys Chem 60:217–239

Mishra AK, Prajapati RK, Verma S (2010) Probing structural consequences of N9-alkylation in silver-adenine frameworks. Dalton Trans 39:10034–10037. https://doi.org/10.1039/c0dt00557f
Mistry L, El-Zubir O, Dura G et al (2019a) Addressing the properties of "Metallo-DNA" with a Ag(I)-mediated supramolecular duplex. Chem Sci 10:3186–3195. https://doi.org/10.1039/c8sc05103h
Mistry L, Waddell PG, Wright NG et al (2019b) Transoid and cisoid conformations in silver-mediated cytosine base pairs: hydrogen bonding dictates argentophilic interactions in the solid state. Inorg Chem 58:13346–13352. https://doi.org/10.1021/acs.inorgchem.9b02228
Mistry L, El-Zubir O, Pope T et al (2021) Silver–cytidine coordination polymer: electrical properties, modulating intrachain Ag…Ag distance, and MOF–nanosheet transformation. Cryst Growth Des 21:4398–4405. https://doi.org/10.1021/acs.cgd.1c00330
Mohan S, Yathindra N (1991) Studies on the cross strand hydrogen bonds in DNA double helices. J Biomol Struct Dyn 9:113–126. https://doi.org/10.1080/07391102.1991.10507897
Mohapatra B, Pratibha, Verma S (2017) Directed adenine functionalization for creating complex architectures for material and biological applications. Chem Commun 53:4748–4758. https://doi.org/10.1039/c7cc00222j
Mohapatra B, Verma S (2016) Crystal engineering with 2-aminopurine and 2,6-diaminopurine derivatives: dimers, metallaquartets, and halide-bridged clusters. Cryst Growth Des 16:696–704. https://doi.org/10.1021/acs.cgd.5b01257
Müller J (2019) Nucleic acid duplexes with metal-mediated base pairs and their structures. Coord Chem Rev 393:37–47. https://doi.org/10.1016/j.ccr.2019.05.007
Munakata M, Wu LP, Kuroda-Sowa T (1998) Toward the construction of functional solid-state supramolecular metal complexes containing copper(I) and silver(I). In: Sykes AG (ed) Advances in inorganic chemistry. Academic Press, pp 173–303
Naskar S, Guha R, Müller J (2020) Metal-modified nucleic acids: metal-mediated base pairs, triples, and tetrads. Angew Chem Int Ed 59:1397–1406. https://doi.org/10.1002/anie.201905913
Nelson HCM, Finch JT, Luisi BF, Klug A (1987) The structure of an oligo(dA·oligo(dT) tract and its biological implications. Nature 330:221–226
New SY, Lee ST, Su XD (2016) DNA-templated silver nanoclusters: structural correlation and fluorescence modulation. Nanoscale 8:17729–17746. https://doi.org/10.1039/C6NR05872H
Nishiyabu R, Hashimoto N, Cho T et al (2009) Nanoparticles of adaptive supramolecular networks self-assembled from nucleotides and lanthanide ions. J Am Chem Soc 131:2151–2158. https://doi.org/10.1021/ja8058843
Nordén B, Matsuoka Y, Kurucsev T (1986) Nucleic acid-metal interactions: V. The effect of silver(I) on the structures of A-DNA and B-DNA forms. Biopolymers 25:1531–1545. https://doi.org/10.1002/bip.360250811
O'Neill PR, Velazquez LR, Dunn DG et al (2009) Hairpins with poly-C loops stabilize four types of fluorescent Ag_n:DNA. J Phys Chem C 113:4229–4233. https://doi.org/10.1021/jp809274m
Olson WK, Bansal M, Burley SK et al (2001) A standard reference frame for the description of nucleic acid base-pair geometry. J Mol Biol 313:229–237
Ono A, Cao S, Togashi H et al (2008) Specific interactions between silver(I) ions and cytosine-cytosine pairs in DNA duplexes. Chem Commun 4825–4827. https://doi.org/10.1039/b808686a
Ono A, Kanazawa H, Ito H et al (2019) A novel DNA helical wire containing Hg^{II}-mediated T:T and T:G pairs. Angew Chem Int Ed 58:16835–16838. https://doi.org/10.1002/anie.201910029
Pal S, Deng Z, Ding B et al (2010) DNA-origami-directed self-assembly of discrete silver-nanoparticle architectures. Angew Chem Int Ed 49:2700–2704. https://doi.org/10.1002/anie.201000330
Park KS, Jung C, Park HG (2010) "Illusionary" polymerase activity triggered by metal ions: use for molecular logic-gate operations. Angew Chem Int Ed 49:9757–9760. https://doi.org/10.1002/anie.201004406
Pauling L, Corey RB (1953) A proposed structure for the nucleic acids. Proc Natl Acad Sci 39:84–97. https://doi.org/10.1073/pnas.39.2.84

Peters GM, Davis JT (2016) Supramolecular gels made from nucleobase, nucleoside and nucleotide analogs. Chem Soc Rev 45:3188–3206. https://doi.org/10.1039/c6cs00183a
Petropoulos V, Uboldi L, Maiuri M et al (2023) Effect of the DNA polarity on the relaxation of its electronic excited states. J Phys Chem Lett 14:10219–10224. https://doi.org/10.1021/acs.jpclett.3c02580
Pettersen EF, Goddard TD, Huang CC et al (2021) UCSF ChimeraX: structure visualization for researchers, educators, and developers. Protein Sci 30:70–82. https://doi.org/10.1002/pro.3943
Petty JT, Zheng J, Hud NV, Dickson RM (2004) DNA-templated Ag nanocluster formation. J Am Chem Soc 126:5207–5212. https://doi.org/10.1021/ja031931o
Polonius F-A, Müller J (2007) An artificial base pair, mediated by hydrogen bonding and metal-ion binding. Angew Chem Int Ed 46:5602–5604. https://doi.org/10.1002/anie.200700315
Purohit CS, Verma S (2006) A luminescent silver-adenine metallamacrocyclic quartet. J Am Chem Soc 128:400–401. https://doi.org/10.1021/ja056452z
Purohit CS, Mishra AK, Verma S (2007) Four-stranded coordination helices containing silver-adenine (purine) metallaquartets. Inorg Chem 46:8493–8495. https://doi.org/10.1021/ic701465d
Rahn RO, Landry LC (1973) Ultraviolet irradiation of nucleic acids complexed with heavy atoms-II. Phosphorescence and photodimerization of DNA complexed with Ag. Photochem Photobiol 18:29–38
Ramasanoff RR, Sokolov PA (2022) Intersystem crossing rates of violet-, green- and red-emitting DNA stabilized silver luminescent clusters. Chem Phys Lett 807:140081. https://doi.org/10.1016/j.cplett.2022.140081
Roberts GM, Marroux HJB, Grubb MP et al (2014) On the participation of photoinduced N-H bond fission in aqueous adenine at 266 and 220 nm: a combined ultrafast transient electronic and vibrational absorption spectroscopy study. J Phys Chem A 118:11211–11225. https://doi.org/10.1021/jp508501w
Röttger K, Marroux HJB, Grubb MP et al (2015) Ultraviolet absorption induces hydrogen-atom transfer in G·C Watson–Crick DNA base pairs in solution. Angew Chem Int Ed 54:14719–14722. https://doi.org/10.1002/anie.201506940
Sasikumar D, John AT, Sunny J, Hariharan M (2020) Access to the triplet excited states of organic chromophores. Chem Soc Rev 49:6122–6140. https://doi.org/10.1039/D0CS00484G
Schmidbaur H, Schier A (2015) Argentophilic Interactions. Angew Chem Int Ed 54:746–784. https://doi.org/10.1002/anie.201405936
Schultz D, Gwinn EG (2012) Silver atom and strand numbers in fluorescent and dark Ag:DNAs. Chem Commun 48:5748–5750. https://doi.org/10.1039/c2cc17675k
Serrano-Andrés L, Merchán M, Borin AC (2006) Adenine and 2-aminopurine: paradigms of modern theoretical photochemistry. Proc Natl Acad Sci USA 103:8691–8696
Sharma B, Mahata A, Mandani S et al (2016) Coordination polymer hydrogels through Ag(I)-mediated spontaneous self-assembly of unsubstituted nucleobases and their antimicrobial activity. RSC Adv 6:62968–62973. https://doi.org/10.1039/c6ra11137h
Shi Y, Sun H, Xiang J et al (2015) i-Motif-modulated fluorescence detection of silver(I) with an ultrahigh specificity. Anal Chim Acta 857:79–84. https://doi.org/10.1016/j.aca.2014.12.002
Shibata M, Zielinski TJ (1994) Unusual cross-strand hydrogen bonds in oligopurine·oligopyrimidine duplexes. In: Modeling the hydrogen bond. American Chemical Society, pp 194–208
Snyder JA, Charnay AP, Kohl FR et al (2019) DNA-like photophysics in self-assembled silver(I)–nucleobase nanofibers. J Phys Chem B 123:5985–5994. https://doi.org/10.1021/acs.jpcb.9b00660
Studdert DS, Davis RC (1974a) Calculations of the circular dichroism of double-helical nucleic acids. I. Effects involving $\pi \rightarrow \pi^*$ transitions. Biopolymers 13:1377–1389. https://doi.org/10.1002/bip.1974.360130709
Studdert DS, Davis RC (1974b) Calculations of the circular dichroism of double-helical nucleic acids. II. Effects involving $\pi \rightarrow \pi^*$ transitions. Biopolymers 13:1391–1403

Stylianou KC, Warren JE, Chong SY et al (2011) CO_2 selectivity of a 1D microporous adenine-based metal–organic framework synthesised in water. Chem Commun 47:3389–3391. https://doi.org/10.1039/C0CC05559J

Su C, Middleton CT, Kohler B (2012) Base-stacking disorder and excited-state dynamics in single-stranded adenine homo-oligonucleotides. J Phys Chem B 116:10266–10274. https://doi.org/10.1021/jp305350t

Sun Q, Xie X, Song Y, Sun L (2022) A review on silver-mediated DNA base pairs: methodology and application. Biomater Res 26:9. https://doi.org/10.1186/s40824-022-00254-w

Swasey SM, Gwinn EG (2016) Silver-mediated base pairings: towards dynamic DNA nanostructures with enhanced chemical and thermal stability. New J Phys 18:045008

Swasey SM, Leal LE, Lopez-Acevedo O et al (2015) Silver (I) as DNA glue: Ag^+-mediated guanine pairing revealed by removing Watson-Crick constraints. Sci Rep 5:9. https://doi.org/10.1038/srep10163

Swasey SM, Rosu F, Copp SM et al (2018) Parallel guanine duplex and cytosine duplex DNA with uninterrupted spines of Ag^I-mediated base pairs. J Phys Chem Lett 9:6605–6610. https://doi.org/10.1021/acs.jpclett.8b02851

Takaya T, Su C, de La Harpe K et al (2008) UV excitation of single DNA and RNA strands produces high yields of exciplex states between two stacked bases. Proc Natl Acad Sci USA 105:10285–10290

Terrón A, Moreno-Vachiano B, Bauzá A et al (2017a) X-ray crystal structure of a metalled double-helix generated by infinite and consecutive C*-Ag^I-C* (C*:N^1-hexylcytosine) base pairs through argentophilic and hydrogen bond interactions. Chem Eur J 23:2103–2108. https://doi.org/10.1002/chem.201604331

Terrón A, Tomas L, Bauza A et al (2017b) The first X-ray structure of a silver-nucleotide complex: interaction of ion Ag(I) with cytidine-5′-monophosphate. CrystEngComm 19:5830–5834. https://doi.org/10.1039/c7ce01400g

Thomas CA (1954) The interaction of $HgCl_2$ with sodium thymonucleate. J Am Chem Soc 76:6032–6034. https://doi.org/10.1021/ja01652a049

Thomas-Gipson J, Pérez-Aguirre R, Beobide G et al (2015) Unravelling the growth of supramolecular metal–organic frameworks based on metal-nucleobase entities. Cryst Growth Des 15:975–983. https://doi.org/10.1021/cg501804h

Tu AT, Reinosa JA (1966) The interaction of silver ion with guanosine, guanosine monophosphate, and related compounds. Determination of possible sites of complexing. Biochemistry 5:3375–3383. https://doi.org/10.1021/bi00874a042

Urata H, Yamaguchi E, Funai T et al (2010) Incorporation of thymine nucleotides by DNA polymerases through T–Hg^{II}–T base pairing. Angew Chem Int Ed 49:6516–6519. https://doi.org/10.1002/anie.201002142

Vecchioni S, Capece MC, Toomey E et al (2019) Construction and characterization of metal ion-containing DNA nanowires for synthetic biology and nanotechnology. Sci Rep 9:15. https://doi.org/10.1038/s41598-019-43316-1

Venkatesh V, Kumaran MDB, Saravanan RK et al (2016) Luminescent silver-purine double helicate: synthesis, self-assembly and antibacterial action. ChemPlusChem 81:1266–1271. https://doi.org/10.1002/cplu.201600293

Verma S, Mishra AK, Kumar J (2010) The many facets of adenine: coordination, crystal patterns, and catalysis. Acc Chem Res 43:79–91. https://doi.org/10.1021/ar9001334

Vohlídal J, Hissler M (2019) Metallo-supramolecular polymers. In: Smart inorganic polymers. Wiley, pp 141–162

Walter A, Luck G (1977) Interactions of Hg(II) ions with DNA as revealed by CD measurements. Nucleic Acids Res 4:539–550. https://doi.org/10.1093/nar/4.3.539

Wang Y, Zhao H, Yang C et al (2019) Degradation of cytosine radical cations in 2′-deoxycytidine and in i-motif DNA: hydrogen-bonding guided pathways. J Am Chem Soc 141:1970–1979

Watson JD, Crick FHC (1953) Molecular structure of nucleic acids. A structure for deoxyribose nucleic acid. Nature 171:737–738

Wei H, Li B, Du Y et al (2007) Nucleobase–metal hybrid materials: preparation of submicrometer-scale, spherical colloidal particles of adenine–Gold(III) via a supramolecular hierarchical self-assembly approach. Chem Mater 19:2987–2993. https://doi.org/10.1021/cm070028a
Weser U, Dönnicke M (1970) Stabilitätskonstanten einiger Co^{2+}-, Ni^{2+}- und Ag^{+}-nucleotide / stability constants of some Co^{2+}-, Ni^{2+}- and Ag^{+}-nucleotides. Z Für Naturforschung B 25:592–595. https://doi.org/10.1515/znb-1970-0608
wwPDB Consortium (2019) Protein Data Bank: the single global archive for 3D macromolecular structure data. Nucleic Acids Res 47:D520–D528. https://doi.org/10.1093/nar/gky949
Xie YP, Shen YL, Duan GX et al (2020) Silver nanoclusters: synthesis, structures and photoluminescence. Mater Chem Front 4:2205–2222. https://doi.org/10.1039/d0qm00117a
Xie Y, Yang M, Zhu L et al (2023) Research progress of DNA aptamer-based silver ions detection. Adv Agrochem 2:231–235. https://doi.org/10.1016/j.aac.2023.06.006
Xu L, Shen X, Hong S et al (2015) Thiazole orange as a fluorescent light-up probe for the i-motif and its application to the development of a molecular logic system. Asian J Org Chem 4:1375–1378. https://doi.org/10.1002/ajoc.201500347
Xu L, Zhang Z, Fang X et al (2018) Robust hydrogels from lanthanide nucleotide coordination with evolving nanostructures for a highly stable protein encapsulation. ACS Appl Mater Interfaces 10:14321–14330. https://doi.org/10.1021/acsami.7b18005
Yamane T, Davidson N (1961) On the complexing of desoxyribonucleic acid (DNA) by mercuric ion. J Am Chem Soc 83:2599–2607. https://doi.org/10.1021/ja01473a001
Yamane T, Davidson N (1962) On complexing of deoxyribonucleic acid by silver(I). Biochim Biophys Acta 55:609–621
Yang JT, Samejima T (1968) Effect of base tilting on the optical activity of nucleic acids: a hypothesis. Biochem Biophys Res Commun 33:739–745. https://doi.org/10.1016/0006-291x(68)90221-0
Yang M, Zhu L, Yang W, Xu W (2023) Nucleic acid-templated silver nanoclusters: a review of structures, properties, and biosensing applications. Coord Chem Rev 491:215247. https://doi.org/10.1016/j.ccr.2023.215247
Zavriev SK, Minchenkova LE, Vorlíčková M et al (1979) Circular dichroism anisotropy of DNA with different modifications at N7 of guanine. Biochim Biophys Acta BBA - Nucleic Acids Protein Synth 564:212–224. https://doi.org/10.1016/0005-2787(79)90220-X
Zhang M, Lu W, Li J-R et al (2014) Design and synthesis of nucleobase-incorporated metal–organic materials. Inorg Chem Front 1:159–162. https://doi.org/10.1039/C3QI00042G
Zhang Y, de La Harpe K, Beckstead AA et al (2015) UV-induced proton transfer between DNA strands. J Am Chem Soc 137:7059–7062. https://doi.org/10.1021/jacs.5b03914
Zhang Y, Li X-B, Fleming AM et al (2016) UV-induced proton-coupled electron transfer in cyclic DNA miniduplexes. J Am Chem Soc 138:7395–7401. https://doi.org/10.1021/jacs.6b03216
Zhang Y, de La Harpe K, Kohl FR, Kohler B (2019) Isotopic substitution affects excited state branching in a DNA duplex in aqueous solution. Chem Commun 55:4174–4177. https://doi.org/10.1039/C9CC01105F
Zhang Y, Qu Z, Jiang C et al (2021) Prescribing silver chirality with DNA origami. J Am Chem Soc 143:8639–8646. https://doi.org/10.1021/jacs.1c00363
Zhou WH, Saran R, Liu JW (2017) Metal sensing by DNA. Chem Rev 117:8272–8325. https://doi.org/10.1021/acs.chemrev.7b00063

Chapter 6
Fluorescence-Detected Circular Dichroism Spectroscopy for Analysis of Nucleic Acids: A Practical Guide and Review

Julia R. Widom

Abstract Fluorescence-detected circular dichroism (FDCD) spectroscopy combines the structural sensitivity of circular dichroism spectroscopy, which is widely used to study the structures of biological macromolecules, with the sensitivity and selectivity of fluorescence detection. It has been used to study a variety of nucleic acid-based systems, including short RNA and DNA oligonucleotides, transfer RNA, DNA–dye complexes, and DNA-templated nanostructures. This chapter surveys both theoretical and experimental aspects of FDCD, with a focus on the features that are most relevant to the study of nucleic acids. The evolution of FDCD theory, instrumentation, and applications is surveyed, and foundational and recent works applying FDCD to nucleic acids are reviewed.

Keywords Fluorescence · Circular dichroism · Polarization · Spectroscopy · Base analogues

6.1 Introduction

As chiral, multichromophoric molecules, nucleic acids interact with light in distinctive ways. Chiroptical spectroscopies such as circular dichroism (CD) have been used for many decades and were the basis for much early knowledge about the global structures that can be adopted by nucleic acids in solution. To this day, however, it remains difficult to draw quantitative structural conclusions from the broad CD spectra that are yielded by large numbers of bases absorbing in the ultraviolet region of the spectrum. As a result, quantitative structural modeling has been mostly limited to small model systems such as dinucleotides (Beyerle et al. 2021), or near-ultraviolet or visible regions of the spectrum where a probe can be excited

J. R. Widom (✉)
Department of Chemistry and Biochemistry, University of Oregon, Eugene, OR, USA
e-mail: jwidom@uoregon.edu

S. Matsika, A. H. Marcus (eds.), *Nucleic Acid Photophysics and Photochemistry*, Nucleic Acids and Molecular Biology 36,
https://doi.org/10.1007/978-3-031-68807-2_6

selectively (Ji et al. 2019; Kringle et al. 2018). As DNA and RNA are almost entirely nonfluorescent, the problem can be simplified considerably by inserting a fluorescent moiety into the nucleic acid of interest. Fluorescence detection then yields signals that are specific to the probe and its local environment, even when excitation is performed at wavelengths at which the native bases absorb. Many modalities of fluorescence spectroscopy can be applied to nucleic acids, such as steady-state and time-resolved emission spectroscopy, fluorescence anisotropy, and fluorescence-detected nonlinear spectroscopy, and the resulting signals can be used to infer information about the structure and dynamics of the native nucleic acid in the vicinity of the probe.

A powerful, though not widely used, exploitation of fluorescence detection is found in the method of fluorescence-detected circular dichroism (FDCD) spectroscopy. While standard CD spectroscopy measures the difference in the transmitted intensity when a sample is illuminated with left- or right-handed circularly polarized light (LCP or RCP), FDCD measures the difference in fluorescence intensity that results from excitation with LCP or RCP (Fig. 6.1). The method of FDCD was first reported in the 1970s (Turner et al. 1974) and the first reported applications were to the study of transfer RNA (Turner et al. 1975; Yoon et al. 1975). However, it fell out of wide use after a flurry of studies of nucleic acids and proteins in the 1980s (Dahl et al. 1982; Lamos and Turner 1985; Lamos et al. 1986a, b; Muto et al. 1986; Thomas et al. 1987, 1988; Lobenstine et al. 1981) and applications to nucleic acids have appeared only sporadically in the literature since then (Fig. 6.2). In most recent FDCD studies, nucleic acids (primarily DNA) have been a means rather than an end, being used to template the supramolecular assembly of dyes (Hofsäß et al. 2018; Ucar and Wagenknecht 2021) or to assemble chiral nanolayers on surfaces (Solomon et al. 2020). This chapter reviews both recent and foundational

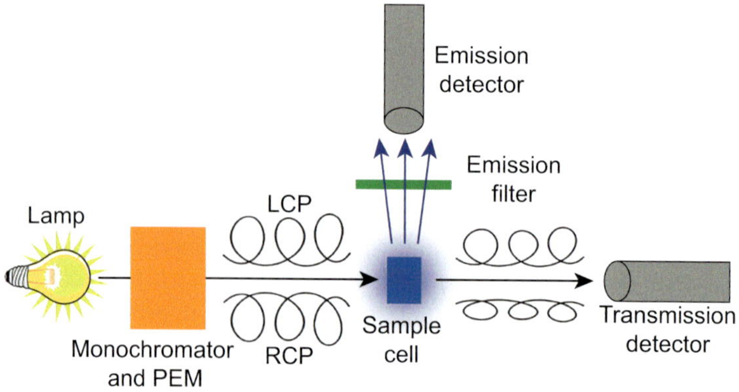

Fig. 6.1 Typical FDCD experimental layout. Light from a lamp is sent through a monochromator and photoelastic modulator (PEM) to generate alternating left-handed and right-handed circularly polarized light (LCP and RCP). An emission detector is placed at 90° from excitation with a filter to remove scattered light. LCP and RCP are depleted to different extents as they propagate through an optically active sample, and a transmission detector records standard CD and absorption spectra

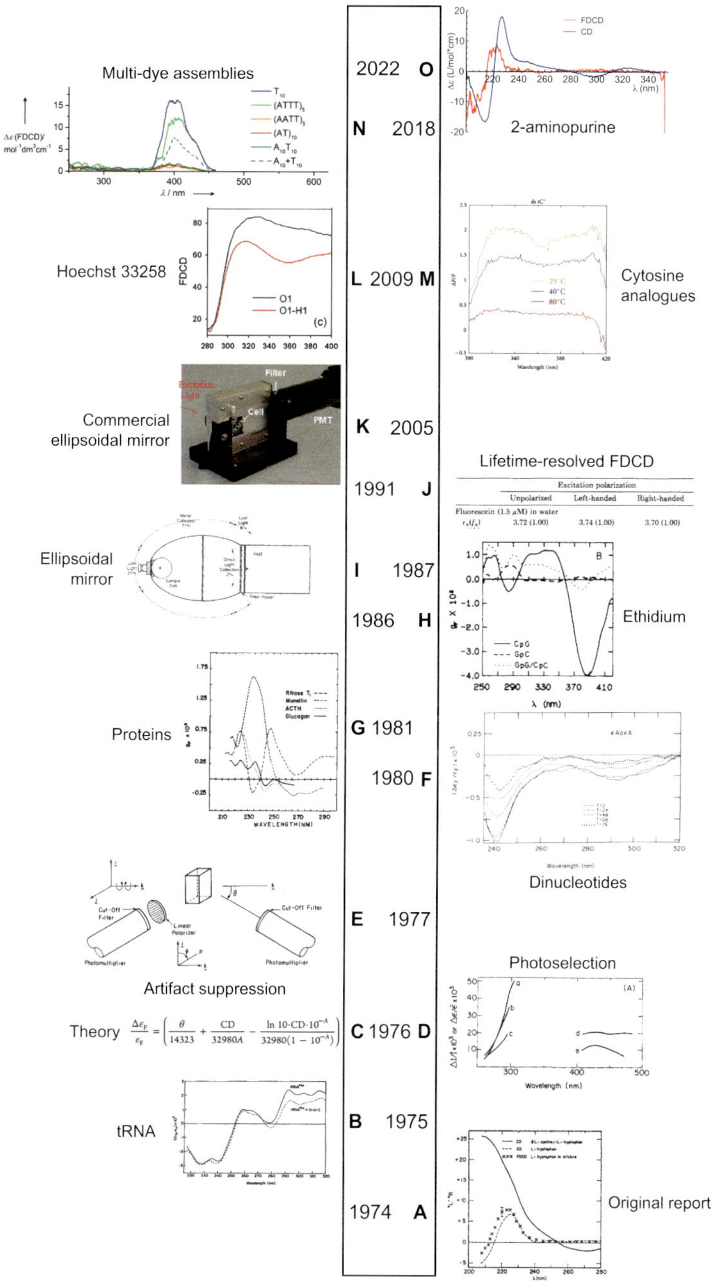

Fig. 6.2 Timeline (not to scale) of FDCD development and applications. (**A**) The original report of FDCD demonstrated that the CD spectrum of tryptophan (dashed line) could be recovered through FDCD (xxx) even in a background of excess L-cysteine (standard CD in solid line) (Turner et al. 1974). (**B**) A study on tRNA demonstrated a reduction in the long-wavelength FDCD signal

applications of FDCD to nucleic acids, presents important experimental and theoretical considerations that must be accounted for, and discusses the potential for deeper and broader applications of FDCD in the analysis of nucleic acids.

6.2 Key Theoretical and Experimental Considerations

Rather than exhaustively reviewing the theory and experimental procedures underlying FDCD, this section discusses the most distinctive aspects of both with a focus on applications to nucleic acids. For more extensive discussion of FDCD theory, the reader is directed to references (Tinoco and Turner 1976; White et al. 1975; Tinoco et al. 1977), and for more extensive discussion of general experimental considerations (Prabodh et al. 2021).

Fig. 6.2 (continued) from the modified base wybutine when an oligonucleotide mimicking the phenylalanine codon was added (Yoon et al. 1975). (**C**) The theoretical basis of much subsequent FDCD work was published in 1976 (Tinoco and Turner 1976). (**D**) The first demonstration of photoselection artifacts in FDCD. Good agreement is seen between CD (curve a) and FDCD (curve b) spectra of D-camphor recorded in methanol or cyclohexane, while the FDCD spectrum recorded in glycerol (curve c) deviates significantly. Curves d and e are the FDCD signals as a function of emission wavelength in methanol or cyclohexane (d) and glycerol (e) (Ehrenberg and Steinberg 1976). (**E**) It was shown that photoselection artifacts can be suppressed by controlling the polarization of the detected emission, or the detector's angle relative to excitation (Tinoco et al. 1977). (**F**) The FDCD spectrum of ethenoadenosine dinucleotide (εApεA) was found to be temperature-dependent, indicating the presence of multiple bright, unstacked conformations (Reich and Tinoco 1980). (**G**) FDCD was used to extract the CD signals specific to the lone tryptophan residues in several proteins (Lobenstine et al. 1981). (**H**) Ethidium exhibits a particularly strong FDCD signal when complexed with dinucleotides consisting of a 5′ pyrimidine and 3′ purine (Lamos et al. 1986b). (**I**) A cell holder based on an ellipsoidal mirror was found to eliminate photoselection artifacts while greatly improving sensitivity (Bicknese and Maestre 1987). (**J**) Lifetime-resolved FDCD was demonstrated, allowing the fluorescence lifetimes following excitation with LCP or RCP to be distinguished (Wu and McGown 1991). (**K**) An ellipsoidal mirror device compatible with commercial instrumentation was developed (Nehira et al. 2005). (**L**) FDCD showed that the local environment of Hoechst 33258 bound to dsDNA (black) was altered upon complexation with histones (red) (Banerjee et al. 2009). (**M**) FDCD spectra of the new cytosine analog tC° were shown to be sensitive to denaturation of dsDNA (Bhavsar et al. 2011). (**N**) FDCD was applied to an assembly of pyrene and Nile Red-labeled nucleosides, showing that the observed fluorescence results from excitation of pyrene in a chiral geometry (Hofsäß et al. 2018). (**O**) FDCD (red) and CD (blue) spectra of 2-aminopurine dinucleotide showed that while stacked structures are prevalent, nearly all fluorescence emission comes from a small unstacked subpopulation (Widom and Hoeher 2022)

6.2.1 FDCD-Specific Theoretical Considerations

An FDCD measurement records the quantity θ_F as a function of excitation wavelength, where θ_F is the difference in fluorescence intensity under excitation with LCP (F_L) or RCP (F_R), divided by the total intensity (Eq. 6.1):

$$\theta_F = \frac{F_L - F_R}{F_L + F_R} \tag{6.1}$$

Due to its normalization by the total detected fluorescence intensity, this quantity is independent of overall sample concentration and brightness, excitation light source intensity, and detector sensitivity. The power of FDCD lies in its ability to extract spectra specific to the fluorescent species in a complex sample. A key desired quantity is $\Delta\varepsilon_F$, the difference in the extinction coefficient for LCP and RCP of the fluorescent species in the sample. As demonstrated below, measurements of θ_F can be processed to extract $\Delta\varepsilon_F$, which can be compared to $\Delta\varepsilon$ spectra obtained through standard transmission-detected CD (Tinoco and Turner 1976; White et al. 1975). $\Delta\varepsilon_F$ can also be compared to the predictions of quantitative structural models without needing to model components of the system that do not affect the photophysics of the fluorophore, such as distal native bases. Alternatively, the "dissymmetry factors" $g_F = \Delta\varepsilon_F/\varepsilon_F$ and $g = \Delta\varepsilon/\varepsilon$ can be directly compared (Fig. 6.3A), which requires fewer assumptions to be made when processing the FDCD data. The fluorescence- and transmission-derived quantities are expected to agree for samples containing a single optically active species (a species that exhibits nonzero $\Delta\varepsilon$) that is also fluorescent, as shown for *d*-10-camphor sulfonic acid and *l*-tryptophan in the original report of FDCD (Turner et al. 1974) (Fig. 6.2A).

To extract g_F from raw FDCD data, application of Beer's law yields the following equation for θ_F in millidegrees per volt (Tinoco and Turner 1976) (Eq. 6.2):

$$\theta_F = 14323\left(\frac{\Delta\varepsilon_F}{\varepsilon_F} - \frac{\Delta A}{A} + \frac{\ln 10 \cdot \Delta A \cdot 10^{-A}}{\left(1 - 10^{-A}\right)}\right) \tag{6.2}$$

where $\Delta\varepsilon_F$ and ε_F are as defined above, A is the overall absorbance of the sample in absorbance units, including both fluorescent and nonfluorescent species, and $\Delta A = A_{LCP}\text{-}A_{RCP}$. Note that this equation differs from that in Tinoco and Turner (1976) by a factor of −1 because that work defines the CD signal "S" as proportional to ΔI (the difference in intensities transmitted by the sample) instead of ΔA. The second and third terms compensate for the differential depletion of LCP and RCP as they propagate through the sample and are obtained by measuring its CD and absorption spectra. The absorbance (in absorbance units) and CD (in millidegrees) signals recorded by the transmission detector on the spectrometer are incorporated into the equation as follows:

$$\theta_{\mathrm{F}} = 14323\left(\frac{\Delta\varepsilon_{\mathrm{F}}}{\varepsilon_{\mathrm{F}}} - \frac{CD}{32980 \cdot A} + \frac{\ln 10 \cdot CD \cdot 10^{-A}}{32980\left(1-10^{-A}\right)}\right) \tag{6.3}$$

These terms are frequently abbreviated to yield the following form:

$$\theta_{\mathrm{F}} = 14323\left(\frac{\Delta\varepsilon_{\mathrm{F}}}{\varepsilon_{\mathrm{F}}} - 2R\right) \tag{6.4}$$

This equation can be rearranged to yield the excitation wavelength-dependent fluorescence-detected dissymmetry factor g_{F}, or, if the extinction coefficient spectrum for the fluorophore (ε_{F}) can be determined, $\Delta\varepsilon_{\mathrm{F}}$ itself. This quantifies the CD signal originating specifically from a fluorescent species, even if it exists in a non-fluorescent and potentially optically active background such as the native bases in a nucleic acid, the majority of the amino acids in a protein, or a quenched conformation of the fluorescent species itself (Fig. 6.3B).

6.2.2 *Special Cases Relevant to Nucleic Acids*

Multiple Fluorescent Species

Many systems cannot be assumed to contain only a single fluorescent species, and as a result are not adequately characterized by a single value of $\Delta\varepsilon_{\mathrm{F}}$. If multiple fluorescent species are present, the detected fluorescence is a sum over all species weighted by their concentrations and quantum yields, and Eq. (6.4) is adjusted to:

$$\theta_{\mathrm{F}} = 14323\left(\frac{\sum_i \phi_i c_i \Delta\varepsilon_i}{\sum_i \phi_i c_i \varepsilon_i} - 2R\right) \tag{6.5}$$

where ϕ_i is the fluorescence quantum yield of species i, c_i is its concentration, and $\Delta\varepsilon_i$ and ε_i are species i's values of $\Delta\varepsilon_{\mathrm{F}}$ and ε_{F}, respectively (Fig. 6.3C). The different fluorescent species i could be distinct conformations of a molecule that place a fluorophore in environments with different optical activity and fluorescence quantum yield, bound and free subpopulations of interaction partners, or different chemical species altogether. The first scenario is frequently observed for nucleic acids labeled with fluorescent base analogs (FBAs), which have fluorescence quantum yields that depend on their local base-stacking and/or hydrogen bonding environment (for example, see (Hardman et al. 2008; Hardman and Thompson 2006; Liang and Matsika 2011; Rachofsky et al. 2001) for discussion of the structure–photophysics relationships of the adenine analog 2-aminopurine [2-AP] and the cytosine analog pyrrolo-cytosine [pC]) (Fig. 6.4A–B). In both early and recent work, the above equation was used to analyze base-stacking heterogeneity in short

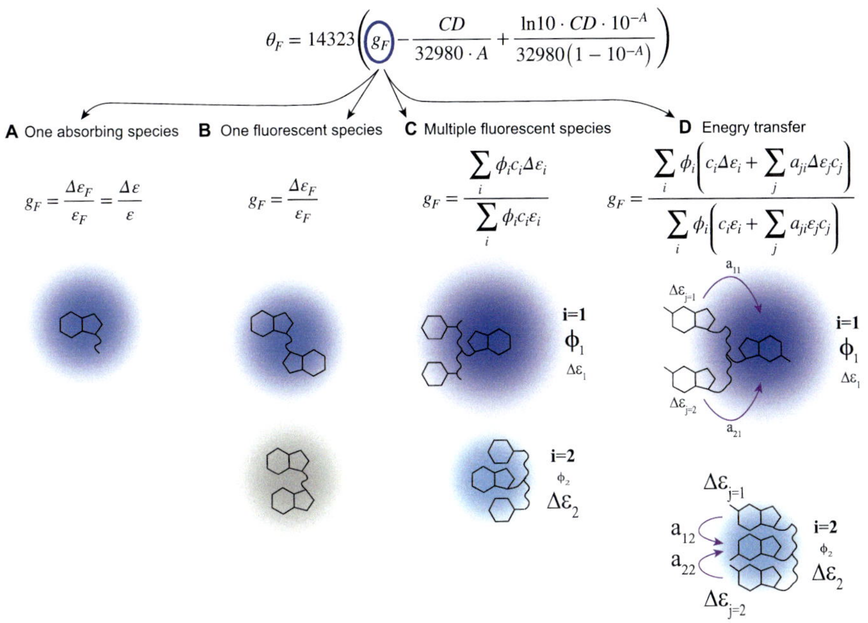

Fig. 6.3 Illustration of scenarios discussed in the FDCD theory section. The examples in A–C are derived from the behavior of 2-aminopurine reported in (Widom and Hoeher 2022). The details of D are a hypothetical scenario based on the known behavior of energy transfer from A to 2-AP (Xu and Nordlund 2000). (**A**) One absorbing species, such as 2-AP riboside. The fluorescence- and absorbance-detected dissymmetry factors are equal. (**B**) One fluorescent species and multiple absorbing species, such as a mixture of unstacked and severely quenched stacked conformations of 2-AP-modified dinucleotides. The observed dissymmetry factor is simply that of the fluorescent species. (**C**) Multiple fluorescent species, such as unstacked ($i = 1$) and stacked ($i = 2$) conformations of a 2-AP-labeled trinucleotide. The observed dissymmetry is weighted by each species' quantum yield Φ_i (higher for an unstacked conformation), CD signal $\Delta\varepsilon_i$ (higher for a stacked conformation), and concentration c_i (dependent on conditions) of each species. The relative magnitudes of these quantities are indicated by the size of their text. (**D**) Energy transfer to detected fluorophore with the potential for multiple energy transfer donors and multiple fluorescent species. The observed dissymmetry is as in panel C, but each donor ($j = 1$ and $j = 2$, depicting the bases to the 5′ and 3′ sides of the fluorophore) contributes a term containing its energy transfer efficiency a (higher for a stacked conformation; similar for the bases to the 5′ and 3′ sides), its CD signal $\Delta\varepsilon$ (higher for a stacked conformation), and concentration c (in this example, equal to that of the detected fluorophore)

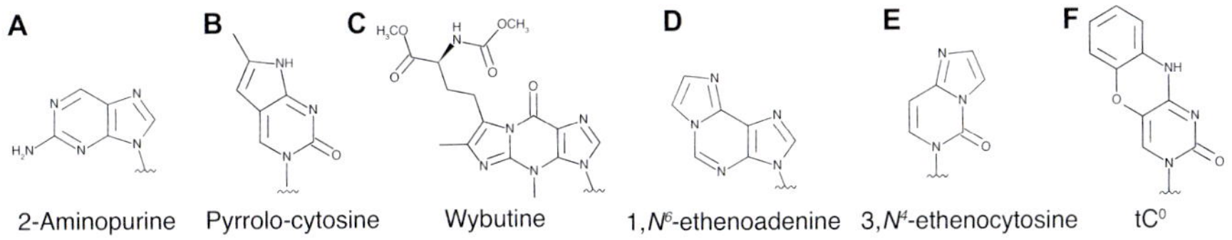

Fig. 6.4 Fluorescent base analogs utilized in the studies discussed in the text

FBA-labeled oligonucleotides (Reich and Tinoco 1980; Widom and Hoeher 2022). It can be rearranged to obtain the quantum yield-weighted average dissymmetry:

$$g_{\mathrm{F}} = \frac{\sum_i \phi_i c_i \Delta\varepsilon_i}{\sum_i \phi_i c_i \varepsilon_i} \tag{6.6}$$

for comparison to the analogous, unweighted quantity obtained through standard CD, which contains contributions from all absorbing species:

$$g = \frac{\sum_i c_i \Delta\varepsilon_i}{\sum_i c_i \varepsilon_i} \tag{6.7}$$

Consider a region of the spectrum in which only the fluorescent moiety absorbs, such as >300 nm in the case of an FBA in a nucleic acid background. If all species have the same fluorescence quantum yield, their values of ϕ_i cancel, and g_{F} and g are expected to be equal. Disagreement between these two quantities thus indicates the presence of multiple species with different fluorescence quantum yields from one another (potentially including nonfluorescent species and non-optically active fluorescent species), as seen previously in FBA-labeled oligonucleotides (Reich and Tinoco 1980; Widom and Hoeher 2022).

Energy Transfer

The relationship between the FDCD signal and the optical activity of the fluorescent species is further altered by energy transfer to the fluorophore whose emission is detected (Fig. 6.3D). This factor can be important in studies of nucleic acids, given that FBAs have been used as energy transfer donors and acceptors (Wranne et al. 2017), and certain native bases can act as donors for certain FBAs (Xu and Nordlund 2000). In addition, the commonly used donor and acceptor fluorophores Cy3 and Cy5 have been introduced into DNA in an "internal" labeling modality that renders them optically active (Kringle et al. 2018; Lee et al. 2014; Heussman et al. 2022), a scheme amenable to investigation using FDCD. Energy transfer should be suspected if features appear in the fluorescence excitation spectrum of the complete system at wavelengths not seen when the fluorophore is free in solution, particularly if those wavelengths coincide with known absorption features of the nonfluorescent components (the latter criterion can be applied even if the molecule is nonfluorescent when free in solution). For example, the fluorescence excitation spectra of oligonucleotides in which 2-AP is flanked by adenosine residues exhibit a pronounced feature centered at the 259 nm absorption maximum of adenine, a feature that does not appear in free 2-AP base or nucleosides (Xu and Nordlund 2000).

Accounting for energy transfer, the observed FDCD signal is given by:

$$\theta_{\mathrm{F}} = 14323\left(\frac{\sum_i \phi_i \left(c_i \Delta\varepsilon_i + \sum_j a_{ji} \Delta\varepsilon_j c_j\right)}{\sum_i \phi_i \left(c_i \varepsilon_i + \sum_j a_{ji} \varepsilon_j c_j\right)} - 2R\right) \tag{6.8}$$

where j counts over all the species that can transfer energy to the fluorescent species i, a_{ji} is the efficiency of energy transfer from species j to species i, and ε_j, $\Delta\varepsilon_j$, and c_j are the extinction coefficient of species j, $\varepsilon_{j,L}$-$\varepsilon_{j,R}$, and the concentration of species j, respectively.

By processing FDCD spectra with this equation, one obtains the quantum yield-weighted and energy transfer-adjusted dissymmetry:

$$g_F = \frac{\sum_i \phi_i \left(c_i \Delta\varepsilon_i + \sum_j a_{ji} \Delta\varepsilon_j c_j \right)}{\sum_i \phi_i \left(c_i \varepsilon_i + \sum_j a_{ji} \varepsilon_j c_j \right)} \tag{6.9}$$

with Eq. (6.7) still providing the analogous expression for standard CD. The different donor species j could be different conformations of a macromolecule with variable energy transfer efficiency and donor optical activity, across which the acceptor optical activity and fluorescence quantum yield are constant. A different fluorescent species i could have a different ensemble of donor optical activities and energy transfer efficiencies. One can see that the observed dissymmetry is decreased by energy transfer from donors that are not optically active ($\Delta\varepsilon_j = 0$ while a_{ji} and $\varepsilon_j \neq 0$), and that g_F can be nonzero in the presence of energy transfer even if the acceptor is not optically active ($\Delta\varepsilon_i = 0$). In the latter scenario, the acceptor acts as an "antenna" that allows detection of dissymmetry in donor excitation, with the FDCD signal being specific to species that exhibit efficient energy transfer (such as compact conformations or bound complexes).

6.2.3 FDCD-Specific Experimental Considerations

Measurement Channels

FDCD is typically measured using the excitation path and electronics of a standard CD spectrometer, with an additional detector at 90° to detect emission (Fig. 6.1). Two data channels must be recorded: "FDCD," which is proportional to $F_L - F_R$, and "DC," which is proportional to $F_L + F_R$ (using the terminology of the Jasco J-1500 CD spectrometer). Following baseline correction of the DC channel, θ_F is computed as FDCD/DC. The gain of the fluorescence detector is typically fixed at a setting that maximizes sensitivity while avoiding saturation, which in practice means setting the gain to yield a DC value of 1 V when exciting at the wavelength that yields the maximum emission intensity. The optimal gain setting will be different for samples of different brightness, so the magnitudes of their $F_L - F_R$ and $F_L + F_R$ signals can not be directly compared. A different approach is advantageous if the goal of an FDCD measurement is to quantify the concentration of an analyte. In that situation, the absolute sizes of FDCD and DC are the desired quantities, so one should either perform all measurements at a single gain setting, or construct a signal vs. gain calibration curve so that the intensities of spectra recorded at different gain values can be directly compared (Prabodh et al. 2021).

Baseline Correction

To construct θ_F from the output of the raw FDCD and DC channels, baseline correction must be performed. The DC spectrum is baseline-corrected using a solvent scan recorded under the same instrument settings as used for the sample. Baseline correction of the FDCD spectrum is more nuanced. Like standard CD, the FDCD channel records the amplitude of modulation ($F_L - F_R$ or $I_L - I_R$) around a nonzero mean $(F_L + F_R)/2$ or $(I_L - I_R)/2$. The modulated (AC) component contains contributions from true unequal absorbance of LCP and RCP by the analyte, as well as spurious contributions from factors like unequal intensity of the incident LCP and RCP. If the analyte is removed from the sample for a CD measurement, the DC signal remains, and any AC signal is the result of contamination or instrument imperfections. If the analyte is removed for an FDCD measurement, however, the fluorescence detector records only dark current and light scatter, and any spurious modulations that would be present in a fluorescence signal (were it present) are missed. As a consequence, background correction in FDCD requires a measurement to be performed on an achiral fluorophore or racemic mixture with similar spectral properties to the analyte. Because the background fluorophore or mixture has no net optical activity, any AC signal produced by its emission is spurious and can be subtracted from the spectrum of the analyte.

Ideally, the baseline fluorophore should have an excitation spectrum similar to that of the analyte. This is easily accomplished for some samples. For example, the fluorescence excitation spectrum of free 2-AP nucleobase is very similar to its spectrum when incorporated into most oligonucleotides (the exception is A-rich sequence contexts in which, as discussed above, energy transfer from A significantly alters the excitation spectrum (Xu and Nordlund 2000)). In recent work, 2-AP nucleobase was thus used as a baseline fluorophore to correct the FDCD spectra of oligonucleotides containing 2-AP and C (Widom and Hoeher 2022). The situation is more complex for chromophores whose excitation spectra are drastically altered in the context of a nucleic acid or other chiral scaffold. In this situation, a different fluorophore must be selected that can be excited at the wavelengths of interest and emits adequately at wavelengths transmitted by the emission filter used for the sample of interest. For example, α-naphthylamine and fluorescein have been used as baseline fluorophores for studies of ethidium (Fig. 6.5A) (Dahl et al. 1982) and the modified base wybutine (Fig. 6.4C) in transfer RNA (Turner et al. 1975). In situations such as this where the DC spectra are not perfectly matched, the sample and baseline fluorophore's DC spectra are background corrected using solvent, then the quantity FDCD/DC for the baseline fluorophore is subtracted from FDCD/DC for the sample.

Fluorescence Polarization and Photoselection Artifacts

It was noted shortly after the original reports on FDCD that the relationship between the measured FDCD signal and the dissymmetry in absorbance of LCP and RCP can be distorted greatly if the fluorophore is unable to rotate freely within its excited state lifetime (Fig. 6.2D) (Ehrenberg and Steinberg 1976). This is often the case for nucleic acids labeled with FBAs such as 2-AP (Barbieri et al. 2007; Guest et al. 1991; Nordlund et al. 1989). In even short DNA duplexes such as a 7-mer, 2-AP

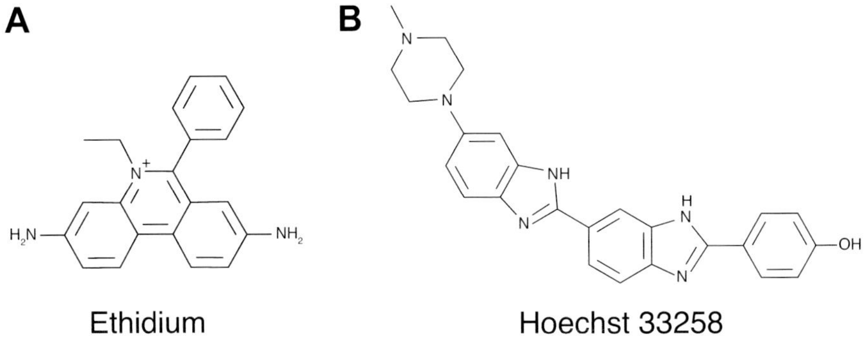

Fig. 6.5 Noncovalent nucleic acid binding dyes utilized in the studies discussed in the text

exhibits rotational correlation times of approximately 100 ps (attributed to fast local motion that leads to partial depolarization of its emission) and 1 ns (attributed to global tumbling), while its most prominent excited state lifetime is approximately 50 ps (Guest et al. 1991). This situation results in photoselection, in which molecules with different orientations relative to the polarization of the incident light are excited with different efficiencies. If their orientations do not have time to randomize before emission, the resulting fluorescence is polarized and spatially anisotropic. As a result, the quantity $F_L - F_R$ is not simply proportional to the dissymmetry in absorbance of LCP and RCP by the fluorescent species, as assumed by the preceding theory. The potential for artifacts should be considered when examining FDCD work that does not state whether they were checked for (typically by measuring fluorescence polarization or fluorescence-detected linear dichroism) or whether measures like those described below were taken to manage them.

Several methods have been devised to enable artifact-free FDCD spectra to be collected. The fluorescence detector can be placed at the "magic angle" of 54.75° relative to the direction of excitation light propagation (Tinoco et al. 1977), or an emission detector at the standard 90° can be outfitted with a polarizer at the magic angle relative to the plane of light propagation (Tinoco et al. 1977; Lobenstine and Turner 1979), or the signals from two detectors oriented at 90° from the excitation path and from each other can be averaged (Nehira 2005) (Fig. 6.2E). However, these methods suffer from either low sensitivity or lack of compatibility with widespread instrumentation. It was later shown that artifacts can be suppressed by increasing the solid angle over which emission is collected, which was accomplished by placing a cylindrical cell inside an ellipsoidal mirror oriented perpendicular to the beam path, with the cell and the detector at the two foci of the mirror (Bicknese and Maestre 1987) (Fig. 6.2I). This apparatus is available commercially from Jasco, currently as the "FDCD551" accessory (previously the FDCD465) (Nehira et al. 2005; Pandiscia n.d.) (Fig. 6.2K). While it requires a large sample volume (5 mL with the provided cylindrical cell), the sensitivity is greatly improved due to a larger fraction of the emitted light being directed to the detector (Nehira 2005; Nehira et al. 2005, 2008).

6.3 Review of Applications of FDCD to Nucleic Acids

6.3.1 *Naturally Occurring Fluorescent Nucleobases*

Measurements on the intrinsically fluorescent "Y" base (wybutine, a hypermodified guanine, Fig. 6.4C) that occurs naturally in tRNA were among the earliest applications of FDCD (Turner et al. 1975; Yoon et al. 1975). These measurements immediately revealed the sensitivity of FDCD and its ability to report on the local environment of a fluorescent moiety within a macromolecule. Y exhibits a low-energy electronic transition that yields an absorbance feature at >300 nm, and this transition exhibited detectable FDCD in yeast $tRNA^{Phe}$ at far lower (~100-fold) concentrations than were required to observe detectable CD (Turner et al. 1975). The FDCD signal in a short wavelength feature (centered around 240 nm) was significantly smaller in magnitude in the absence of magnesium, despite the fluorescence excitation spectrum of Y being almost independent of Mg^{2+}. Thermal denaturation curves obtained by detecting low-energy CD and FDCD exhibited the same two transitions, an unassigned transition just below 20 °C and a transition around 80 °C corresponding to melting of the tRNA secondary structure. Another early study investigated changes in Y fluorescence upon binding of short oligonucleotides, UUC and UUCA, that mimic the phenylalanine codon that this tRNA recognizes (Fig. 6.2B) (Yoon et al. 1975). Binding of both oligonucleotides induced decreases in the FDCD of the Y base in the long-wavelength region, with the more strongly binding UUCA inducing changes throughout the UV. This decrease in FDCD indicated that either the Y base had become more disordered, or that its environment had become less asymmetric. It should be noted that the potential for photoselection artifacts in FDCD was noted only later, and Y exhibits significant fluorescence polarization in $tRNA^{Phe}$. (Ehrlich et al. 1980) Still, these early studies on Y highlighted the potential of FDCD as a highly sensitive method for measuring the local environments of fluorescent moieties in complex biological macromolecules.

6.3.2 *Artificial Fluorescent Nucleobases*

The fortuitous presence of the Y base in tRNA presented a convenient target for early FDCD studies, but the ability to incorporate artificial fluorophores into nucleic acids offered the promise of far greater control and flexibility. This was first explored by using chloroacetaldehyde modification of adenosine and cytidine to generate dinucleotides containing 1,N^6-ethenoadenosine (εA) and 3,N^4-ethenocytidine (εC) (Reich and Tinoco 1980) (Fig. 6.4D–E). Temperature-dependent FDCD and CD spectra of εApεA, εApεC, and εApU were used to assess whether the dinucleotides exhibit a two-state stacking–unstacking equilibrium. The fluorescence of εA and εC is quenched by base-stacking, so a starting assumption was made that any stacked structures would be completely dark and only unstacked structures would contribute to the FDCD signal. As $\Delta\varepsilon_F/\varepsilon_F$ is concentration-independent, a two-state system in which only a single conformation is fluorescent should show a signal that is independent of temperature. This is what was observed for εApεC and εApU. In contrast, εApεA exhibited an FDCD signal that was strongly temperature-dependent above 35 °C (Fig. 6.2F). It was concluded that εApεA exhibits a two-state equilibrium between a stacked dark state and an unstacked bright state below 35 °C, and that a second unstacked species with a different CD spectrum and fluorescence quantum yield emerges above that temperature (Reich and Tinoco 1980).

In the decades since, there has been an explosion in the use of artificially introduced FBAs to probe RNA and DNA structure in a sensitive and site-specific manner. A large library of FBAs has been assembled, spanning the entire RNA and DNA alphabets (Wranne et al. 2017; Shin et al. 2011; Rovira et al. 2015, 2017; Bood et al. 2018; Dierckx 2014; Füchtbauer et al. 2020; Sandin et al. 2005, 2007; Wilhelmsson et al. 2001). They have been used to study a huge array of nucleic acid targets using a huge array of different measurement techniques, extending as far as multidimensional spectroscopy (Widom et al. 2013) and intracellular imaging (Wang et al. 2022). FBAs typically exhibit redshifted absorption spectra relative to the native bases, a property that has enabled them to be used as site-specific probes in absorbance-based spectroscopy such as standard CD (Datta et al. 2012; Johnson et al. 2004, 2005). Surprisingly, the use of FDCD to study FBA-labeled nucleic acids has been quite limited. It was reported that the tricyclic cytosine analog tC° (Fig. 6.4F) was a promising candidate for FDCD studies of DNA. It was found to exhibit a robust FDCD signal in the near-ultraviolet (300–420 nm) that decreased with increasing temperature and was stronger in dsDNA than ssDNA (Bhavsar et al. 2011) (Fig. 6.2M), suggesting that FDCD on this base analog could be a useful probe of structure. This work noted that no FDCD signal was observed from pyrrolo-cytosine (pC, Fig. 6.4B) in the same DNA constructs. Interestingly, recent work from our lab has shown robust signals from pC in short RNA oligonucleotides in the 200–300 nm range (unpublished data). Small but clearly resolved signals were observed in the 300–420 nm range, originating from the longest-wavelength electronic transition of pC.

The widely used FBA 2-AP (Fig. 6.4A) was recently investigated using FDCD as well (Widom and Hoeher 2022). It was found that in the dinucleotides (2-AP)(2-AP), C(2-AP) and (2-AP)C, 2-AP is almost completely dark in the stacked conformations that give rise to the majority of the standard CD signal. Furthermore, the bright species are unstacked to the point where electronic coupling between the bases is not detectable, indicated by a lack of FDCD signal arising from 2-AP's longest-wavelength electronic transition (Fig. 6.2O). A two-state model (base-stacked and unstacked) was constructed that accounted for the relative fluorescence intensities of an oligonucleotide and free 2-AP riboside, and the CD and FDCD spectra of the oligonucleotide. These three measurements, combined with the above observation that 2-AP exhibits no detectable long-wavelength CD in unstacked conformations, were sufficient to solve for the populations of the stacked and unstacked states as well as the degree of fluorescence quenching between them. This revealed that 2-AP is much more severely quenched in stacked conformations than would be assumed based on its emission intensity alone (Widom and Hoeher 2022).

6.3.3 Nucleic Acid–Dye Complexes

FDCD has also found use in the study of complexes formed through noncovalent interactions between nucleic acids and fluorophores or fluorogenic molecules. Early work explored complexes of the intercalating dye ethidium (Fig. 6.5A) and di- and trinucleotides (Dahl et al. 1982), polynucleotides (Lamos et al. 1986a, b), genomic DNA, and cells (Lamos and Turner 1985). In the earliest of these studies (Dahl et al. 1982), ethidium was present in the samples at concentrations of at most 50% as high as di- or trinucleotide dimers, and often far less than that. The CD spectrum below 300 nm therefore originated mostly from electronic transitions on the bases, including both bound and unbound dimers, precluding the use of that spectral region for analyzing binding. In contrast, FDCD spectra arose only from complexes of DNA and ethidium across the entire analyzed range of 220–400 nm (Dahl et al. 1982) (Fig. 6.2H). The shorter wavelength transitions that were obscured in the standard CD spectra proved to be very sensitive to the sequence of the target DNA, perhaps in part due to energy transfer from bases near the binding sites (Lamos et al. 1986a, b). In studies of ethidium in complex with polynucleotides, the presence of multiple binding modes was established through comparison of the induced CD and FDCD in the unobscured long-wavelength region where only ethidium absorbs (Lamos and Turner 1985; Lamos et al. 1986a). It was found that a 385 nm feature was more intense in the FDCD spectrum than in the CD spectrum, indicating that the most highly fluorescent binding modes have stronger dissymmetry (Lamos and Turner 1985). Another study compared the spectra of ethidium bound to poly(dG-dC) and poly(dG-m^{5}dC) under salt conditions that favored right-handed B- or left-handed Z-form dsDNA structures. Excellent agreement between CD and FDCD was observed under B-form conditions. In contrast, under Z-form conditions [4.4 M NaCl for poly(dG-dC) or ≥2 mM $MgCl_2$ for poly (dG-m^{5}dC)], the standard CD spectra showed the expected shape for a Z-form structure, while the FDCD spectrum was unchanged between B-form and Z-form conditions. It was concluded that ethidium must be binding in clusters that induce a locally right-handed helical structure even under Z-form conditions (Lamos et al. 1986a).

A later study investigated the binding of the minor groove binder Hoechst 33258 (Fig. 6.5B) to dsDNA with and without histones assembled on it (Banerjee et al. 2009). Standard CD spectroscopy revealed that addition of histones induced a more significant change in an dsDNA dodecamer containing two consecutive A-T base-pairs as its Hoechst binding site (O1), than a site consisting of three consecutive A-T base-pairs (O_2). Femtosecond time-resolved emission spectroscopy and FDCD confirmed that the local environment of the dye was more perturbed upon addition of histones to O1 as well (Fig. 6.2L).

6.3.4 Nucleic Acid-Based Nanostructures

In recent years, attention has turned to the use of nucleic acids as templates for programmable nanostructures. For example, nucleic acids have been used as chemical scaffolds to promote assembly of dyes to achieve specific optical properties, and FDCD has emerged as a useful method for characterizing the resulting assemblies (Hofsäß et al. 2018; Ucar and Wagenknecht 2021). In one such application, a pyrene-modified diaminopurine deoxynucleoside (Py-dAp) and a Nile Red-modified deoxyuridine (NrdU) were synthesized and hybridized to ssDNA containing A and T in various sequences (Hofsäß et al. 2018). NrdU assemblies were completely quenched, and interfaces between NrdU and Py-dAP in mixed A-T sequences led to quenching of Py-dAp fluorescence and replacement of pyrene excimer emission features with monomer features. FDCD confirmed that the observed fluorescence originated from excitation of the pyrene moieties, and that these moieties were in a chiral geometry due to templating by the ssDNA (Fig. 6.2N). The monomers were noted to have no CD signal in the spectral range in which pyrene absorbs (Hofsäß et al. 2018). A nontemplated dye assembly was additionally created using tetraphenylethylene-modified deoxyuridine (Ucar and Wagenknecht 2021). FDCD on these assemblies revealed that the emissive species adopts a helical structure with chirality controlled by that of the sugar.

6.4 Conclusion and Future Outlook

FDCD has the potential to play a far greater role in the analysis of nucleic acids than recent levels of usage would suggest. The library of FBAs is constantly expanding, and understanding of their structure–photophysics relationships would be greatly enhanced by wider application of FDCD. Likewise, a new generation of noncovalent nucleic acid binding dyes such as the Sybr series (Tuma et al. 1999; Kolbeck et al. 2021) has largely supplanted ethidium for laboratory usage, and new intracellular sensors have been constructed using fluorogenic RNA aptamers that bind to small molecules and enhance their fluorescence (Paige et al. 2011; You et al. 2015). These systems are natural targets for analysis by FDCD, which could be used to investigate how the local environment of the dye binding site modulates its photophysical properties. The landscape is ripe for more sophisticated and sensitive variants of FDCD measurements, such as lifetime-resolved FDCD and high-sensitivity FDCD recorded through single-photon counting. These possibilities were noted and in some cases demonstrated in early reports of FDCD (Wu and McGown 1991; Thomas et al. 1986), but their potential has yet to be fully realized. Furthermore, fluorescence detection appears to offer the most promising route toward chiroptical spectroscopy of single molecules, though efforts toward that goal have been complicated by the difficulty of avoiding or correcting for the strong *linear* dichroism that single molecules intrinsically exhibit even without bulk orientational ordering (Tang et al. 2009; Hassey et al. 2006, 2008). As a result, researchers have been working to establish methods of signal enhancement that can facilitate detection of CD from single molecules. Signal enhancements have been reported using superchiral light (Tang and Cohen 2011), plasmonic nanostructures (Solomon et al. 2020), and excitation via 2-photon absorption (Pniakowska et al. 2023). These enhancements of the asymmetry and/or strength of light–matter interactions have the potential to offer not only detection approaching the single molecule level, but also control over chiral photochemical reaction pathways.

In summary, FDCD has already yielded significant insight into nucleic acid structure and interactions, and into the photophysics of the fluorescent probes that are used to study it. The scope of its potential applications is constantly growing as new fluorescent probes are developed, new biologically significant roles of nucleic acid structure are discovered, and experimental methods are enhanced.

References

Banerjee D, Makhal A, Pal SK (2009) Sequence dependent femtosecond-resolved hydration dynamics in the minor groove of DNA and histone—DNA complexes. J Fluoresc 19(6):1111–1118. https://doi.org/10.1007/s10895-009-0512-9

Barbieri CM, Kaul M, Pilch DS (2007) Use of 2-aminopurine as a fluorescent tool for characterizing antibiotic recognition of the bacterial rRNA A-site. Tetrahedron 63(17):3567–3574. https://doi.org/10.1016/j.tet.2006.08.107

Beyerle ER, Dinpajooh M, Ji H, von Hippel PH, Marcus AH, Guenza MG (2021) Dinucleotides as simple models of the base stacking-unstacking component of DNA 'breathing' mechanisms. Nucleic Acids Res 49(4):1872–1885. https://doi.org/10.1093/nar/gkab015

Bhavsar YP, Reilly SM, Wadkins RM (2011) Evaluation of fluorescent analogs of deoxycytidine for monitoring DNA transitions from duplex to functional structures. J Nucleic Acids 2011:e986820. https://doi.org/10.4061/2011/986820

Bicknese SE, Maestre MF (1987) Correction of fluorescence detected circular dichroic photoselection artifacts using light collecting ellipsoidal mirrors. Rev Sci Instrum 58(11):2060–2063. https://doi.org/10.1063/1.1139463

Bood M, Füchtbauer AF, Wranne MS, Ro JJ, Sarangamath S, El-Sagheer AH, Rupert DLM, Fisher RS, Magennis SW, Jones AC, Höök F, Brown T, Kim BH, Dahlén A, Wilhelmsson LM, Grøtli M (2018) Pentacyclic adenine: a versatile and exceptionally bright fluorescent DNA base analogue. Chem Biol 9(14):3494–3502. https://doi.org/10.1039/C7SC05448C

Dahl KS, Pardi A, Tinoco I Jr (1982) Structural effects on the circular dichroism of ethidium ion nucleic acid complexes. Biochemistry 21(11):2730–2737. https://doi.org/10.1021/bi00540a024

Datta K, Johnson NP, Villani G, Marcus AH, von Hippel PH (2012) Characterization of the 6-methyl isoxanthopterin (6-MI) base analog dimer, a spectroscopic probe for monitoring guanine base conformations at specific sites in nucleic acids. Nucleic Acids Res 40(3):1191–1202. https://doi.org/10.1093/nar/gkr858

Dierckx A (2014) Fluorescent nucleobase analogues and their use for investigating DNA interactions. Chalmers Univ Technol:1–74

Ehrenberg B, Steinberg IZ (1976) Effect of photoselection on fluorescence-detected circular dichroism. J Am Chem Soc 98(5):1293–1295. https://doi.org/10.1021/ja00421a057

Ehrlich R, Lefevre J-F, Remy P (1980) Fluorimetric study of the complex between yeast phenylalanyl-tRNA synthetase and tRNAPhe. Eur J Biochem 103(1):145–153. https://doi.org/10.1111/j.1432-1033.1980.tb04298.x

Füchtbauer AF, Wranne MS, Sarangamath S, Bood M, El-Sagheer AH, Brown T, Gradén H, Grøtli M, Wilhelmsson LM (2020) Lighting up DNA with the environment-sensitive bright adenine analogue qAN4. ChemPlusChem 85(2):319–326. https://doi.org/10.1002/cplu.201900712

Guest CR, Hochstrasser RA, Sowers LC, Millar DP (1991) Dynamics of mismatched base pairs in DNA. Biochemistry 30(13):3271–3279. https://doi.org/10.1021/bi00227a015

Hardman SJO, Thompson KC (2006) Influence of base stacking and hydrogen bonding on the fluorescence of 2-aminopurine and pyrrolocytosine in nucleic acids. Biochemistry 45(30):9145–9155. https://doi.org/10.1021/bi060479t

Hardman SJO, Botchway SW, Thompson KC (2008) Evidence for a nonbase stacking effect for the environment-sensitive fluorescent base pyrrolocytosine—comparison with 2-aminopurine. Photochem Photobiol 84(6):1473–1479. https://doi.org/10.1111/j.1751-1097.2008.00368.x

Hassey R, Swain EJ, Hammer NI, Venkataraman D, Barnes MD (2006) Probing the chiroptical response of a single molecule. Science 314(5804):1437–1439. https://doi.org/10.1126/science.1134231

Hassey R, McCarthy KD, Swain E, Basak D, Venkataraman D, Barnes MD (2008) Single-molecule chiroptical spectroscopy: fluorescence excitation of individual helicene molecules in polymer-supported thin-films. Chirality 20(9):1039–1046. https://doi.org/10.1002/chir.20584

Heussman D, Kittell J, von Hippel PH, Marcus AH (2022) Temperature-dependent local conformations and conformational distributions of cyanine dimer labeled single-stranded–double-stranded DNA junctions by 2D fluorescence spectroscopy. J Chem Phys 156(4):045101. https://doi.org/10.1063/5.0076261

Hofsäß R, Sinn S, Biedermann F, Wagenknecht H-A (2018) Programmable and sequence-selective supramolecular assembly of two different chromophores along DNA templates. Chem Eur J 24(61):16257–16261. https://doi.org/10.1002/chem.201804314

Ji H, Johnson NP, von Hippel PH, Marcus AH (2019) Local DNA base conformations and ligand intercalation in DNA constructs containing optical probes. Biophys J 117(6):1101–1115. https://doi.org/10.1016/j.bpj.2019.08.006

Johnson NP, Baase WA, von Hippel PH (2004) Low-energy circular dichroism of 2-aminopurine dinucleotide as a probe of local conformation of DNA and RNA. Proc Natl Acad Sci USA 101(10):3426–3431

Johnson NP, Baase WA, von Hippel PH (2005) Investigating local conformations of double-stranded DNA by low-energy circular dichroism of pyrrolo-cytosine. Proc Natl Acad Sci USA 102(20):7169–7173. https://doi.org/10.1073/pnas.0502359102

Kolbeck PJ, Vanderlinden W, Gemmecker G, Gebhardt C, Lehmann M, Lak A, Nicolaus T, Cordes T, Lipfert J (2021) Molecular structure, DNA binding mode, photophysical properties and recommendations for use of SYBR gold. Nucleic Acids Res 49(9):5143–5158. https://doi.org/10.1093/nar/gkab265

Kringle L, Sawaya NPD, Widom J, Adams C, Raymer MG, Aspuru-Guzik A, Marcus AH (2018) Temperature-dependent conformations of exciton-coupled Cy3 dimers in double-stranded DNA. J Chem Phys 148(8):085101. https://doi.org/10.1063/1.5020084

Lamos ML, Turner DH (1985) Fluorescence-detected circular dichroism of ethidium in vivo and bound to deoxyribonucleic acid in vitro. Biochemistry 24(12):2819–2822. https://doi.org/10.1021/bi00333a001

Lamos ML, Walker GT, Krugh TR, Turner DH (1986a) Fluorescence-detected circular dichroism of ethidium bound to poly(dG-dC) and poly(dG-m5dC) under B- and Z-form conditions. Biochemistry 25(3):687–691. https://doi.org/10.1021/bi00351a027

Lamos ML, Lobenstine EW, Turner DH (1986b) Fluorescence-detected circular dichroism of ethidium bound to nucleic acids. J Am Chem Soc 108(15):4278–4284. https://doi.org/10.1021/ja00275a007

Lee W, von Hippel PH, Marcus AH (2014) Internally labeled Cy3/Cy5 DNA constructs show greatly enhanced photo-stability in single-molecule FRET experiments. Nucleic Acids Res 42(9):5967–5977. https://doi.org/10.1093/nar/gku199

Liang J, Matsika S (2011) Pathways for fluorescence quenching in 2-aminopurine π-stacked with pyrimidine nucleobases. J Am Chem Soc 133(17):6799–6808. https://doi.org/10.1021/ja2007998

Lobenstine EW, Turner DH (1979) Photoselected fluorescence detected circular dichroism. J Am Chem Soc 101(8):2205–2207. https://doi.org/10.1021/ja00502a048

Lobenstine EW, Schaefer WC, Turner DH (1981) Fluorescence detected circular dichroism of proteins with single fluorescent tryptophans. J Am Chem Soc 103(16):4936–4940. https://doi.org/10.1021/ja00406a044

Muto K, Yoshida R, Yashida R, Ishii T, Handa T (1986) Circular dichroism of the backbone and side chains separated from natural circular dichroism of poly-L-tryptophan by the fluorescence detected circular dichroism method. J Am Chem Soc 108(20):6416–6417. https://doi.org/10.1021/ja00280a064

Nehira T (2005) Fluorescence detected circular dichroism for highly sensitive observation of exciton coupled CD. Monatshefte Für Chem 136(3):477–487. https://doi.org/10.1007/s00706-004-0271-x

Nehira T, Tanaka K, Takakuwa T, Ohshima C, Masago H, Pescitelli G, Wada A, Berova N (2005) Development of a universal ellipsoidal mirror device for fluorescence detected circular dichroism: elimination of polarization artifacts. J Appl Spectrosc 59(1):121–125. https://doi.org/10.1366/0003702052940459

Nehira T, Boiadjiev SE, Lightner DA (2008) Fluorescence detected circular dichroism of a red-shifted exciton-coupling chromophore N,N′-carbonyl-bridged dipyrrinone derivative using an ellipsoidal device. Monatshefte Für Chem - Chem Mon 139(6):591–595. https://doi.org/10.1007/s00706-007-0808-x

Nordlund TM, Andersson S, Nilsson L, Rigler R, Graeslund A, McLaughlin LW (1989) Structure and dynamics of a fluorescent DNA oligomer containing the EcoRI recognition sequence: fluorescence, molecular dynamics, and NMR studies. Biochemistry 28(23):9095–9103. https://doi.org/10.1021/bi00449a021

Paige JS, Wu KY, Jaffrey SR (2011) RNA mimics of green fluorescent protein. Science 333(6042):642–646. https://doi.org/10.1126/science.1205295

Pandiscia L (n.d.) High-sensitivity fluorescence-detected circular dichroism (FDCD) measurement - JASCO. https://jascoinc.com/applications/high-sensitivity-fluorescence-detected-circular-dichroism-fdcd-measurement/. Accessed 2024-02-18

Pniakowska A, Samoć M, Olesiak-Bańska J (2023) Strong fluorescence-detected two-photon circular dichroism of chiral gold nanoclusters. Nanoscale 15(19):8597–8602. https://doi.org/10.1039/D3NR01091K

Prabodh A, Wang Y, Sinn S, Albertini P, Spies C, Spuling E, Yang L-P, Jiang W, Bräse S, Biedermann F (2021) Fluorescence detected circular dichroism (FDCD) for supramolecular host–guest complexes. Chem Sci 12(27):9420–9431. https://doi.org/10.1039/D1SC01411K

Rachofsky EL, Osman R, Ross JBA (2001) Probing structure and dynamics of DNA with 2-aminopurine: effects of local environment on fluorescence. Biochemistry 40:946–956. https://doi.org/10.1021/bi001664o

Reich C, Tinoco I (1980) Fluorescence-detected circular dichroism of dinucleoside phosphates. A study of solution conformations and the two-state model. Biopolymers 19(4):833–848. https://doi.org/10.1002/bip.1980.360190409

Rovira AR, Fin A, Tor Y (2015) Chemical mutagenesis of an emissive RNA alphabet. J Am Chem Soc 137:14602–14605. https://doi.org/10.1021/jacs.5b10420

Rovira AR, Fin A, Tor Y (2017) Expanding a fluorescent RNA alphabet: synthesis, photophysics and utility of isothiazole-derived purine nucleoside surrogates. Chem Sci 8(4):2983–2993. https://doi.org/10.1039/C6SC05354H

Sandin P, Wilhelmsson LM, Lincoln P, Powers VEC, Brown T, Albinsson B (2005) Fluorescent properties of DNA base analogue tC upon incorporation into DNA—negligible influence of neighbouring bases on fluorescence quantum yield. Nucleic Acids Res 33(16):5019–5025. https://doi.org/10.1093/nar/gki790

Sandin P, Börjesson K, Li H, Martensson J, Brown T, Wilhelmsson LM, Albinsson B (2007) Characterization and use of an unprecedentedly bright and structurally non-perturbing fluorescent DNA base analogue. Nucleic Acids Res 36(1):157–167. https://doi.org/10.1093/nar/gkm1006

Shin D, Sinkeldam RW, Tor Y (2011) Emissive RNA alphabet. J Am Chem Soc 133(38):14912–14915. https://doi.org/10.1021/ja206095a

Solomon ML, Abendroth JM, Poulikakos LV, Hu J, Dionne JA (2020) Fluorescence-detected circular dichroism of a chiral molecular monolayer with dielectric metasurfaces. J Am Chem Soc 142(43):18304–18309. https://doi.org/10.1021/jacs.0c07140

Tang Y, Cohen AE (2011) Enhanced enantioselectivity in excitation of chiral molecules by superchiral light. Science 332(6027):333–336. https://doi.org/10.1126/science.1202817

Tang Y, Cook TA, Cohen AE (2009) Limits on fluorescence detected circular dichroism of single helicene molecules. J Phys Chem A 113(22):6213–6216. https://doi.org/10.1021/jp903598t

Thomas M, Patonay G, Warner I (1986) Novel instrument for measurement of multidimensional fluorescence detected circular dichroism. Rev Sci Instrum 57(7):1308–1313. https://doi.org/10.1063/1.1138594

Thomas MP, Patonay G, Warner IM (1987) Fluorescence-detected circular dichroism studies of serum albumins. Anal Biochem 164(2):466–473. https://doi.org/10.1016/0003-2697(87)90520-3

Thomas MP, Nelson G, Patonay G, Warner IM (1988) Analysis of drug binding sites on human serum albumin using multidimensional fluorescence measurements. Spectrochim Acta Part B At Spectrosc 43(4):651–660. https://doi.org/10.1016/0584-8547(88)80087-9

Tinoco I Jr, Ehrenberg B, Steinberg IZ (1977) Fluorescence detected circular dichroism and circular polarization of luminescence in rigid media: direction dependent optical activity obtained by photoselection. J Chem Phys 66(3):916–920. https://doi.org/10.1063/1.433998

Tinoco I, Turner DH (1976) Fluorescence detected circular dichroism. Theory J Am Chem Soc 98(21):6453–6456. https://doi.org/10.1021/ja00437a003

Tuma RS, Beaudet MP, Jin X, Jones LJ, Cheung C-Y, Yue S, Singer VL (1999) Characterization of SYBR gold nucleic acid gel stain: a dye optimized for use with 300-nm ultraviolet transilluminators. Anal Biochem 268(2):278–288. https://doi.org/10.1006/abio.1998.3067

Turner DH, Tinoco I, Maestre M (1974) Fluorescence detected circular dichroism. J Am Chem Soc 96(13):4340–4342. https://doi.org/10.1021/ja00820a057

Turner DH, Tinoco I, Maestre MF (1975) Fluorescence detected circular dichroism study of the anticodon loop of yeast tRNAPhe. Biochemistry 14(17):3794–3799. https://doi.org/10.1021/bi00688a011

Ucar H, Wagenknecht H-A (2021) DNA-templated control of chirality and efficient energy transport in supramolecular DNA architectures with aggregation-induced emission. Chem Sci 12(29):10048–10053. https://doi.org/10.1039/D1SC02351A

Wang D, Shalamberidze A, Arguello AE, Purse BW, Kleiner RE (2022) Live-cell RNA imaging with metabolically incorporated fluorescent nucleosides. J Am Chem Soc 144(32):14647–14656. https://doi.org/10.1021/jacs.2c04142

White TG, Pao Y-H, Tang MM (1975) Interpretation of fluorescence detected circular dichroism data. J Am Chem Soc 97(16):4751–4753. https://doi.org/10.1021/ja00849a046

Widom JR, Hoeher JE (2022) Base-stacking heterogeneity in RNA resolved by fluorescence-detected circular dichroism spectroscopy. J Phys Chem Lett 13(34):8010–8018. https://doi.org/10.1021/acs.jpclett.2c01778

Widom JR, Johnson NP, von Hippel PH, Marcus AH (2013) Solution conformation of 2-aminopurine (2-AP) dinucleotide determined by ultraviolet 2D fluorescence spectroscopy (UV-2D FS). New J Phys 15:15. https://doi.org/10.1088/1367-2630/15/2/025028

Wilhelmsson LM, Holmén A, Lincoln P, Nielsen PE, Nordén B (2001) A highly fluorescent DNA base analogue that forms Watson–Crick base pairs with guanine. J Am Chem Soc 123(10):2434–2435. https://doi.org/10.1021/ja0025797

Wranne MS, Füchtbauer AF, Dumat B, Bood M, El-Sagheer AH, Brown T, Gradén H, Grøtli M, Wilhelmsson LM (2017) Toward complete sequence flexibility of nucleic acid base analogue FRET. J Am Chem Soc 139(27):9271–9280. https://doi.org/10.1021/jacs.7b04517

Wu K, McGown LB (1991) A modified phase-modulation spectrofluorometer for lifetime-resolved fluorescence-detected circular dichroism. Appl Spectrosc 45(1):1–3. https://doi.org/10.1366/0003702914337858

Xu D-G, Nordlund TM (2000) Sequence dependence of energy transfer in DNA oligonucleotides. Biophys J 78(2):1042–1058. https://doi.org/10.1016/S0006-3495(00)76663-X

Yoon K, Turner DH, Tinoco I (1975) The kinetics of codon-anticodon interaction in yeast phenylalanine transfer RNA. J Mol Biol 99(4):507–518. https://doi.org/10.1016/S0022-2836(75)80169-0

You M, Litke JL, Jaffrey SR (2015) Imaging metabolite dynamics in living cells using a spinach-based riboswitch. Proc Natl Acad Sci USA 112(21):E2756–E2765. https://doi.org/10.1073/pnas.1504354112

Chapter 7
Spectroscopic Approaches for Studies of Site-Specific DNA Base and Backbone "Breathing" Using Exciton-Coupled Dimer-Labeled DNA

Andrew H. Marcus, Spiridoula Matsika, Dylan Heussman, Mohammed I. Sorour, Jack Maurer, Claire S. Albrecht, Lulu Enkhbaatar, Patrick Herbert, Kurt A. Kistler, and Peter H. von Hippel

Abstract DNA regulation and repair processes require direct interactions between proteins and DNA at specific sites. Local fluctuations of the sugar-phosphate backbones and bases of DNA (a form of DNA "breathing") play a central role in such processes. Here we review the development and application of novel spectroscopic methods and analyses—both at the ensemble and single-molecule levels—to study structural and dynamic properties of exciton-coupled cyanine and fluorescent nucleobase analog dimer-labeled DNA constructs at key positions involved in protein–DNA complex assembly and function. The exciton-coupled dimer probes act

---

A. H. Marcus (✉) · D. Heussman · J. Maurer · L. Enkhbaatar · P. Herbert
Center for Optical, Molecular and Quantum Science, Department of Chemistry and Biochemistry, University of Oregon, Eugene, OR, USA

Department of Chemistry and Biochemistry, Institute of Molecular Biology, University of Oregon, Eugene, OR, USA
e-mail: ahmarcus@uoregon.edu

S. Matsika · M. I. Sorour
Department of Chemistry, Temple University, Philadelphia, PA, USA

C. S. Albrecht
Center for Optical, Molecular and Quantum Science, Department of Physics, University of Oregon, Eugene, OR, USA

Institute of Molecular Biology, Department of Physics, University of Oregon, Eugene, OR, USA

K. A. Kistler
Department of Chemistry, Brandywine Campus, The Pennsylvania State University, Media, PA, USA

P. H. von Hippel
Department of Chemistry and Biochemistry, Institute of Molecular Biology, University of Oregon, Eugene, OR, USA

S. Matsika, A. H. Marcus (eds.), *Nucleic Acid Photophysics and Photochemistry*, Nucleic Acids and Molecular Biology 36,
https://doi.org/10.1007/978-3-031-68807-2_7

as "sensors" of the local conformations adopted by the sugar-phosphate backbones and bases immediately surrounding the dimer probes. These methods can be used to study the mechanisms of protein binding and function at these sites.

Keywords DNA "breathing" · Protein–DNA interactions · Free energy surface · Exciton-coupled dimer probe · Exciton-coupled probe spectroscopy · Fluorescent base analog · Polarization-sweep single-molecule fluorescence (PS-SMF) microscopy · Fourier transform spectroscopy · Two-photon excitation · Two-dimensional coherent spectroscopy · Two-dimensional fluorescence spectroscopy (2DFS)

7.1 Biological Significance of DNA "Breathing" and Its Connection to Functional Protein–DNA Interactions and Site-Specific Free Energy Surfaces

Biochemical processes central to gene regulation and expression are often described in terms of *local*, site-specific interactions between nucleic acids and proteins. For example, during DNA replication, proteins recognize and bind to key sites at and near single-stranded (ss)–double-stranded (ds) DNA junctions to establish the "trombone"-shaped framework of the functional DNA replication-elongation complex (or replisome) (Alberts 1987). Because the various sub-assemblies of the DNA replisome must copy the entire paternal genome, regardless of specific DNA base sequences, the binding affinities and interactions with the DNA scaffold must be controlled primarily by the local secondary structures of the bases, base pairs, and sugar-phosphate backbones at the ss–dsDNA fork junctions at which replication proteins bind and function.

The local conformations of DNA near ss–dsDNA junctions undergo thermally activated fluctuations (termed DNA "breathing") within an unknown distribution of macrostates to permit the proper binding of replication proteins. These macrostates are perturbed relative to the canonical Watson–Crick (W-C) duplex structure (von Hippel et al. 2013), exhibiting one or more of the following characteristics: (1) flanking bases may be "twisted" relative to their neighbors and thus somewhat more or less stacked; (2) inter-base hydrogen bonds may be disrupted; and/or (3) flanking bases may be fully unstacked or un-hydrogen-bonded, thus exposing the base sequences to the external environment (McConnell and von Hippel 1970a, b). Because non-canonical conformations are unstable relative to the W-C structure, they occur transiently at trace concentrations. Nevertheless, such non-canonical conformations likely act as rate-limiting intermediates for the initial chemical steps of recognition, assembly, and function of the protein–DNA complexes involved in core biological processes (von Hippel 2007).

To understand the mechanisms by which multicomponent complexes such as the DNA replisome self-assembles and functions, it is important to identify key features of the underlying macromolecular interactions. In Fig. 7.1 is shown a hypothetical free energy surface (FES) that depicts a possible cooperative assembly mechanism of the T4 bacteriophage helicase-primase (primosome)–DNA fork complex (Benkovic and Spiering 2017; Lee et al. 2016). The primosome is a sub-assembly of the T4 replisome, which is a useful model system to understand the replication complexes of higher organisms (Barry et al. 2018). The FES describes variations in the stability of protein and nucleic acid components bound at specific interaction sites and as a function of the relevant chemical reaction coordinates (labeled Q_1, Q_2, …). Points on the FES corresponding to free energy minima and maxima define the local conformations of stable macrostates and unstable transition states, respectively, which lie along the functional assembly pathway of the protein–DNA complex. "Basins" that surround each local minimum of the FES are associated with the distributions of quasi-degenerate states, which define the conformational disorder of the macrostate.

The FES shown in Fig. 7.1 emphasizes the role of DNA "breathing" at a specific site near the ss–dsDNA fork junction where the helicase and primase must preferentially bind and assemble in a defined sequence of events. In the absence of

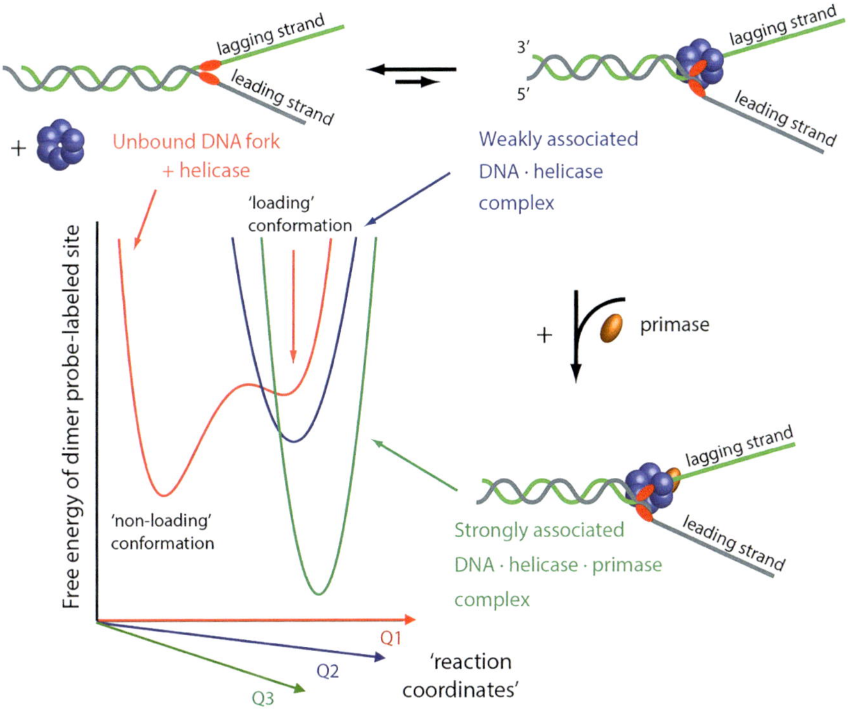

Fig. 7.1 Hypothetical free energy surface (FES) for the assembly steps of the T4 helicase-primase (primosome)–DNA fork sub-assembly as monitored by exciton-coupled $(iCy3)_2$ dimer backbone probes (indicated by red ovals). Successive reaction steps are indicated by red, blue, and green 2D cross-sections of the multidimensional FES. Figure adapted from von Hippel et al. (2013)

protein, the DNA bases and sugar-phosphate backbones at a specific position near the ss–dsDNA fork junction may fluctuate between a stable conformation that does not favor functional helicase loading and an unstable conformation that does favor such loading (von Hippel et al. 2013). The association of a weakly bound helicase with the subsequent formation of a stable and fully functional primosome–DNA fork complex is represented as a series of successive "free-energy-surface-crossing" events.

Many of the molecular details currently known about site-specific protein–DNA interactions—and how they participate in functional assembly mechanisms as depicted in Fig. 7.1—have been inferred by applying powerful structural methods such as nuclear magnetic resonance (NMR), X-ray crystallography, and cryo-electron microscopy (Kelch et al. 2011; Acton et al. 2011; Brown and Wimperis 1995; Delius et al. 1972; Feng et al. 2023). However, such methods are not well suited to study the DNA "breathing" of protein–DNA complexes in situ at "physiological" concentrations and solution conditions that simulate the intracellular environment.[1] In the remainder of this chapter, we review novel optical spectroscopic approaches that can be applied to study the dynamics of protein–DNA interactions under biologically relevant conditions in which such complexes function.

7.2 Exciton-Coupled Probe Spectroscopy for Studies of In Situ Site-Specific DNA "Breathing"

An approach that is well suited for in situ studies of protein–DNA complexes at "physiological" concentrations is to perform spectroscopic experiments on a DNA molecule that has been site-specifically labeled with fluorescent optical probes (Levitus and Ranjit 2011; Roy et al. 2008; Lee et al. 2014; Algar et al. 2019). For example, the fluorescent cyanine dyes Cy3 and Cy5 can be chemically attached to a nucleic acid base or to the distal end of a nucleic acid chain using a chemical linker. Such cyanine dye-labeled DNA constructs are useful—due largely to their relatively high absorption cross-sections and fluorescence quantum yields—for fluorescence microscopy, gene sequencing technologies, and other bio-analytical applications. Spectroscopic measurements based on Förster resonance energy transfer (FRET) of a DNA strand labeled with a Cy3 "donor" and a Cy5 "acceptor" can monitor global conformation changes on the length scale of a few nanometers (Algar et al. 2019; Screenivasan et al. 2020; Murphy et al. 2004; Israels et al. 2021; Phelps et al. 2013). Alternatively, pairs of Cy3 dye molecules can be attached "internally" (termed "iCy3") and rigidly at defined opposing positions within the sugar-phosphate

[1] The "physiological" salt concentrations referred to in this chapter are typically [NaCl] = 100 mM and [$MgCl_2$] = 6 mM for in vitro studies. These are not the actual intra-cellular salt concentrations, which fluctuate according to changing environmental conditions. Rather these values correspond to a set of salt concentrations that result in the same protein–DNA binding constants in in vitro model systems as those measured in living *E. coli* cells (Kao-Huang et al. 1977).

backbones of complementary DNA strands to create an exciton-coupled $(iCy3)_2$ dimer-probe-labeled DNA duplex or ss–dsDNA construct. Such exciton-coupled $(iCy3)_2$ dimer-labeled DNA constructs can be used to "sense" the local environment of the DNA bases and backbones within the immediate vicinity of the dimer probe.

In Fig. 7.2a, we show how iCy3 is attached internally within the framework of a DNA strand using phosphoramidite chemistry. The iCy3 label acts as a relatively stiff molecular bridge between DNA bases and as part of the sugar-phosphate backbones connecting adjacent nucleotide residues. The electric dipole transition moment (EDTM, shown as a double-headed green arrow) lies along the longitudinal direction of the π-conjugated trimethine group. By annealing two complementary single strands of DNA with opposite iCy3 labeling positions, an exciton-coupled $(iCy3)_2$ dimer-labeled DNA backbone probe can be formed at a predetermined position within a DNA fork construct (see Fig. 7.2b). As discussed in the following sections, the absorption and circular dichroism (CD) spectra of an $(iCy3)_2$ dimer-labeled DNA construct are highly sensitive to the conformation of the dimer probe, which in turn depends sensitively on both the labeling position within the DNA construct and on environmental conditions (e.g., salt concentration, temperature, etc.) that affect the stabilities of the local DNA bases and sugar-phosphate backbones immediately adjacent to the probe (Heussman et al. 2019, 2022, 2024).

Another probe-labeling strategy that we discuss in later sections uses pairs of fluorescent DNA base analogs, which can be substituted for native DNA bases as a flanking dimer within the native DNA duplex at sites close to a ss–dsDNA junction (Datta et al. 2010, 2012; Camel et al. 2021). In Fig. 7.2c, d, respectively, are shown molecular structures of the fluorescent base analogs 6-methylisoxanthopterin (6-MI, an analog of guanine) and 2-aminopurine (2-AP, an analog of adenine) (Hoeher et al. 2023; Kodali et al. 2012; Widom et al. 2012; Hawkins 2001; Jose et al. 2009; Joyce et al. 2008; Ji et al. 2019). Like their canonical nucleobase counterparts, 6-MI and 2-AP can form W-C base pairs with their complementary cytosine and thymine bases, respectively. An example structure of a $(6\text{-MI})_2$ dimer-substituted ss–dsDNA fork construct is shown in Fig. 7.2e. 6-MI and 2-AP absorb light at lower energy than the absorption of native nucleic acids (i.e., at wavelengths longer than ~300 nm, where the native bases are transparent) and they exhibit significant fluorescence quantum yields. Example absorption and fluorescence spectra of a 6-MI-substituted ss–dsDNA construct are shown in Fig. 7.2f. Two closely spaced fluorescent base analogs—incorporated as flanking bases within the same strand—can form an exciton-coupled dimer probe. Like the $(iCy3)_2$ dimer backbone probes discussed above, local conformational changes of the $(6\text{-MI})_2$ dimer probe can be monitored using fluorescence, absorption, and CD spectroscopy, as discussed in further detail below (Datta et al. 2012; Camel et al. 2021; Ji et al. 2019).

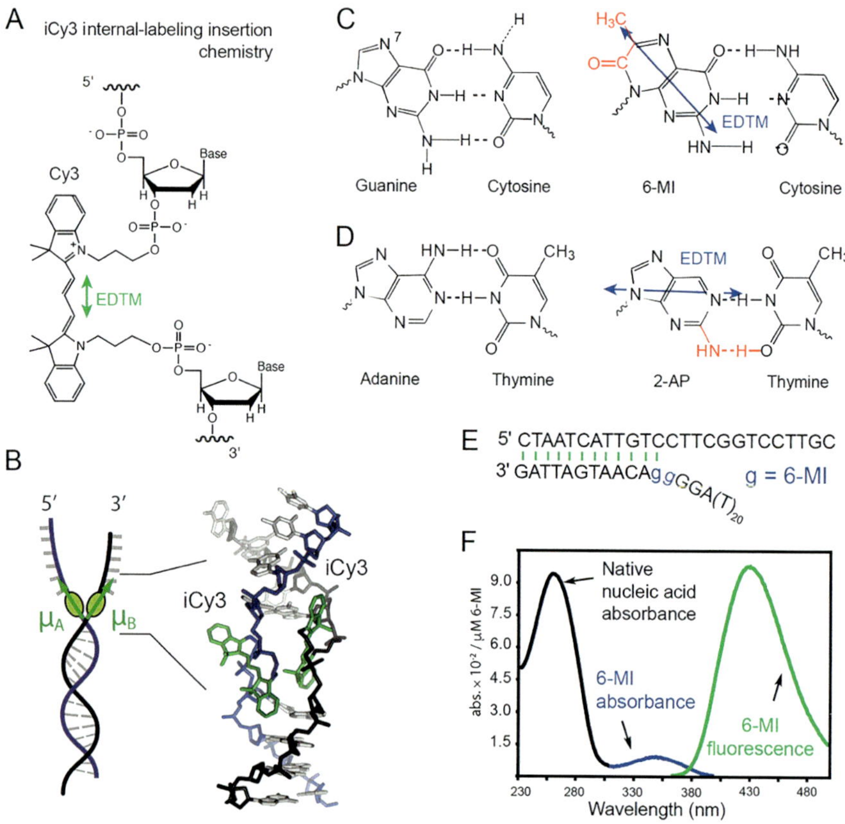

Fig. 7.2 (**a**) Chemical structure of the internally labeled iCy3 chromophore, indicating the direction of its lowest energy electric dipole transition moment (EDTM) and its internal-labeling insertion chemistry. (**b**) Example of a DNA fork construct with an exciton-coupled $(iCy3)_2$ dimer probe labeling a specific position relative to the ss–dsDNA junction. (**c**) Structure of the G:C (left) versus the 6-MI:C (right) Watson–Crick base pair. (**d**) Structure of the A:T (left) versus the 2-AP:T base pair (right). The difference between the natural bases and the analogs is shown in red. The directions of the lowest energy EDTMs are shown in blue. (**e**) An example of a DNA fork construct labeled with two adjacent 6-MI nucleotide analogs (indicated by gg in blue) positioned at specific positions relative to the fork junction. (**f**) Absorption and fluorescence spectra of 6-MI substituted DNA

7.3 Exciton-Coupled $(iCy3)_2$ Dimer-Labeled ss–dsDNA Constructs as Probes of Local DNA Base and Backbone Conformations

An annealed $(iCy3)_2$ dimer-labeled DNA fork construct contains both double-stranded and single-stranded DNA regions, and the dimer probe can be selectively inserted relative to the ss–dsDNA fork junction. In Fig. 7.3a is shown the nomenclature that we use to specify the probe-labeling position relative to the ss–dsDNA fork

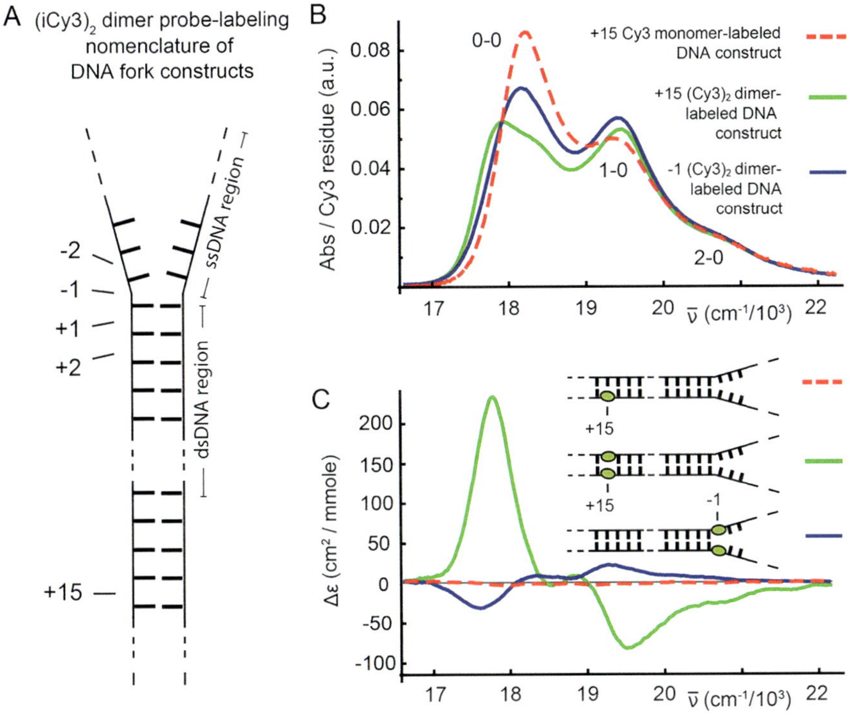

Fig. 7.3 (**a**) The insertion site position of the $(iCy3)_2$ dimer probes is indicated relative to the ss–dsDNA fork junction using positive integers in the direction toward the double-stranded region and negative integers in the direction toward the single-stranded region. (**b**) Room temperature (23 °C) absorption and (**c**) CD spectra for iCy3 monomer-labeled +15 duplex (dashed red curve), $(iCy3)_2$ dimer-labeled +15 duplex (green), and $(iCy3)_2$ dimer-labeled −1 fork (blue) DNA constructs. Measurements were performed at 23 °C using 100 mM NaCl and 6 mM $MgCl_2$, and 10 mM Tris at pH 8.0. Figure adapted from Heussman et al. (2019)

junction with progressively increasing positive integers indicating positions toward the double-stranded region and negative integers indicating positions toward the single-stranded regions. An iCy3 monomer-labeled ss–dsDNA construct can also be prepared by including a thymine (T) base in the complementary strand at the position directly opposite to the probe chromophore. The addition of the T base partially compensates for the misalignment of complementary base pairs introduced by the presence of the iCy3 monomer probe within one strand.

In Fig. 7.3b, c are shown experimental (room temperature, 23 °C) absorption and CD spectra of iCy3 monomer- and $(iCy3)_2$ dimer-labeled DNA constructs with probes inserted deep within the duplex region (+15 position), and an $(iCy3)_2$ dimer-labeled DNA construct with probes inserted near the ss–dsDNA fork junction (−1 position). The linear absorption spectrum of the iCy3 monomer (red dashed curve) exhibits a pronounced electronic-vibrational (vibronic) progression with progressively increasing energy features (labeled 0–0, 1–0, and 2–0), while its CD spectrum is very weak. The absorption and CD of the iCy3 monomer are largely insensitive to factors that affect the local environment of the chromophore, such as temperature, solution salt concentrations, and probe-labeling position relative to the ss–dsDNA fork junction (Heussman et al. 2019, 2022; Kringle et al. 2018; Sorour et al. 2022). In contrast, the absorption and CD spectra of the $(iCy3)_2$ dimer-labeled DNA constructs are highly sensitive to probe-labeling position and environmental factors that affect the local probe conformation. The vibronic progression is still present in the spectra of both the +15 and −1 $(iCy3)_2$ dimer-labeled ss–dsDNA constructs (solid green and blue curves, respectively). However, the individual vibronic features of the $(iCy3)_2$ dimers are broadened relative to those of the iCy3 monomer, and the ratios of the 0–0 to 1–0 vibronic peak intensities are smaller compared to the monomer [$I_{\mathrm{mon}}^{(0-0)} / I_{\mathrm{mon}}^{(1-0)} = 1.60$] (Kringle et al. 2018). The CD of both the +15 and −1 $(iCy3)_2$ dimer-labeled ss–dsDNA constructs exhibit a progression of bisignate lineshapes (i.e., a change of sign within a given vibronic band), which is a signature of vibronic excitons in a chiral dimer (Harada and Berova 2012). Most notably, the CD spectra of the +15 and −1 $(iCy3)_2$ dimer-labeled ss–dsDNA constructs have opposite signs, indicating that the two systems have opposite chiral symmetries.

7.3.1 Characterization of Mean Local Conformational Parameters from Absorption and CD Spectra of Exciton-Coupled (iCy3)₂ Dimer-Labeled ss–dsDNA Fork Constructs

The Cy3 chromophore consists of a conjugated trimethine bridge, which cojoins two indole-like groups, as shown in Fig. 7.2a. The trimethine bridge of the Cy3 monomer exists in an *all-trans* configuration, and the lowest energy electronic transition is a $\pi \rightarrow \pi^*$ transition (Jia et al. 2007). Like many π-conjugated molecules, Cy3 exhibits numerous Franck–Condon active modes, which range in energy from tens-to-several hundreds of wavenumbers (Aydin et al. 2011). Nevertheless, the homogeneous spectral lineshape of the Cy3 monomer in solution is broadened due to rapid electronic dephasing and thermal occupation of low energy vibrational levels in the electronic ground state. The lowest energy electronic transition from ground state $|g\rangle$ to excited state $|e\rangle$ is predominantly coupled to one low-frequency (~30 cm^{-1}) symmetric bending vibration along the trimethine bridge, which in turn couples anharmonically to a cluster of relatively high-frequency modes in the vicinity of ~1300 cm^{-1} (Sorour et al. 2022), as described in further detail in Sect. 7.3.2.

The spectrum of the Cy3 monomer in solution—and that of the iCy3 monomer inserted within the DNA sugar-phosphate backbone—can be simulated using a relatively simple quantum mechanical model Hamiltonian, $\hat{H}_M$, where the $|g\rangle \rightarrow |e\rangle$ transition with energy ε_{eg} is coupled to a single "effective" harmonic mode with energy $\hbar\omega_{vib}$, as shown in Fig. 7.4a, b (Heussman et al. 2019, 2022, 2024; Kringle et al. 2018). The monomer Hamiltonian is written as:

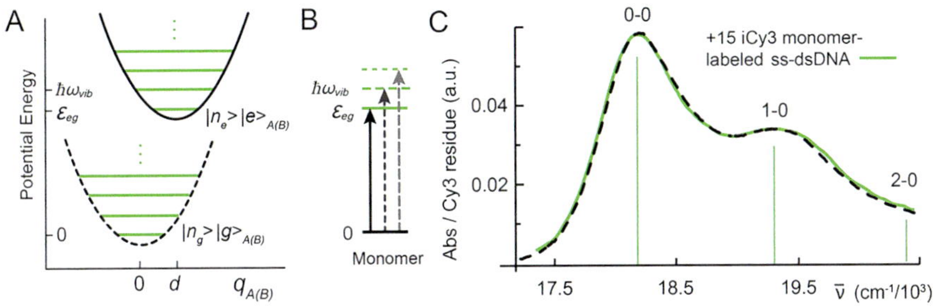

Fig. 7.4 (**a**) Electronic-vibrational (vibronic) potential energy diagram for the iCy3 monomer [labeled $A(B)$] ground and excited electronic state levels [labeled $|g\rangle_{A(B)}$ and $|e\rangle_{A(B)}$, respectively], which are assumed to be coupled to a single "effective" high-frequency vibrational mode with ground and excited vibrational states [labeled $|n_g\rangle_{A(B)}$ and $|n_e\rangle_{A(B)}$, respectively]. (**b**) Simplified energy level diagram of the iCy3 monomer. (**c**) Experimental and simulated absorption spectrum of +15 iCy3 monomer-labeled ss–dsDNA construct. Individual vibronic sub-bands are labeled 0–0, 1–0, and 2–0, as shown, indicating the vibrational levels in the ground and excited electronic states. Figure adapted from Marcus et al. (2023)

$$\hat{H}_M = \varepsilon_{eg}\,|e\rangle\langle e| + \hbar\omega_{vib}\hat{b}_M^{\dagger}\hat{b}_M + \hbar\omega_{vib}\left\{\lambda\left(\hat{b}_M^{\dagger} + \hat{b}_M\right) + \lambda^2\right\}|e\rangle\langle e| \tag{7.1}$$

where $\hat{b}_M^{\dagger}$ and $\hat{b}_M$ are, respectively, the operators for creating and annihilating a vibrational excitation in the ground and excited electronic potential energy surfaces. These operators obey the commutation relation $\left[\hat{b}_{M'}, \hat{b}_M^{\dagger}\right] = \delta_{M',M}$, where $\delta_{M',M}$ is the Kronecker delta function and the monomer labels $M', M \in \{A, B\}$.

The strength of the vibronic coupling is characterized by Franck–Condon (FC) factors, $\left|\langle n_e|0\rangle\right|^2 = e^{-\lambda^2}\lambda^{2n_e}/n_e!$, where $|0\rangle = |n_g = 0\rangle\,|g\rangle$ is the zero vibrational level of the ground electronic state and $|n_e\rangle$ is the vibrational quantum number of the electronically excited state with vibrational quantum number n_e. The FC factors depend on the Huang–Rhys parameter, $\lambda^2 = d^2\omega_{vib}m/2\hbar$ where d is the displacement of the electronically excited vibrational potential energy surface relative to the ground state (see Fig. 7.4a) surface and m is the reduced mass of the effective mode (Heussman et al. 2024). The Huang–Rhys parameter also determines the vibrational relaxation energy, $\lambda^2\hbar\omega_{vib}$, which is the amount of optical transition energy that is converted into heat.

In Fig. 7.4c are shown the experimental (green curve) and simulated (dashed black) absorption spectrum of the +15 iCy3 monomer-labeled ss–dsDNA construct at room temperature (23 °C). The simulated monomer spectrum accounts for the presence of both homogeneous and inhomogeneous line broadening. The homogeneous spectrum, $\sigma_H^M(\varepsilon)$, is the sum of homogeneous lineshapes associated with the individual vibronic transitions (Heussman et al. 2022).

$$\sigma_H^M(\varepsilon) = |\mu_{\text{eg}}|^2 \sum_{n_e=0}^{\infty} |\langle n_e | 0 \rangle|^2 L_H(\varepsilon - \varepsilon_{eg} - n_e \hbar\omega_{\text{vib}}) \tag{7.2}$$

Individual homogeneous lineshapes are modeled as Lorentzian functions, $L_H(\varepsilon) = \frac{1}{2}\Gamma_H / \left[\varepsilon^2 + \left(\frac{1}{2}\Gamma_H\right)^2\right]$, with full-width-at-half-maximum (FWHM) equal to Γ_H. The magnitude of the homogeneous linewidth is due to coupling between the optical transition and the phonon bath, which rapidly modulates the monomer transition energy. The value of Γ_H = 186 cm^{-1} for the iCy3 monomer was determined using two-dimensional fluorescence spectroscopy (2DFS) (Heussman et al. 2022; Kringle et al. 2018). The magnitude of the iCy3 monomer EDTM, μ_{eg} = ~12.8 D, was determined by integrating numerically the experimental absorption lineshape (Kringle et al. 2018).

Although each monomer is chemically identical, variations in the local environment can lead to the presence of static inhomogeneity of the transition energy, ε_{eg} (Heussman et al. 2019, 2022). We modeled static inhomogeneity using a Gaussian distribution, $G_{I,\text{mon}}(\varepsilon_{\text{eg}}) = \exp\left[-(\varepsilon_{\text{eg}} - \bar{\varepsilon}_{\text{eg}})^2 / 2\sigma_{I,\text{mon}}^2\right]$, with mean transition energy $\bar{\varepsilon}_{\text{eg}}$ and spectral inhomogeneity standard deviation $\sigma_{I,\text{ mon}}$. The simulated absorption spectrum is thus a convolution integral of homogeneous and inhomogeneous line shape functions (Heussman et al. 2022):

$$\sigma_I^M(\varepsilon) = \int_{-\infty}^{\infty} \sigma_H^M(\varepsilon - \varepsilon') G_{I,\text{mon}}(\varepsilon') d\varepsilon' \tag{7.3}$$

As discussed above, the simulated absorption spectrum of the iCy3 monomer depends on four independent model parameters: (i) the monomer electronic transition energy, ε_{eg}; (ii) the Huang–Rhys electronic-vibrational coupling parameter, λ^2; (iii) the effective vibrational energy, $\hbar\omega_{\text{vib}}$; and (iv) the spectral inhomogeneity parameter $\sigma_{I,\text{ mon}}$. These parameters were determined by fitting the above model to experimental spectra of the +15 iCy3 monomer-labeled DNA construct from which we determined the following values: ε_{eg} = ~18,277 cm^{-1}, λ^2 = ~0.56, $\hbar\omega_{\text{vib}}$ = ~1109 cm^{-1}, and $\sigma_{I,\text{ mon}}$ = ~347 cm^{-1} (Kringle et al. 2018). We note that the vibrational relaxation energy corresponding to these values is $\lambda^2\hbar\omega_{\text{vib}}$ = ~605 cm^{-1}. As mentioned above, these values for iCy3 monomer-labeled ss–dsDNA constructs are relatively insensitive to labeling position and other environmental factors.

For the $(\text{iCy3})_2$ dimer-labeled ss–dsDNA constructs, the closely spaced iCy3 monomers can couple through a transition density interaction, J (Heussman et al. 2022, 2024; Kringle et al. 2018; Sorour et al. 2021, 2023). The value of J and the resulting spectroscopic properties of the $(\text{iCy3})_2$ dimer-labeled DNA construct depend sensitively on the relative orientation and spacing between the iCy3 monomers on the length scale of a few angstroms (see, for example, Fig. 7.3b, c), which in turn depends on the local stabilities and dynamics of the sugar-phosphate backbones and bases immediately adjacent to the probes.

The Holstein–Frenkel (H-F) Hamiltonian of the coupled dimer is written as (Spano 2010):

$$\hat{H}_A + \hat{H}_B + J\left[|eg\rangle\langle ge| + |ge\rangle\langle eg|\right] \tag{7.4}$$

where $|eg\rangle$ is the product state in which monomer A is electronically excited and monomer B is unexcited. For non-zero electrostatic coupling J, the absorption spectrum of the $(iCy3)_2$ dimer has a shape roughly like that of the monomer. However, the electrostatic interaction leads to each of the vibronic features (i.e., 0–0, 0–1, …) being split and additionally broadened into symmetric (+) and anti-symmetric (−) sub-bands, as illustrated in Fig. 7.5a. The relative peak intensities of the absorption

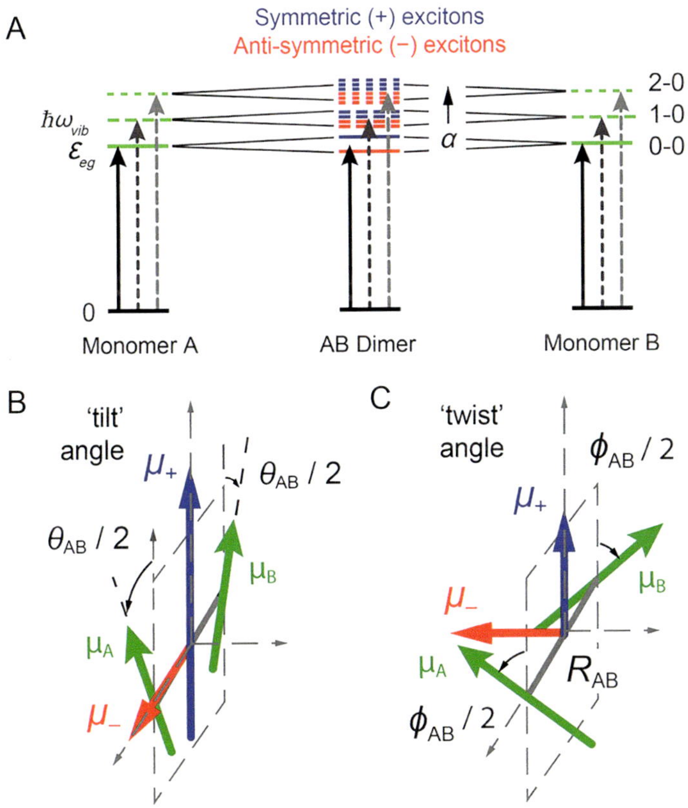

Fig. 7.5 (**a**) Energy level diagrams of the separated iCy3 monomers [labeled $A(B)$] and of the coupled $(iCy3)_2$ dimer. The electronic coupling (J) between closely spaced monomers of the $(iCy3)_2$ dimer results in delocalized symmetric (+, blue) and anti-symmetric (−, red) excitons. The dimer structural parameters are (**b**) the "tilt" angle, θ_{AB}, (**c**) the "twist" angle, ϕ_{AB}, and the inter-chromophore separation, R_{AB}. The electric dipole transition moments of the monomers [$\mu_{A(B)}$] are indicated by green arrows, and those of the symmetric (μ_+) and anti-symmetric (μ_-) excitons are indicated by blue and red arrows, respectively. Figure adapted from Marcus et al. (2023)

spectrum depend on—in addition to the Franck–Condon factors that affect the vibronic features of the iCy3 monomer—the local conformation of the dimer, which determines the magnitudes of the symmetric and anti-symmetric transition dipole moments $\mu_{\pm} = \frac{\mu_{eg}}{\sqrt{2}}\left[\hat{\boldsymbol{d}}_A \pm \hat{\boldsymbol{d}}_B\right]$, where $\hat{\boldsymbol{d}}_A$ and $\hat{\boldsymbol{d}}_B$ are unit vectors that specify the monomer EDTM directions (see Fig. 7.5b, c). The symmetric and anti-symmetric excitons each consist of a manifold of delocalized electronically coupled vibronic states, which are superpositions of electronic-vibrational products of the *A* and *B* monomer states given by $\left|e_{\pm}^{(\alpha)}\right\rangle = \sum_{n_e,n_g} c_{\pm}^{(\alpha)}\left[u_{n_e,n_g}\,|\,eg\rangle \pm u_{n_g,n_e}\,|\,ge\rangle\right]$ (Kistler et al. 2011, 2013). The coefficients, $c_{\pm}^{(\alpha)}$, depend on the vibrational coordinates of the monomers, $u_{n_{e(g)},n_{g(e)}} = |\,n_{e(g)}n_{g(e)}\rangle$ is the vibrational product state, $n_{e(g)}$ is the vibrational quantum number in the electronic excited (ground) state, and $\alpha = (0, 1, \ldots)$ is a state index in order of increasing energy.

The $(\text{iCy3})_2$ dimer homogeneous absorption spectrum is the sum of symmetric (+) and anti-symmetric (−) excitons:

$$\sigma_H^D(\varepsilon) = \sigma_+(\varepsilon) + \sigma_-(\varepsilon) \tag{7.5}$$

where $\sigma_{\pm}(\varepsilon) = \sum_{\alpha}\left|\langle 0\,|\,\mu^{\text{tot}}\,|\,e_{\pm}^{(\alpha)}\rangle\right|^2 L_H\left(\varepsilon - \varepsilon_{\pm}^{(\alpha)}\right)$, $\mu^{\text{tot}} = \mu_{eg}\left(\hat{\boldsymbol{d}}_A + \hat{\boldsymbol{d}}_B\right)$ is the collective EDTM and $L_H(\varepsilon) = \frac{1}{2}\Gamma_H \Big/ \left[\varepsilon^2 + \left(\frac{1}{2}\Gamma_H\right)^2\right]$ is a Lorentzian homogeneous lineshape with eigen-energy, $\varepsilon_{\pm}^{(\alpha)}$, and FWHM line width, $\Gamma_H = 186\ \text{cm}^{-1}$. Similarly, the dimer homogeneous CD spectrum is the sum of symmetric and anti-symmetric rotational strengths (Nordén et al. 2010)

$$CD_H^D(\varepsilon) = \sum_{\alpha}\left[RS_+^{(\alpha)} L_H\left(\varepsilon - \varepsilon_+^{(\alpha)}\right) + RS_-^{(\alpha)} L_H\left(\varepsilon - \varepsilon_-^{(\alpha)}\right)\right] \tag{7.6}$$

where $RS_{\pm}^{(\alpha)} = \frac{\varepsilon_{eg}}{4\hbar c}\langle 0\,|\,\hat{\boldsymbol{d}}_A\,|\,e_{\pm}^{(\alpha)}\rangle \times \langle e_{\pm}^{(\alpha)}\,|\,\hat{\boldsymbol{d}}_B\,|\,0\rangle \bullet \boldsymbol{R}_{AB}$. In the above expressions, the ground electronic-vibrational state of the *AB* dimer is $|\,0\rangle = |n_A = 0, n_B = 0\rangle|gg\rangle$.

In solution, the $(\text{iCy3})_2$ dimer conformation may vary from molecule to molecule due to local (and position-specific) DNA “breathing” fluctuations. To account for spectral inhomogeneity, the homogeneous dimer absorption and CD line shapes are convolved with an inhomogeneous distribution function, $G\left(\varepsilon_{\pm}^{(\alpha)}\right) = \left(2\pi\sigma_I^2\right)^{-\frac{1}{2}} \exp\left[-\left(\varepsilon_{\pm}^{(\alpha)} - \bar{\varepsilon}_{\pm}^{(\alpha)}\right)^2 / 2\sigma_I^2\right]$, which is centered at the average transition energy, $\bar{\varepsilon}_{+}^{(\alpha)}$, and has standard deviation, σ_I. The final expressions for the absorption and CD spectra are given by the convolution integrals:

$$\sigma_I^D(\varepsilon) = \int_{-\infty}^{\infty} \sigma_H^D(\varepsilon - \varepsilon') G(\varepsilon') d\varepsilon' \tag{7.7}$$

$$\mathrm{CD}_I^D(\varepsilon) = \int_{-\infty}^{\infty} \mathrm{CD}_H^D(\varepsilon - \varepsilon') G(\varepsilon') d\varepsilon' \tag{7.8}$$

According to the H-F model, the simulated absorption and CD spectra of the $(\text{iCy3})_2$ dimer-labeled ss–dsDNA constructs depend on, in addition to the monomer Hamiltonian parameters (i)–(iv) discussed above, (v) the electrostatic coupling, J, and (vi) the spectral inhomogeneity parameter, σ_I. The value of J depends on structural parameters that specify the conformation of the $(\text{iCy3})_2$ dimer while the value of σ_I depends on the conformational disorder, which can be determined experimentally using two-dimensional fluorescence spectroscopy (2DFS), as discussed in Sect. 7.3.2.

The electrostatic coupling, J, can be calculated using quantum chemical models of the Coulomb interaction between the iCy3 monomer electronic transition density (Heussman et al. 2024; Sorour et al. 2021, 2023).

$$J = \frac{1}{4\pi\epsilon} \int_{-\infty}^{\infty} d\boldsymbol{r}_A \int_{-\infty}^{\infty} d\boldsymbol{r}_B \frac{\rho_A^{ge}(\boldsymbol{r}_A)\rho_B^{eg}(\boldsymbol{r}_B)}{|\boldsymbol{R}_{AB}|} \tag{7.9}$$

where $\boldsymbol{R}_{AB} = \boldsymbol{r}_B - \boldsymbol{r}_A$ is the inter-chromophore center-of-mass separation vector, $\rho_M^{ge}(\boldsymbol{r}_M) = \langle g|_M \, \hat{\rho}(\boldsymbol{r}_M) | e_M$ is the monomer transition density matrix element, and ϵ is the electric permittivity. The ability of these calculations to determine conformational parameters of $(\text{iCy3})_2$ dimer-labeled ss–dsDNA constructs by analyzing experimental optical spectra has been assessed by comparing transition density models of varying levels of approximation (Heussman et al. 2024). In the point-dipole (PD) model, the electrostatic coupling is given by

$$J^{PD} = \frac{|\mu_{eg}|^2}{4\pi\epsilon |R_{AB}|^3} \left[\hat{\boldsymbol{d}}_A \cdot \hat{\boldsymbol{d}}_B - 3\left(\hat{\boldsymbol{d}}_A \cdot \hat{\boldsymbol{R}}_{AB}\right)\left(\hat{\boldsymbol{R}}_{AB} \cdot \hat{\boldsymbol{d}}_B\right) \right] \tag{7.10}$$

The PD model assumes that the transition density can be approximated as a point-dipole located at the molecular center-of-mass, which is an unrealistic approximation for intermolecular separations smaller than the iCy3 monomer dimension. The finite dimension of the molecule can be accounted for by using the "extended dipole" (ED) model, which represents the transition density as a line segment that is oriented parallel to the monomer EDTM with equal and opposite charges on its endpoints (Czikklely et al. 1970; Didraga et al. 2004).

$$J^{\mathrm{ED}} = \frac{\left|\mu_{\mathrm{eg}}\right|^2}{4\pi\varepsilon l^2}\left[\frac{1}{R_{AB}^{++}} - \frac{1}{R_{AB}^{-+}} - \frac{1}{R_{AB}^{+-}} + \frac{1}{R_{AB}^{--}}\right] \tag{7.11}$$

Equation (7.11) approximates the transition density for each monomer as two separate point charges of equal magnitude (q) and opposite sign, which lie separated by the distance l such that $ql = \mu_{eg}$. The distances between point charges are given by $R_{AB}^{\pm\pm} = \left[R_{AB} \pm l\left(\hat{d}_A - \hat{d}_B\right)/2\right]$ and $R_{AB}^{\mp\pm} = \left[R_{AB} \mp l\left(\hat{d}_A + \hat{d}_B\right)/2\right]$. The three-dimensional atomic structure of the iCy3 chromophore can be accounted for by applying an atomistically detailed electrostatic coupling model that is based on "transition charges" (TQs) derived from ab initio quantum chemical calculations. In the TQ model, the electrostatic interaction is approximated as a discrete sum over individual transition electrostatic potential charges (q_i), which are assigned to the centers of each atomic coordinate of the monomer (Sorour et al. 2021).

$$J^{TQ} = \frac{1}{4\pi\epsilon}\sum_a^{N_A}\sum_b^{N_B}\frac{q_a q_b}{\left|\boldsymbol{r}_a^A - \boldsymbol{r}_b^B\right|} \tag{7.12}$$

In Eq. (7.12), $\boldsymbol{r}_i^M$ is the position of the ith atom of chromophore M ($\in\{A, B\}$) and N_M is the number of atoms in chromophore M.

A minimal set of structural parameters are used to characterize the conformations of $(\text{iCy3})_2$ dimer-labeled ss–dsDNA constructs. For the PD and ED models, the conformation is fully specified by the "tilt," θ_{AB}, "twist," ϕ_{AB}, and "inter-chromophore separation," R_{AB} (Fig. 7.5b, c). However, for TQ model calculations, which include higher dimensionality, additional structural parameters are needed to distinguish between relative orientations and translational displacements of the iCy3 monomers. These are the "roll" angle, η_{AB}, to specify the degree of chromophore stacking, and the vertical "shear," δ_{AB}, and the horizontal "shift," ξ_{AB}, displacements (Heussman et al. 2024). For a specific set of conformational parameters, and depending on the choice of coupling model, the value of J and simulated optical spectra are calculated and compared to experimental data. "Optimized values" for the conformational parameters are thus obtained by applying the above procedure iteratively while rejecting sterically overlapping dimer conformations using a multiparameter optimization procedure.

The above procedure provides nearly identical results for the three different coupling models in terms of the twist angle, ϕ_{AB}, tilt, θ_{AB}, roll, η_{AB} and inter-chromophore separation R_{AB}. In principle, information about the roll angle, η_{AB}, is only available using the atomistically detailed TQ model (Heussman et al. 2024). However, by systematically rejecting sterically overlapping dimer conformations (approximating the van der Waals radii of the major component atoms of the iCy3 monomers as r_w ~1.5 Å), corroborating information between the three models for all structural parameters are obtained. Thus, the reliability of the above approach for determining structural information from optical spectra is established.

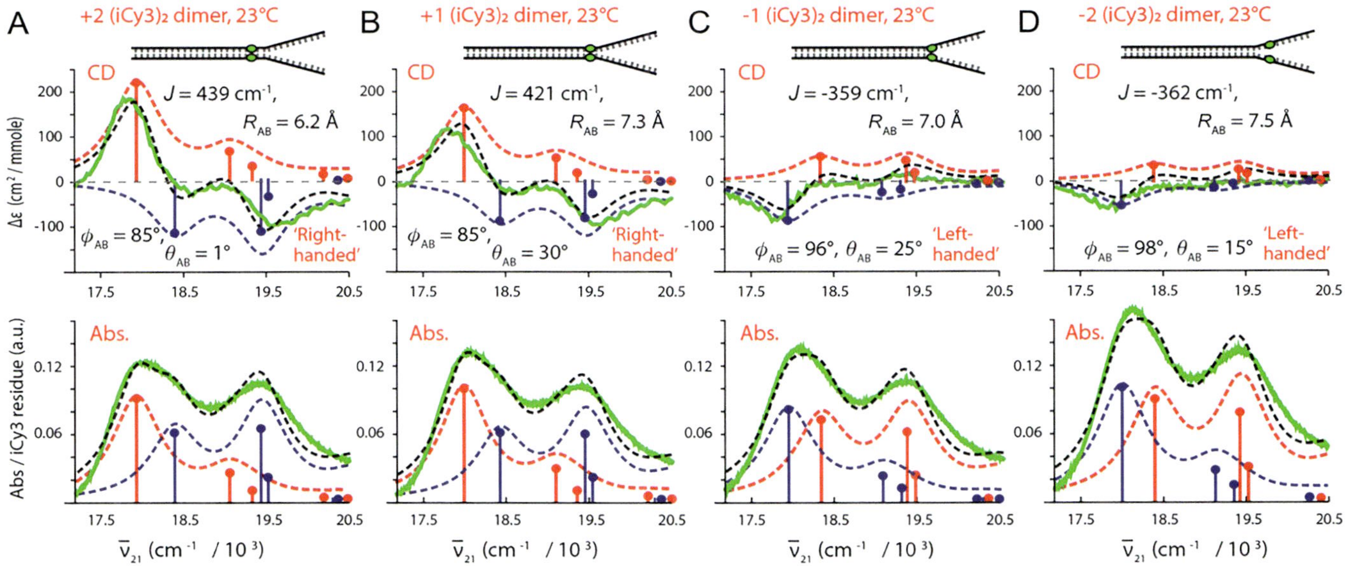

Fig. 7.6 Experimental and simulated CD and absorption measurements performed at room temperature (25 °C) for $(iCy3)_2$ dimer ss–dsDNA constructs as a function of probe-labeling position. (**a**) +2; (**b**) +1; (**c**) −1; and (**d**) −2. Samples were prepared using 100 mM NaCl and 6 mM $MgCl_2$, and 10 mM Tris at pH 8.0. The experimental CD (top row) and absorption spectra (bottom row) are shown (in green) overlaid with vibronic spectral features (black dashed curves) obtained from optimized fits to the H-F model. The symmetric (+) and anti-symmetric (−) exciton state manifold contributions to the spectra are shown separately in blue and red, respectively. Values of the optimized parameters are shown in the insets of the corresponding panels. The mean local conformation changes abruptly from a right-handed to a left-handed twist symmetry as the label position is changed from the +1 to the −1 position. Figure adapted from Heussman et al. (2022, 2024)

In Fig. 7.6 are shown comparisons between experimental and simulated room temperature optical spectra of $(iCy3)_2$ dimer-labeled ss–dsDNA fork constructs labeled at the +2, +1, −1, and −2 positions. The experimental CD and absorption spectra (green curves) are shown overlaid with simulated spectra resulting from the H-F model analysis. Optimized values for the corresponding structural parameters are presented in the insets. The results indicate that the electrostatic coupling, J, undergoes a sign inversion as the $(iCy3)_2$ dimer-probe position is changed from the +1 to −1 position across the ss–dsDNA junction. This is accompanied by non-continuous changes of the local conformational coordinates: the inter-chromophore separation R_{AB} [6.2 Å (+2), 7.3 Å (+1), 7.0 Å (−1), and 7.5 Å (−2)], the twist angle ϕ_{AB} [85° (+2), 85° (+1), 96° (−1), and 98° (−2)], and the tilt angle θ_{AB} [1° (+2), 30° (+1), 25° (−1), and 15 (−2)]. The sign inversion of the electrostatic coupling between the +1 and −1 positions indicates that the conformation of the $(iCy3)_2$ dimer probe, and presumably that of the sugar-phosphate backbones labeled at these sites, changes from right-handed to left-handed. This change of handedness is correlated to a change of the twist angle ϕ_{AB} from values that are less than 90° (right-handed) to values that are greater than 90° (left-handed). In addition, the tilt angle θ_{AB} undergoes an abrupt increase from 1° to 30° between the +2 and +1 positions, followed by a decrease to ~25° at the −1 position, and again to ~15° at the −2 position. This indicates that the local conformation of the sugar-phosphate backbones within the ss–dsDNA fork constructs undergoes an abrupt loss of cylindrical symmetry at the interface between the +2 and +1 positions. However, some of the cylindrical symmetry is recovered at the −1 and −2 positions.

7.3.2 Adiabatic Franck–Condon Models of iCy3 Monomer and $(iCy3)_2$ Dimers

While "effective mode" models for simulating the optical spectra of iCy3 monomer and $(iCy3)_2$ dimer-labeled DNA constructs are useful for their relative simplicity of implementation, more accurate computational approaches are also available to test the validity of such utilitarian models. Of particular interest is the Franck–Condon (FC) approach, which considers all the vibrational degrees of freedom of the iCy3 molecule rather than a single effective mode. The FC approach is particularly valuable for simulating the spectra of molecules with strong electron-vibrational coupling. Unlike the minimal state "effective mode" models, the adiabatic FC approach is a full ab initio method that accounts for excited state processes and all normal modes on an equal footing (Gozem and Krylov 2022; Barone et al. 2009).

Within the FC approach a harmonic potential is assumed for both the ground and excited states from which the normal modes are obtained. The probability of electronic transitions is calculated using the Franck–Condon factors, which describe the overlap between the vibrational wave functions of the ground and excited (final) electronic states at their respective equilibrium positions, as described by (Gozem and Krylov 2022; Barone et al. 2009).

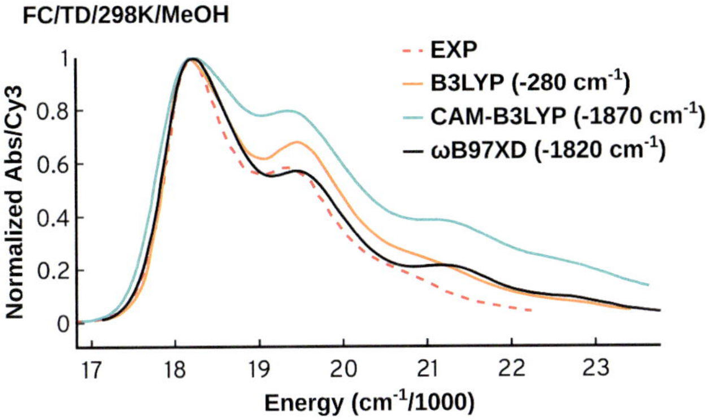

Fig. 7.7 Time-dependent Franck–Condon (FC) density functional theory (DFT) calculations of the Cy3 monomer. Comparisons are shown between the B3LYP, CAM-B3LYP, and ωB97XD functionals at 298 K. The calculations were performed using implicit methanol (MeOH) solvent via a polarizable continuum model. The calculated spectra were shifted to match the experimental absorption maximum with the shift value shown in parentheses. From Sorour et al. (2022)

$$\langle \Psi' | \hat{\mu} | \Psi'' \rangle = \langle \chi_v' | \chi_v'' \rangle \langle \psi_e' | \hat{\mu} | \psi_e'' \rangle \tag{7.13}$$

where Ψ and ψ_e are the total and electronic wavefunctions, respectively, of the initial (″) and final (′) states; χ_v is the vibrational wavefunction, and μ is the electronic transition dipole operator.[2] The FC approach has been used to simulate the absorption spectra of several members of the cyanine family with reasonable accuracy (Sorour et al. 2022; Alia and Flack 2020). The FC approach is highly sensitive to the choice of quantum mechanical method (i.e., the choice of functional) used for the structural optimization and calculation of the spectrum. In Fig. 7.7 are shown comparisons between the experimental spectrum of the Cy3 chromophore in methanol and FC calculations using the B3LYP, CAM-B3LYP, and the ωB97XD functionals, with the latter providing the best agreement with experiment.

The FC approach calculates the overlap terms between vibrational wavefunctions of all the normal modes, and thus provides detailed information about the vibrational excitations contributing to the different bands of the spectrum. In Fig. 7.8 are shown the vibronic excitations contributing to the Cy3 monomer spectrum obtained from the FC approach. The first (lowest energy) spectral band is dominated by low-frequency modes, e.g. mode 3 (with frequency = 35 cm^{-1}) and combined vibronic excitations from the same and other normal modes. The labels in Fig. 7.8 indicate the mode numbers of specific vibronic transitions with significant contributions to the spectrum. For example, 0–0 indicates the vertical FC transition without vibrational excitation, 0-3^n indicates the *nth* FC transition (in order of increasing

[2] Note that in the current section, we adopt a slightly different nomenclature for vibrational and electronic wavefunctions and transition dipole operators than in previous sections to be consistent with the cited literature.

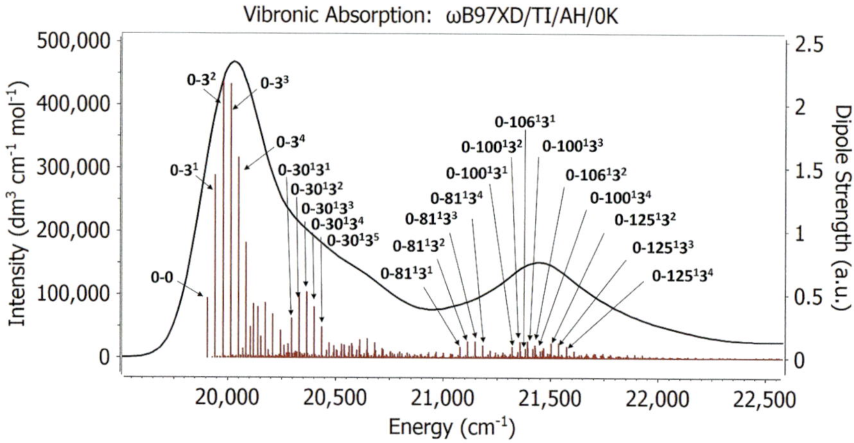

Fig. 7.8 The vibronic spectrum of the Cy3 monomer obtained from structures optimized using the ωB97XD functional and the FC/Adiabatic Hessian (AH) approach at 0 K. The red sticks correspond to the vibronic excitations. The labels on the sticks are selected transitions with significant contributions. TI: Time-independent. From Sorour et al. (2022)

energy) that is coupled to mode 3, and $0\text{-}30^1 3^n$ indicates the FC combined transition, which is coupled to mode 30 through its anharmonic coupling to mode 3^n. The higher energy bands are dominated by combined vibrational excitations between normal mode number 3 and higher energy modes with frequencies ranging from 1100 cm^{-1} to 1550 cm^{-1}. Thus, the broadly defined shape of the Cy3 monomer absorption spectrum—with "effective mode" vibrational spacing ~1100 cm^{-1}—can be understood in terms of the direct coupling of the optical transition to mode 3, which in turn couples indirectly (anharmonically) to higher frequency modes. These findings emphasize the importance of accounting for all normal modes to accurately model and interpret the Cy3 spectrum. Moreover, they provide an understanding that justifies the application of minimal state "effective mode" models for simulations of iCy3 monomer and $(iCy3)_2$ dimer-labeled DNA spectra, as discussed in Sect. 7.3.1.

Full quantum mechanical modeling of the optical spectra of $(iCy3)_2$ dimer-labeled ss–dsDNA constructs is less straightforward than for the iCy3 monomer due to the need to optimize the minima of the ground and excited electronic state surfaces of such multi-subunit flexible molecules. As discussed in Sect. 7.3.1, approximation methods for calculating the electrostatic coupling, J, within an $(iCy3)_2$ dimer can be obtained using a "transition charge" TQ model derived from quantum chemical calculations (Sorour et al. 2021). The TQ approach was applied successfully to model the optical spectra of $(iCy3)_2$ dimer-labeled DNA constructs to obtain structural information (Heussman et al. 2024). Furthermore, the FC approach was used to accurately model a well-defined $(Cy3)_2$ dimer in which the Cy3 monomers were directly attached to each other using covalent linkers (Sorour et al. 2023). Such FC calculations provide detailed information about the electronic-vibrational properties of the $(Cy3)_2$ dimer spectral features, which cannot be accurately determined using minimal state "effective mode" models.

7.3.3 Characterization of Conformational Disorder of Exciton-Coupled $(iCy3)_2$ Dimer-Labeled ss–dsDNA Fork Constructs by Two-Dimensional Fluorescence Spectroscopy (2DFS)

While absorption and CD measurements can be used to determine the mean structural parameters of the $(iCy3)_2$ dimer probes within site-specifically labeled ss–dsDNA constructs, two-dimensional fluorescence spectroscopy (2DFS) provides additional information about the distributions of these parameters (Heussman et al. 2022). 2DFS is a fluorescence-detected Fourier transform (FT) optical method that is well suited to measure the electronic-vibrational structure of fluorescent molecular chromophores (Tekavec et al. 2007; Perdomo et al. 2012; Lott et al. 2011; Karki et al. 2014; Widom et al. 2013a). The underlying spectroscopic principles of 2DFS resemble those of 2D NMR (Jonas 2003; Slichter 1996) and 2DIR (Zheng et al. 2007; Hochstrasser 2007), with the latter providing structural and dynamic information about local vibrational modes in proteins (Reppert and Tokmakoff 2016; Ghosh et al. 2017), nucleic acids (Sanstead et al. 2016; Sanstead and Tokmakoff 2018), and biomolecular hydration shells (Laage et al. 2017). By virtue of its ability to sensitively monitor fluorescence against a dark background, 2DFS can detect relatively weak excited state population signals above the background of scattered excitation light. The signals detected by 2DFS on $(iCy3)_2$ dimer-labeled ss–dsDNA constructs directly monitor DNA backbone conformations and conformational disorder, which likely play a central role in protein recognition and binding events.

In 2DFS, a repeating sequence of four collinear optical pulses (see Fig. 7.9a) is generated using interferometric methods. The pulses are characterized by their inter-pulse delays, $t_{ij} = t_i - t_j$, and relative phases, φ_{ij} $[i, j \in \{1, 2, 3, 4\}]$. For a given measurement, the time delays, t_{21}, t_{32} and t_{43}, are held fixed while the relative phases, φ_{21}, φ_{32} and φ_{43}, are swept continuously at tens-of-kilohertz frequencies (Tekavec et al. 2007). The laser is optically resonant with the symmetric and anti-symmetric excitons of the $(iCy3)_2$ dimer probes, such that each of the four pulses can excite a one-photon dipole-allowed optical transition between ground, singly-excited, and doubly-excited states (see Fig. 7.9b). The resulting phase-dependent fluorescence signals are proportional to the nonlinear populations of molecular excited states, which can be detected separately using phase-sensitive methods (Tamimi et al. 2020). "Rephasing" (RP) and "non-rephasing" (NRP) signals are of specific interest, which have "phase signatures" $\varphi_{43} + \varphi_{21}$ and $\varphi_{43} - \varphi_{21}$, respectively. The recorded signals are complex-valued response functions of the inter-pulse delays, $S_{RP}(t_{21}, t_{32} = 0, t_{43})$ and $S_{\mathrm{NRP}}(t_{21}, t_{32} = 0, t_{43})$, which oscillate in time at the transition frequencies of the $(iCy3)_2$ dimer probe (see Fig. 7.9c) (Heussman et al. 2022). The RP response function contains the line shape factor $\exp\left[-\Gamma_H\left(t_{21}+t_{43}\right)\right]\exp\left[-\sigma_I^2\left(t_{21}-t_{43}\right)^2/2\right]$, which decays exponentially at the homogeneous dephasing rate along the diagonal axis $(t_{21} + t_{43})$ and as a Gaussian envelope with inhomogeneous decay constant $\sigma_I^2/2$ along the anti-diagonal axis $(t_{21} - t_{43})$. In contrast, the NRP response function contains the line shape factor

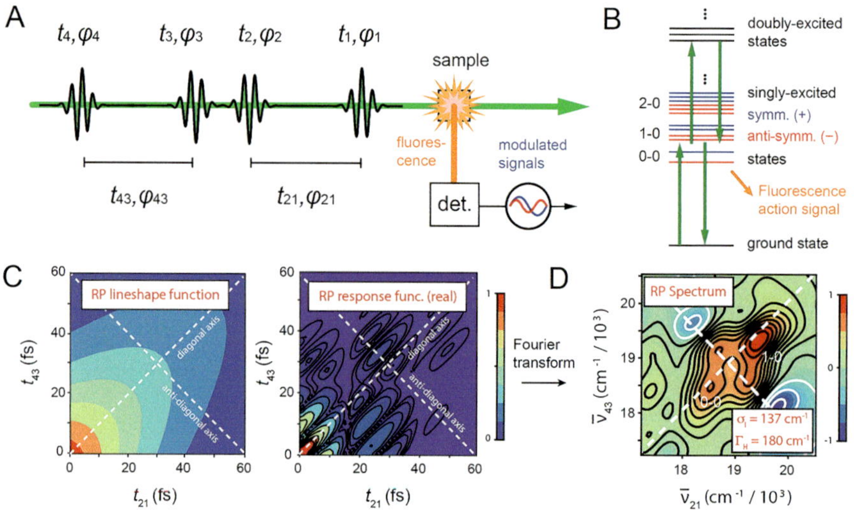

Fig. 7.9 (**a**) The 2DFS instrument prepares a sequence of four collinear optical pulses with relative phases, φ_{21} and φ_{43}, and time delays, t_{21} and t_{43}. The ensuing fluorescence signals are detected as a function of the inter-pulse delays and phases. (**b**) Energy level diagram indicating the laser pulse energy and exciton-coupled $(iCy3)_2$ dimer state energies. Each laser pulse can induce a transition between ground, singly-excited, and doubly-excited exciton states. Fluorescence is spectrally separated from scattered laser excitation light. (**c**) Example calculations of the real parts of the 2DFS rephasing (RP) line shape (left panel) and response (right panel) functions. These functions are displayed as two-dimensional contour plots with diagonal and anti-diagonal axes indicated as white dashed lines. The RP response function, $S_{RP}(t_{21}, t_{32} = 0, t_{43})$, shown in the right panel contains both the line shape function and the transition frequency phase factors. (**d**) The experimental RP 2D fluorescence spectrum, $\hat{S}_{RP}\left(\bar{\nu}_{21}, t_{32} = 0, \bar{\nu}_{43}\right)$, of the +15 $(iCy3)_2$ dimer-labeled ss–dsDNA construct is calculated by performing a two-dimensional FT of the experimental response function with respect to the time variables t_{21} and t_{43}. Comparisons between model calculations and experimental data determine the values of the homogeneous and inhomogeneous line shape parameters, Γ_H and σ_I, respectively. Optimized values of Γ_H and σ_I are given in the inset. Figure adapted from Heussman et al. (2022), Marcus et al. (2023)

$\exp\left[-\Gamma_H\left(t_{21}+t_{43}\right)\right]\exp\left[-\sigma_I^2\left(t_{21}+t_{43}\right)^2/2\right]$, which decays along the diagonal axis with rate constants that depend on both the homogeneous and inhomogeneous parameters (Heussman et al. 2022). Simultaneous comparisons of both RP and NRP simulated and experimental 2D spectra permits accurate determination of the homogeneous and inhomogeneous line shape parameters, Γ_H and σ_I, respectively. FT of the response functions with respect to t_{21} and t_{43} yields the frequency-dependent RP and NRP spectra, $\hat{S}_{RP}\left(\bar{\nu}_{21}, t_{32} = 0, \bar{\nu}_{43}\right)$ and $\hat{S}_{NRP}\left(\bar{\nu}_{21}, t_{32} = 0, \bar{\nu}_{43}\right)$, respectively. An example experimental RP 2DFS spectrum of the +15 $(iCy3)_2$ dimer-labeled ss–dsDNA construct, and the resulting values for Γ_H and σ_I, are shown in Fig. 7.9d.

In Fig. 7.10 are shown temperature-dependent measurements of the +15 $(iCy3)_2$ dimer-labeled ss–dsDNA construct, which includes the absorption, CD, and experimental and simulated RP 2DFS data (Heussman et al. 2022). At the lowest temperatures (5–23 °C), the absorption and CD spectra exhibit, respectively, the intensity

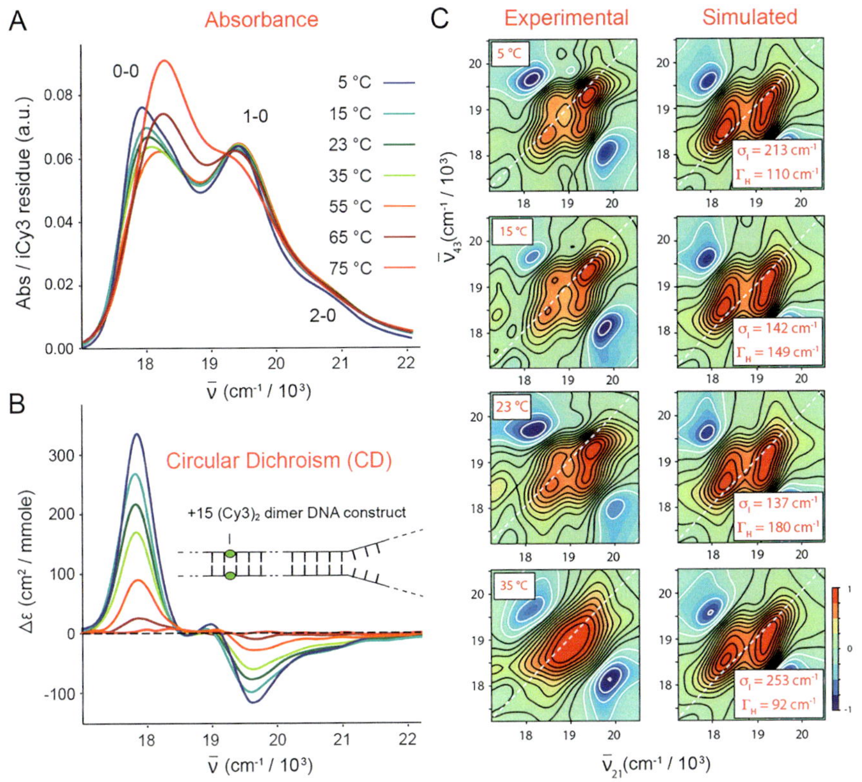

Fig. 7.10 Melting behavior of the +15 $(iCy3)_2$ dimer-labeled ss–dsDNA construct. Temperature-dependent measurements of the (**a**) absorption, (**b**) CD, (**c**) experimental and simulated 2DFS data. Comparisons between model calculations and experimental data determine the values of the homogeneous and inhomogeneous line shape parameters, Γ_H and σ_I, respectively. Optimized values of Γ_H and σ_I are given in the insets. Samples were prepared using 100 mM NaCl and 6 mM $MgCl_2$, and 10 mM Tris at pH 8.0. Figure adapted from Heussman et al. (2022)

borrowing and bisignate lineshapes that are characteristic of exciton-coupled $(iCy3)_2$ dimer probes (Fig. 7.10a, b). The 2DFS data at these relatively low temperatures show well-separated peaks and cross-peaks, which indicate the presence of the spatially delocalized symmetric (+) and anti-symmetric (−) excitons (see discussion in Sect. 7.3.1).

As the temperature is elevated above room temperature toward the melting point (T_m = ~65 °C), the Watson–Crick (W-C) B-form structure of the DNA bases and sugar-phosphate backbones becomes locally unstable. The absorption and CD spectra change systematically with increasing temperature to reflect the reduced coupling strength of the $(iCy3)_2$ dimer probe due to the disruption of structural order. At the highest temperature for which the single strands of the DNA duplex have fully separated (75 °C) the coupling strength is approximately zero such that the absorption and CD spectra resemble those of the iCy3 monomer substituted ss–dsDNA

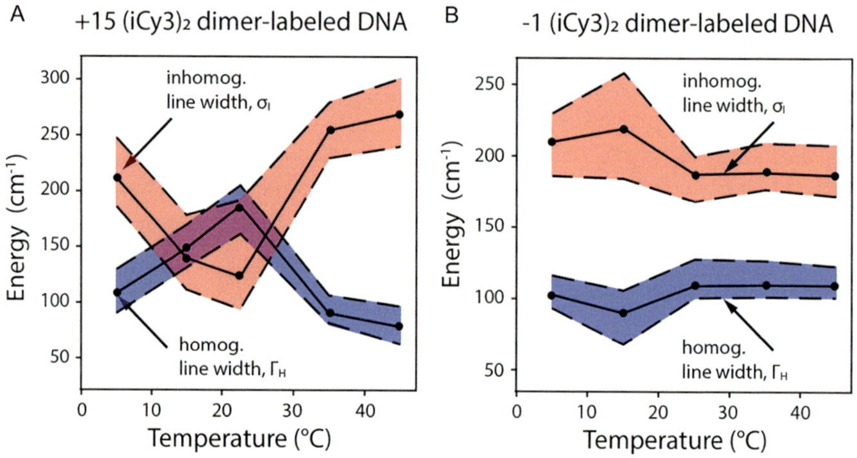

Fig. 7.11 Optimized homogeneous (Γ_H, blue) and inhomogeneous (σ_I, teal) line width parameters as a function of temperature obtained from 2DFS line shape analyses. (**a**) +15 $(iCy3)_2$ dimer-labeled ss–dsDNA construct; (**b**) −1 $(iCy3)_2$ dimer-labeled ss–dsDNA construct. Shaded regions bounded by dashed lines indicate error bars determined from the analysis. The melting temperature, T_m, of the duplex regions of the DNA constructs is 65 °C. Samples were prepared using 100 mM NaCl and 6 mM $MgCl_2$, and 10 mM Tris at pH 8.0. Figure adapted from Heussman et al. (2022), Marcus et al. (2023)

constructs. The 2DFS data reveal the sensitivity of this method to the presence of local conformational disorder that is "sensed" by the $(iCy3)_2$ dimer probe deep within the duplex region (+15 position) of the ss–dsDNA construct. At low temperatures, the spectral features of the 2DFS data are well-defined. However, at elevated temperatures the spectral features become progressively less pronounced and the exciton-split diagonal features merge into a single diffuse feature (Fig. 7.10c).

Analyses of 2DFS line shape data provide temperature-dependent values for the homogeneous and inhomogeneous line shape parameters, Γ_H and σ_I, respectively (Heussman et al. 2022). Such measurements show how conformational disorder of $(iCy3)_2$ dimer-labeled ss–dsDNA constructs—proportional to σ_I—varies with temperature and probe-labeling position. In Fig. 7.11 are shown the results of such studies for the +15 "duplex" and the −1 "fork" $(iCy3)_2$ dimer-labeled ss–dsDNA construct. The values of Γ_H and σ_I are shown as blue- and teal-shaded regions, respectively. The optimized values are presented as points, and the shaded regions bounded by dashed lines indicate error bars.

For the +15 "duplex" $(iCy3)_2$ dimer-labeled ss–dsDNA construct, the local conformation is minimally disordered at room temperature (see Fig. 7.11a). As the temperature is raised or lowered from room temperature (23 °C), the local conformational disorder abruptly increases, while the interactions between the probes and the solvent environment (proportional to the homogeneous linewidth) decreases. This indicates that the W-C B-form conformation of the $(iCy3)_2$ dimer probes at local sites deep within the duplex region is a marginally stable structure at room temperature, which becomes unstable at temperatures just above or below room temperature. These findings are consistent with the notion that the room temperature stability of the local B-form conformation results from a nearly equal balance between opposing thermodynamic forces (entropy–enthalpy compensation), and that small departures from room temperature (in either the positive or negative directions) alter the free energy landscape and serve to populate non-B-form conformations (Privalov et al. 1986; Vaitiekunas et al. 2015; Mikulecky and Feig 2006). Such a picture is sometimes invoked in "DNA breathing and trapping models," in which non-canonical local conformations of the DNA framework are transiently populated under "physiological" conditions and function as activated states required for protein–DNA complex assembly (von Hippel et al. 2013; von Hippel 2007; von Hippel and McGhee 1972; McGhee and von Hippel 1975a, b).

For the −1 "fork" $(iCy3)_2$ dimer-labeled ss–dsDNA construct, in which the average local conformation of the sugar-phosphate backbones at room temperature is left-handed and relatively disordered (in comparison to the +15 position), increasing the temperature above 23 °C does not further increase the local conformational disorder (see Fig. 7.11b). However, decreasing temperature below 23 °C does lead to a significant increase of conformational disorder at the −1 position. These findings indicate that the onset of local conformational disorder at elevated and reduced temperatures observed for the +15 labeled construct is associated with "breathing" fluctuations of the DNA duplex. For the −1 labeled construct, the $(iCy3)_2$ dimer probes "sense" the local conformations of the bases and backbones at and near the ss–dsDNA junction, where the local W-C B-form conformations are less stable and more susceptible to "breathing" fluctuations in comparison to positions deep within the duplex region. Thus, increasing temperature leads to dissociation of the DNA duplex near the ss–dsDNA junction, while decreasing temperature leads to trapping of disordered local conformations near the ss–dsDNA junction.

In Fig. 7.12 are shown the results of 2DFS studies of local conformational disorder in $(iCy3)_2$ dimer-labeled ss–dsDNA constructs at sites traversing the ss–dsDNA junction. Analyses of these data show that the local conformational disorder increases as the dimer-probe position is changed from the +2 to the +1 position. However, the local conformational disorder slightly decreases as the position is changed from the +1 to −1 position, and again from the −1 to −2 position. Thus, the position-dependent behavior of the conformational disorder parameter, σ_I, reflects the same trends observed for the mean local conformational parameters, ϕ_{AB}, θ_{AB} and R_{AB}, which were discussed in Sect. 7.3.1 and summarized in Fig. 7.6.

In Fig. 7.13a is shown a summary of results of position-dependent studies of $(iCy3)_2$ dimer-labeled ss–dsDNA constructs (Heussman et al. 2022). These results suggest a detailed picture of the mean local conformations and conformational disorder of the bases and sugar-phosphate backbones at probe-labeling sites at and near the ss–dsDNA junction. The average local conformation of the $(iCy3)_2$ dimer-probe changes systematically as the labeling site is varied from the +2 to the −2 positions. For positive integer positions, the mean local conformation is right-handed (defined as exhibiting a mean twist angle, $\phi_{AB} < 90°$), and for negative integer positions the local conformation is left-handed ($\phi_{AB} > 90°$). Local conformations deep within the

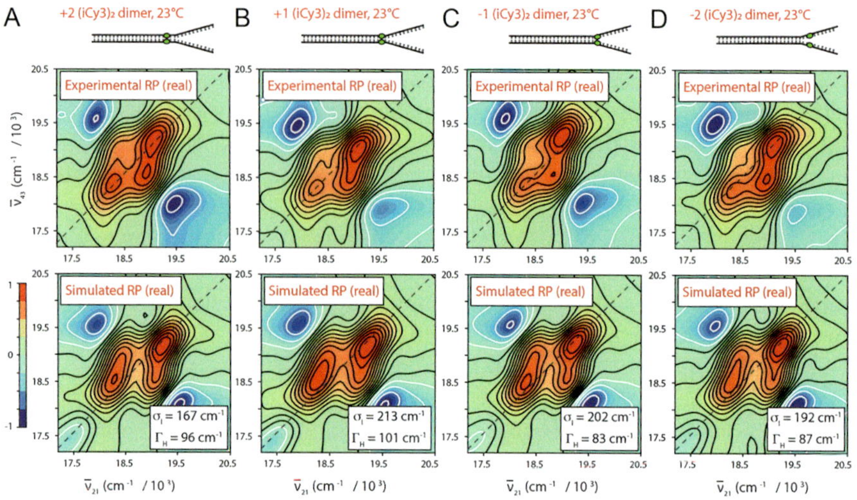

Fig. 7.12 Experimental RP spectra (top row) are compared to the optimized simulated RP spectra (bottom row) at room temperature (23 °C) for $(iCy3)_2$ dimer ss–dsDNA constructs as a function of probe-labeling position. (**a**) +2; (**b**) +1; (**c**) −1; and (**d**) −2. Samples were prepared using 100 mM NaCl and 6 mM $MgCl_2$, and 10 mM Tris at pH 8.0. Simulated spectra are based on the structural parameters obtained from our optimization analyses of the CD and absorbance spectra. The resulting homogeneous and inhomogeneous line shape parameters (Γ_H and σ_I, respectively) are given in the insets. The local conformational disorder increases as the probe-labeling position is varied across the ss–dsDNA junction in the direction from the duplex to the single-stranded DNA region. Figure adapted from Heussman et al. (2022), Marcus et al. (2023)

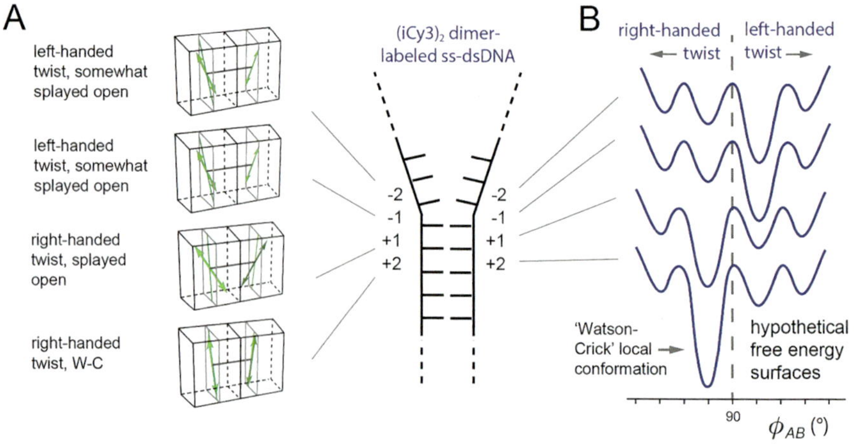

Fig. 7.13 (**a**) Schematic of the position-dependent changes in mean local conformations of the $(iCy3)_2$ dimer-labeled ss–dsDNA constructs. (**b**) Hypothetical free energy landscape specific to the probe-labeling position. Individual surfaces are vertically offset for clarity. Figure adapted from Heussman et al. (2022), Marcus et al. (2023)

duplex region are cylindrically symmetric and minimally disordered. The disorder increases significantly for positive positions approaching the ss–dsDNA fork junction. At the +1 position, there is an abrupt loss of cylindrical symmetry, which coincides with an additional increase in conformational disorder. The left-handed conformations at the −1 and −2 positions exhibit somewhat smaller mean tilt angles and decreasing conformational disorder, suggesting that the peak perturbation to secondary structure within the ss–dsDNA fork junction occurs at the +1 position.

7.3.4 *Studies of the Equilibrium Distributions of Conformational Macrostates and the Kinetics of State-to-State Interconversion by Polarization-Sweep Single-Molecule Fluorescence (PS-SMF) Microscopy*

A significant finding of the studies of $(iCy3)_2$ dimer-labeled ss–dsDNA constructs described in Sect. 7.3.3 is that sites at and near the ss–dsDNA fork junction exhibit only moderate levels of structural disorder, suggesting that the distributions of conformational states at positions moving across the ss–dsDNA junction toward the single-stranded regions are relatively narrow. A possible interpretation of these results is that the $(iCy3)_2$ dimer probe, which is covalently linked to the sugar-phosphate backbones and bases immediately adjacent to the probe, is limited to only a small number of local right- or left-handed conformations with relative stabilities that depend on the probe-labeling position relative to the ss–dsDNA junction. Figure 7.13b depicts hypothetical free energy surfaces to illustrate this point. Each surface contains four possible local conformations of the $(iCy3)_2$ dimer probe, which may exhibit either a right-handed or a left-handed twist. In this picture, the free energy differences between the "Watson-Crick" (B-form) ground state and the other (non-canonical) local conformations are sufficiently high to ensure that the Boltzmann-weighted distribution of available macrostates is dominated by the B-form structure. At the same time, the moderate free energies of activation allow for infrequent transitions between non-canonical local structures, which are present in trace amounts. As the probe-labeling position is varied across the ss–dsDNA junction toward the single-stranded region, one might expect the free energy surface to exhibit local minima with approximately the same coordinate values, but with relative stabilities and barrier heights shifted to reflect the presence of non-canonical structures observed in ensemble measurements (Heussman et al. 2019, 2022, 2024).

To directly study the local conformational fluctuations—and the associated free energy surfaces—of $(iCy3)_2$ dimer-labeled ss–dsDNA fork constructs, a novel single-molecule optical method was developed. Polarization-sweep single-molecule fluorescence (PS-SMF) microscopy was designed to directly monitor the local conformational fluctuations of the $(iCy3)_2$ dimer probe as a function of its position relative to the ss–dsDNA fork junction (Maurer et al. 2023, 2024). These measurements exploit the sensitivity of the conformation-dependent coherent coupling between

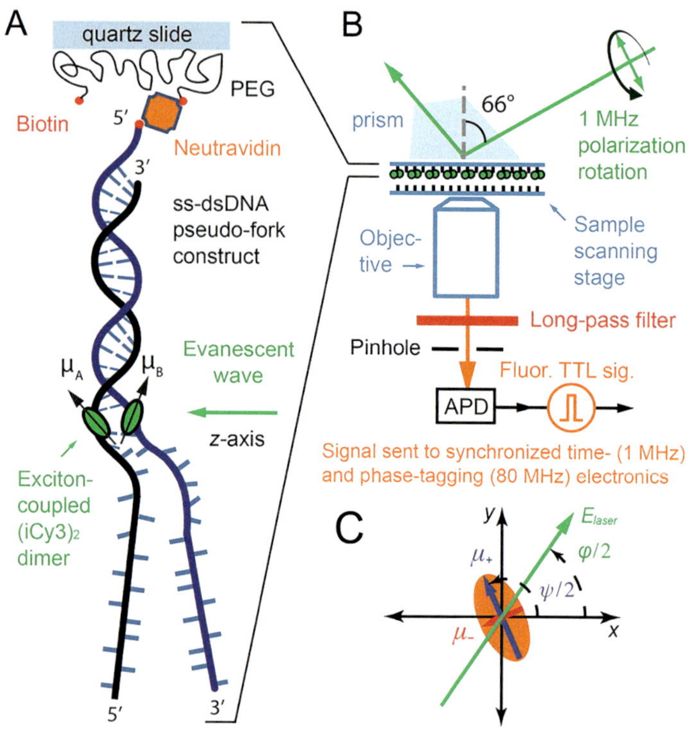

Fig. 7.14 Schematic of PS-SMF microscopy setup. (**a**) $(iCy3)_2$ dimer-labeled ss–dsDNA constructs are attached to a microscope slide using biotin/neutravidin linkers. (**b**) A total internal reflection fluorescence (TIRF) illumination geometry is used to preferentially excite molecules attached near the liquid–solid interface. The cw 532 nm laser generates an evanescent wave that propagates parallel to the sample surface with electric field vector perpendicular to the surface. The plane-polarized electric field vector is rotated continuously at a frequency of 1 MHz by sweeping the phase of an interferometer. (**c**) The symmetric and anti-symmetric EDTMs of the $(iCy3)_2$ dimer probe (labeled $\boldsymbol{\mu}_{\pm}$, and shown as blue and red vectors, respectively) define major and minor axes of a "polarization ellipse" in the transverse cross-sectional area of the incident laser beam. The angles $\psi/2$ and $\varphi/2$ define the orientations of the $(iCy3)_2$ dimer probe and the laser polarization vector, respectively. Reproduced from Maurer et al. (2023)

the closely spaced monomers of the $(iCy3)_2$ dimer probe. Certain structural aspects of DNA "breathing" in duplex DNA were previously revealed using hydrogen exchange and chemical trapping studies (McConnell and von Hippel 1970a, b; McGhee and von Hippel 1974, 1975b; Printz and von Hippel 1964, 1965). PS-SMF experiments provide more detailed, structure-specific information about the relative stabilities and transition barriers that mediate conformational fluctuations near the ss–dsDNA fork junctions that are characteristic of DNA replication complexes.

A schematic of the PS-SMF instrument is shown in Fig. 7.14. An $(iCy3)_2$ dimer-labeled ss–dsDNA fork construct is attached to the inner surface of a quartz sample chamber using a biotin/neutravidin linkage (Chandradoss et al. 2014). The sample is placed on the stage of an inverted TIRF (total internal reflection fluorescence) microscope. The optical source—a continuous wave (cw) laser with wavelength ~532 nm—is used to create an evanescent field, which propagates parallel to the slide surface with its electric field vector contained within a plane normal to the surface. The plane polarization state of the laser is prepared using a polarized Mach–Zehnder interferometer and quarter-wave plate, and the electric field vector is rotated continuously at the frequency 1 MHz by sweeping the phase of the interferometer using acousto-optic Bragg cells (Maurer et al. 2023). The symmetric (+) and anti-symmetric (−) excitons of the $(iCy3)_2$ dimer probe, which are oriented perpendicular to one another, are alternately excited as the laser polarization direction is rotated across the corresponding EDTMs. The modulated fluorescence from a single $(iCy3)_2$ dimer-labeled DNA construct is isolated through a pinhole and detected using an avalanche photodiode (APD). The signal is processed using the phase-tagged photon counting (PTPC) method (Tamimi et al. 2020).

The PS-SMF signal has modulation amplitude that depends on the phase of the polarization rotation cycle. The signal modulation amplitude is termed the *visibility*, v,which is defined as (Maurer et al. 2023):

$$v = \left[|\mu_+|^2 - |\mu_-|^2 \right] / \left[|\mu_+|^2 + |\mu_-|^2 \right] \tag{7.14}$$

In Eq. (7.14), $|\mu_\pm|^2 = |\boldsymbol{E} \cdot \mu_\pm|^2 = |A(\omega)|^2 \sum_\alpha \int d\omega \left| \mu_\pm^{(\alpha)}(\omega) \right|^2$, are the intensities associated with the symmetric (+) and anti-symmetric (−) excitons, $\boldsymbol{E}$ is the plane-polarized electric field vector of the laser, $A(\omega)$ is the (narrow) spectral envelope, $\boldsymbol{\mu}_\pm^{(\alpha)} = \langle 0 | \boldsymbol{\mu}^{\text{tot}} | e_\pm^{(\alpha)} \rangle$ are the symmetric and anti-symmetric EDTMs between electronic ground and excited states of the $(iCy3)_2$ dimer, and $\boldsymbol{\mu}^{\text{tot}} = \boldsymbol{\mu}_+ + \boldsymbol{\mu}_-$, as described in Sect. 7.3.1.

As illustrated in Fig. 7.5b, c, the magnitudes of the $\boldsymbol{\mu}_\pm$ EDTMs depend sensitively on the local conformation of the $(iCy3)_2$ dimer. The orthogonal EDTMs define a "polarization ellipse" in the cross-sectional area in which the laser beam projects onto the molecular frame, as shown in Fig. 7.14c. Thus, the visibility, v, is directly related to the instantaneous local conformation of the $(iCy3)_2$ dimer probe within the ss–dsDNA construct, which can be monitored on microsecond and longer time scales.

Studies of the local conformational fluctuations of $(iCy3)_2$ dimer-labeled ss–dsDNA fork constructs were recently carried out by monitoring time-dependent trajectories of the PS-SMF signal visibility, $v(t)$ (Maurer et al. 2023). The data were analyzed by constructing from a time series of PS-SMF measurements ensemble-averaged probability distribution functions (PDFs), $P(v)$, and the two-point and three-point time-correlation functions (TCFs), $\bar{C}^{(2)}(\tau)$ and $\bar{C}^{(3)}(\tau_1, \tau_2)$, respectively (Israels et al. 2021; Maurer et al. 2023; Phelps et al. 2016, 2017). The two-point TCF is the average product of two successive measurements separated by the interval τ:

$$\bar{C}^{(2)}(\tau) = \langle \delta v(\tau) \delta v(0) \rangle \tag{7.15}$$

The angle brackets in Eq. (7.14) indicate a running average over all possible initial measurement times, and $\delta v(t) = v(t) - \langle v \rangle$ is the signal fluctuation about the mean value, $\langle v \rangle$. The two-point TCF, $\bar{C}^{(2)}$, contains information about the number of quasi-stable macrostates of the system, the mean visibility of each macrostate, and the characteristic time scales of interconversion between macrostates (Phelps et al. 2016). The three-point TCF is the time-averaged product of three consecutive measurements separated by the intervals τ_1 and τ_2.

$$\bar{C}^{(3)}(\tau_1, \tau_2) = \langle \delta v(\tau_1 + \tau_2)\delta v(\tau_1)\delta v(0) \rangle \tag{7.16}$$

The function $\bar{C}^{(3)}$ contains detailed kinetic information about the transition pathways that connect the macrostates of the system. This function is sensitive to the roles of intermediates, whose presence can facilitate or hinder successive transitions between macrostates (Phelps et al. 2016). Unlike even-moment TCFs, the three-point TCF, $\bar{C}^{(3)}$, is devoid of noise-related background contributions (Dhar et al. 2023).

The above functions can be simulated using a kinetic network model, which assumes that the system interconverts between a finite number of Boltzmann-weighted macrostates, and that the rates of interconversion between macrostates are related to Arrhenius activation barriers. The minimum number of macrostates needed to fully describe PS-SMF data is equal to the number of decay components exhibited by the two-point TCF plus one. By comparing experimental results to the predictions of the kinetic network model, it is possible to extract the relative stabilities and activation barriers between distinct macrostates, which may be used to parameterize the free energy surfaces associated with the local environment of DNA bases and sugar-phosphate backbones "sensed" by the $(iCy3)_2$ dimer probes.

In Fig. 7.15 are shown examples of PS-SMF data for the +1 $(iCy3)_2$ dimer-labeled ss–dsDNA fork construct. The mean signal intensity for these measurements was ~7500 s^{-1}. From the PTPC data stream the signal visibility, $v(t)$, was determined using an integration period of ~100 ms (see Fig. 7.15a). The visibility trajectory undergoes discontinuous transitions between a small number of discrete values within the range $0 < v < 0.4$ (dashed lines are guides to the eye). The visibility is a direct measure of the conformational changes of the $(iCy3)_2$ dimer probe. Control measurements of iCy3 monomer-labeled DNA constructs do not exhibit discontinuous transitions, in contrast to $(iCy3)_2$ dimer-labeled ss–dsDNA constructs (Maurer et al. 2023). From ~40 individual PS-SMF measurements, each of ~30 s in duration, the ensemble-averaged functions were constructed. In Fig. 7.15b is shown the probability distribution function (PDF), which was constructed using an integration period ~100 ms. The PDF exhibits multiple "low visibility" and "high visibility" features. There is a prominent "low visibility" feature in the range $0 \lesssim v \lesssim 0.1$ in addition to sparsely populated "high visibility" features in the range $0.1 \lesssim v \lesssim 0.4$. This broad distribution of visibility features indicates the presence of both stable and thermally activated local conformational macrostates of the $(iCy3)_2$ dimer probes.

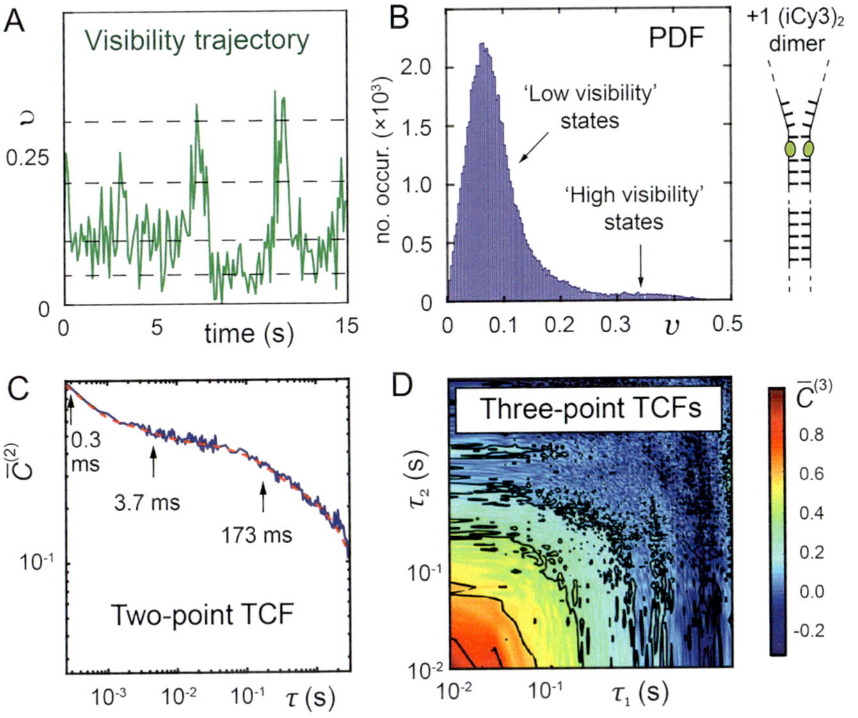

Fig. 7.15 Sample experimental data from PS-SMF measurements of +1 $(iCy3)_2$ dimer-labeled ss–dsDNA construct in 100 mM NaCl, 6 mM $MgCl_2$, and 10 mM Tris at pH 8.0. (**a**) The first 15 seconds of a PS-SMF visibility trajectory, $v(t)$, is shown using a bin period of 100 ms. The mean signal intensity was ~7500 s^{-1}. Horizontal dashed lines are guides to the eye to indicate the presence of multiple discrete conformational states. (**b**) Normalized probability distribution function (PDF) of visibility values, $P(v)$, using a bin period of 100 ms. (**c**) Two-point time-correlation function (TCF) is well modeled as a weighted sum of three exponentially decaying terms (red curve) with decay constants that are well separated in time. (**d**) Three-point TCF is shown as a two-dimensional contour plot. Adapted from Maurer et al. (2024)

The dynamics of the system are reflected by the behaviors of the two-point and three-point TCFs of the signal visibility. In Fig. 7.15c is shown the two-point TCF, which was constructed using an integration period of ~250 μs. The two-point TCF is plotted alongside a model fit to a tri-exponential function, with time constants indicated in the figure. There are three decay components at $t_1 = 0.3$ ms, $t_2 = 3.7$ ms, and $t_3 = 173$ ms. The presence of three well-separated decay components indicates that the underlying dynamics of the system must involve at least four quasi-stable macrostates in dynamic equilibrium. Figure 7.15d shows the corresponding three-point TCF, which was constructed using an integration period of ~10 ms. The three-point TCF contains detailed kinetic information about the transition pathways that connect the macrostates of the system. The three-point TCF is sensitive to the roles of pathway intermediates, whose presence can facilitate or hinder successive transitions between macrostates.

7.3.5 Studies of Local DNA "Breathing" of +1, −1, and −2 $(iCy3)_2$ Dimer-Labeled ss–dsDNA Fork Constructs

In Fig. 7.16 are shown PS-SMF results for the +1, −1, and −2 $(iCy3)_2$ dimer-labeled ss–dsDNA constructs, which were obtained at room temperature (23 °C) and "physiological" buffer salt conditions ([NaCl] = 100 mM, $[MgCl_2]$ = 6 mM). The left column shows the visibility PDFs, $P(v)$, the middle column shows the two-point TCFs, $\bar{C}^{(2)}(\tau)$, which are overlaid with multi-exponential fits, and the right column shows contour plots of the three-point TCFs, $\bar{C}^{(3)}(\tau_1, \tau_2)$. Like the PDF of the +1 $(iCy3)_2$ dimer-labeled ss–dsDNA construct (discussed above and shown in Fig. 7.16a), the PDFs of the −1 and −2 dimer-labeled constructs exhibit a major "low visibility" feature in the region $0 \lesssim v \lesssim 0.1$, and minor "high visibility" features in the region $0.1 \lesssim v \lesssim 0.4$. The relative weights of the "high visibility" features depend on $(iCy3)_2$ dimer-probe-labeling position relative to the ss–dsDNA fork junction, with "high visibility" features notably less prominent for the −1 construct than for the +1 and −2 constructs. Furthermore, the PDF of the −2 construct is significantly different from that of the +1 construct, as expected given the differential stabilities of stacked bases within dsDNA versus ssDNA regions spanning the ss–dsDNA fork junction.

The two-point and three-point TCFs shown in Fig. 7.16 characterize the conformational dynamics of the $(iCy3)_2$ dimer-labeled ss–dsDNA constructs as a function of probe-labeling position. The two-point TCFs are shown overlaid with multi-exponential fits (dashed red curves). In each case, the four decay components are well separated in time: t_1 ~0.3–0.6 ms, t_2 ~2–4 ms, t_3 ~50–200 ms, and t_4 ~1–2 s. The seconds-long decay, t_4, was determined to be due to mechanical room vibrations. Thus, there are three relevant, well-separated decay components present in the +1, −1, and −2 $(iCy3)_2$ dimer-labeled ss–dsDNA constructs under "physiological" buffer salt conditions, suggesting the presence of four quasi-stable macrostates in thermal equilibrium.

The +1 $(iCy3)_2$ dimer-labeled ss–dsDNA construct exhibits the slowest relaxation dynamics of the three constructs examined in Fig. 7.16. As the position of the dimer probe is varied across the ss–dsDNA junction from +1 to −1, and again from −1 to −2, the relaxation dynamics become faster. These observations are consistent with the notion that the conformational fluctuations of the $(iCy3)_2$ dimer probe depend on the stabilities and dynamics of the DNA bases and sugar-phosphate backbones immediately adjacent to the probe. The W-C base pairs within the duplex side of the ss–dsDNA junction are largely stacked, while the stacking interactions between bases on the ssDNA side of the junction are much weaker due to the lack of complementary base pairing. A noteworthy feature of the three-point TCFs, $\bar{C}^{(3)}$

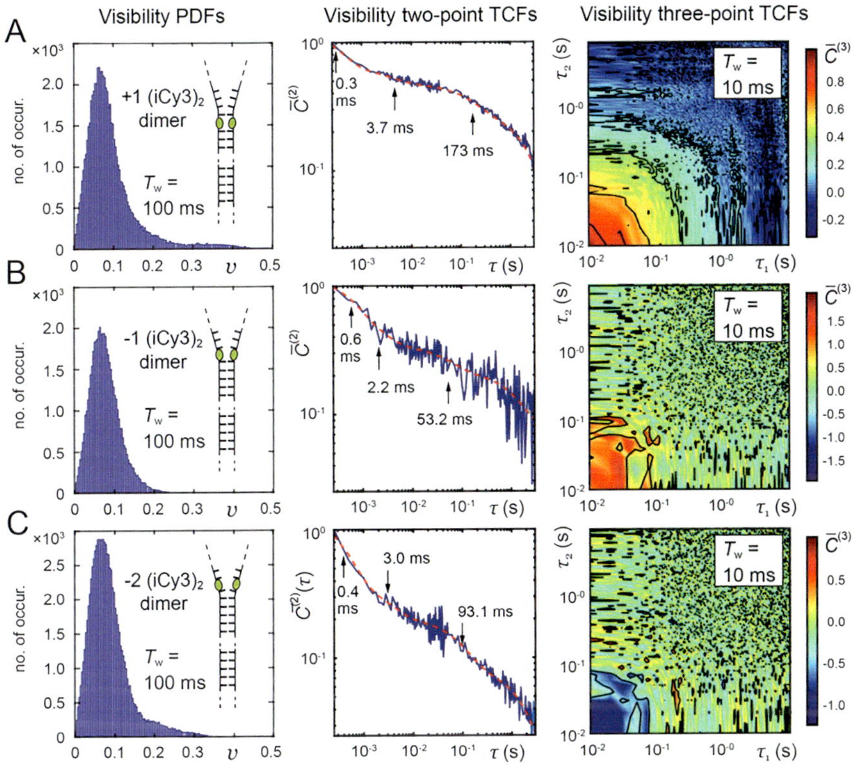

Fig. 7.16 Probability distribution functions (PDFs, left column), two-point time-correlation functions (TCFs, middle column), and three-point TCFs (left column) of the PS-SMF visibility for the (**a**) +1, (**b**) -1, and (**c**) -2 $(iCy3)_2$ dimer-labeled ss–dsDNA fork constructs. The integration periods, T_w, used for the PDFs and three-point TCFs are indicated in the insets. The two-point TCFs were stitched together over the range 250 μs–25 ms using T_w = 250 μs, and from 25 ms to 2.5 s using T_w = 10 ms. The two-point TCFs are shown overlaid with multi-exponential fits (dashed red curves). The three-point TCFs are plotted as two-dimensional contour diagrams. Experiments were performed at room temperature (23 °C) in 100 mM NaCl, 6 mM $MgCl_2$, and 10 mM Tris at pH 8.0. Adapted from Maurer et al. (2024)

, is that they exhibit positive amplitude for the +1 and −1 constructs, and negative amplitude for the −2 construct. These observations suggest that the prevailing kinetic pathways of state-to-state interconversion for the +1 and −1 constructs are significantly different from that of the −2 construct.

The results summarized in Fig. 7.16 indicate that the bases and sugar-phosphate backbones sensed by the $(iCy3)_2$ dimer probes can adopt four quasi-stable local conformations, whose relative stabilities and transition state barriers depend on position relative to the ss–dsDNA fork junction. Under "physiological" buffer salt conditions, the +1 $(iCy3)_2$ ss–dsDNA construct exhibits a relatively broad distribution of local base and backbone conformations within the duplex region of the ss–dsDNA junction. The conformational macrostates of the +1 $(iCy3)_2$ ss–dsDNA construct undergo the slowest dynamics of the three positions investigated, indicating that the most thermodynamically stable states within the duplex side of the fork junction are also mechanically stable with relatively long population lifetimes. In comparison, the −2 $(iCy3)_2$ ss–dsDNA construct exhibits significantly faster dynamics and a distinctly different distribution of conformational macrostates, indicating that the thermodynamically favored conformations of bases and sugar-phosphate backbones within the ssDNA region of the fork junction are distinct from the duplex and mechanically unstable.

Additional insights about the energetics of the local conformational macrostates at these key positions across the ss–dsDNA fork junction were obtained from PS-SMF studies performed at varying salt concentrations (Maurer et al. 2024). For the +1 $(iCy3)_2$ ss–dsDNA construct, increasing or decreasing salt concentrations relative to "physiological" concentrations shifted the equilibrium distribution to favor macrostates that are mechanically unstable with relatively low transition barriers. For the −2 $(iCy3)_2$ ss–dsDNA construct, varying salt concentration had the opposite effect, shifting the equilibrium distribution to favor macrostates that are mechanically stable with relatively high transition barriers. These results indicate that destabilization of the ss–dsDNA junction by varying salt concentrations leads to compensatory effects on the thermodynamic and mechanical stabilities of local conformational macrostates within the duplex and single-stranded DNA regions.

In Fig. 7.17 is shown a hypothetical mechanism to account for the PS-SMF results discussed above. For the +1 $(iCy3)_2$ ss–dsDNA construct, the local base and backbone conformations adjacent to the dimer probes are within the DNA duplex region of the fork junction. At "physiological" salt concentrations, the majority of conformational macrostates sensed by the probe are dominated by W-C base stacking, and these conformations are mechanically stable. Increasing or decreasing salt concentration leads to disruption of the local W-C conformations within the duplex region and a reduction of mechanical stability. For the −2 $(iCy3)_2$ ss–dsDNA construct, the local base and backbone conformations adjacent to the dimer probes are within the ssDNA region. At "physiological" salt concentrations, the majority of conformational macrostates sensed by the dimer in the ssDNA region are dominated by unstacked base conformations, which are distinct from the duplex region and mechanically unstable. Additional evidence for the presence of unstacked and dynamically labile base conformations within the ssDNA region of oligo$(dT)_{15}$ tails of ss–dsDNA fork constructs was determined by microsecond-resolved single-molecule FRET experiments (Israels et al. 2021) and corroborated by small-angle X-ray scattering experiments (Plumridge et al. 2017a, b). At elevated or reduced salt concentrations, the distribution of conformational macrostates sensed by the dimer probes within the ssDNA region of the fork junction is shifted to mechanically stable base-stacked conformations. This hypothetical mechanism is consistent with the findings of previous ensemble studies of $(iCy3)_2$ dimer-labeled ss–dsDNA fork constructs (Heussman et al. 2022) and is further interpreted using kinetic network modeling approaches in a subsequent paper.

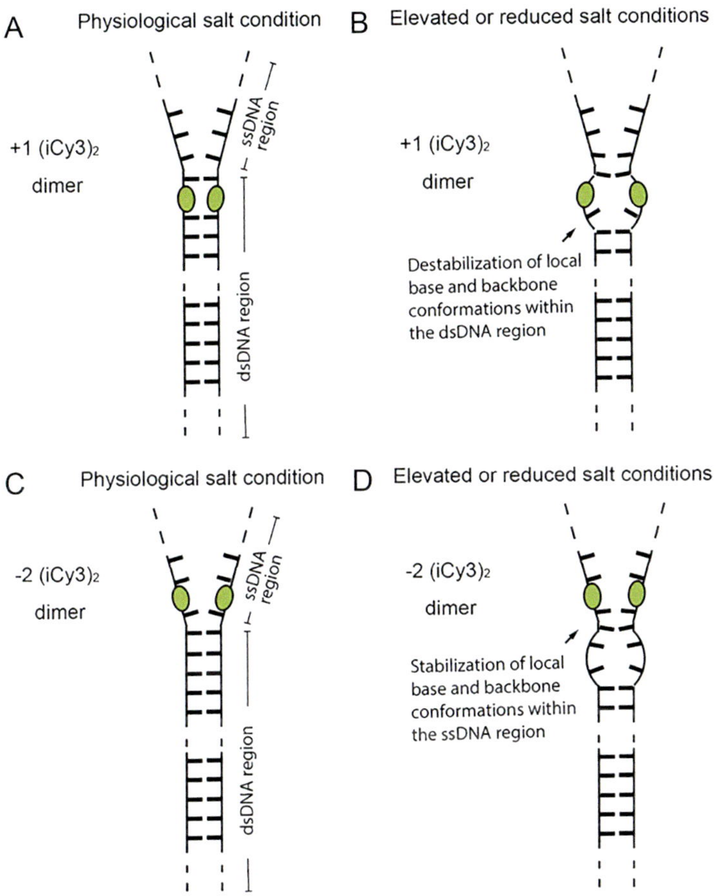

Fig. 7.17 Schematic diagram illustrating the hypothesized mechanism of salt-induced instability of $(iCy3)_2$ dimer-labeled ss–dsDNA fork constructs labeled at (panel **a** and **b**) the +1 position and (panel **c** and **d**) the −2 position. Left column diagrams (panels **a** and **c**) represent the situation at "physiological" salt conditions, and diagrams in the right column (panels **b** and **d**) represent the hypothesized mechanisms at elevated or reduced salt conditions. Adapted from Maurer et al. (2023)

7.3.6 Inference of Conformational Macrostates of $(iCy3)_2$ Dimer-Labeled ss–dsDNA Constructs

The ensemble and single-molecule studies discussed in Sects. 7.3.1–7.3.5 indicate that the local conformations of DNA bases and backbones immediately adjacent to the $(iCy3)_2$ dimer probes fluctuate in thermal equilibrium between four quasi-stable macrostates. The results of these studies suggest that the FESs that govern the equilibrium distributions and interconversion dynamics are sensitive to position relative to the ss–dsDNA junction, as illustrated in Fig. 7.13. At positive integer positions within the duplex side of the ss–dsDNA junction the distribution of macrostates is

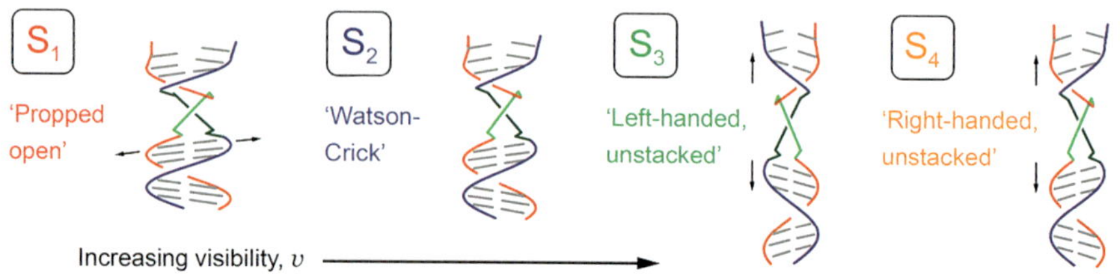

Fig. 7.18 Conformational macrostates of $(iCy3)_2$ dimer-labeled ss–dsDNA constructs arranged in order of increasing visibility. The $(iCy3)_2$ dimer is depicted as light green and dark green line segments attached to the sugar-phosphate backbones (shown in red and blue). The macrostates are color coded: S_1 (red), S_2 (blue), S_3 (green), and S_4 (orange). Adapted from Maurer et al. (2024)

dominated by a right-handed conformation, while at negative integer positions within the single-stranded side of the junction the distribution is dominated by a left-handed conformation. Based on the above observations and considering prior studies of DNA "breathing" using hydrogen exchange methods (McConnell and von Hippel 1970a, b; Printz and von Hippel 1964, 1965), it is possible to hypothesize structural assignments to the four conformational macrostates of the system.

From the work of Printz et al. and McConnell et al., it was concluded that a local "cooperatively exchanging" region of duplex DNA (consisting of a ~5–10 base-pair segment) can undergo thermally activated transitions to populate unstable, "open" conformations (von Hippel et al. 2013). In Fig. 7.18 are illustrated four conformational macrostates (labeled S_1–S_4) in order of increasing visibility, which are consistent with the experimental observations described in previous sections. The lowest visibility state, S_1, represents the so-called propped open macrostate, in which a hydrogen bond within an affected base pair is disrupted (i.e., held open) by the gain or loss of an inter-base hydrogen atom, resulting in two H-bond donors or two H-bond acceptors facing one another within a W-C base pair. The propped open state is right-handed and has stacked flanking bases. Thus, in the S_1 macrostate, the complementary hydrogen bonds of opposing bases are separated by relatively large distances (indicated by horizontal arrows), such that the twist angle ϕ_{AB} approaches 90° and the visibility approaches zero. S_2 represents the canonical W-C conformation, which is right-handed and has stacked flanking bases. The S_2 macrostate has right-handed twist angle $\phi_{AB} \lesssim 90°$ (Heussman et al. 2024). S_3 represents a left-handed conformation in which flanking bases are unstacked by relatively large distances (indicated by vertical arrows). The S_3 macrostate has a left-handed twist angle $\phi_{AB} > 90°$. The S_4 macrostate, like S_3, has its flanking bases unstacked, but here with a right-handed twist angle $\phi_{AB} < 90°$.

7.3.7 DNA Breathing of +1, −1, and −2 $(iCy3)_2$ Dimer-Labeled ss–dsDNA Fork Constructs and Determination of Free Energy Surface Parameters

The combined statistical-dynamical functions shown in Fig. 7.16 [i.e., the visibility PDF, $P(v)$, two-point TCF, $\bar{C}^{(2)}(\tau)$, and three-point TCF, $\bar{C}^{(3)}(\tau_1,\tau_2)$] serve to constrain a multiparameter optimization of solutions to a four-state kinetic network model, which assumes that the four macrostates at equilibrium interconvert stochastically with forward and backward rate constants. Details of this approach, which have been implemented in previous studies (Israels et al. 2021; Maurer et al. 2024; Phelps et al. 2016, 2017), are briefly summarized below.

The time-dependent population of the ith macrostate, $p_i(t)$, is modeled using the master equation (Reichl 1998).

$$\dot{p}_i = \sum_{j \neq i=1}^{4} -k_{ij} p_i + k_{ji} p_j \tag{7.17}$$

where the $k_{ij(ji)}$ are the forward (backward) rate constants between macrostates i and j $[i, j \in \{1, 2, 3, 4\}]$. A solution to Eq. (7.17) is obtained for an assumed set of input rate constants, which are subject to completeness and detailed balance conditions. The output parameters from the solution include: (i) the equilibrium probabilities, p_i^{eq}; and (ii) the time-dependent conditional probabilities, $p_{ij}(\tau)$, that the system undergoes transitions from macrostate i to macrostate j during the time interval τ.

Using the output parameters from a solution to Eq. (7.17), the PDF of the signal visibility is modeled as a sum of four macrostate contributions

$$P(v) = \sum_{i=1}^{4} p_i(v) = \sum_{i=1}^{4} A_i \exp\left[-(v - \bar{v}_i)^2 / 2\sigma_i^2\right] \tag{7.18}$$

where the probability of the ith macrostate, $p_i(v)$, is assumed to have a Gaussian form with mean visibility $\bar{v}_i$, standard deviation σ_i, and amplitude A_i. The probability that the ith macrostate is present at equilibrium is the integrated area $p_i^{\text{eq}} = \int_{-\infty}^{\infty} dv\, p_i(v) = A_i \sigma_i \sqrt{2\pi}$. The two-point and three-point TCFs, respectively, are simulated according to

$$\bar{C}^{(2)}(\tau) = \sum_{i,j=1}^{4} \delta\bar{v}_j p_{ij}(\tau) \delta\bar{v}_i p_i^{\text{eq}} \tag{7.19}$$

and

$$\bar{C}^{(3)}(\tau_1,\tau_2) = \sum_{i,j,k=1}^{4} \delta\bar{v}_k p_{jk}(\tau_2) \delta\bar{v}_j p_{ij}(\tau_1) \delta\bar{v}_i p_i^{\text{eq}} \tag{7.20}$$

where $\delta\overline{v}_i = \overline{v}_i - v$ is the visibility fluctuation of the ith macrostate. The two-point TCF is a time-averaged product of two consecutive observations separated by the interval τ. Similarly, the three-point TCF is the time-averaged product of three consecutive measurements separated by the intervals τ_1 and τ_2, respectively.

For each of the +1, −1, and −2 $(iCy3)_2$ dimer-labeled ss–dsDNA constructs, the statistical functions $P(\upsilon)$ $P(v)]$, $\overline{C}^{(2)}(\tau)$ and $\overline{C}^{(3)}(\tau_1, \tau_2)$ were calculated using Eqs. (7.18)–(7.20), respectively, for a given set of input rate constants, k_{ij}, mean visibility values, $\overline{v}_i$, amplitudes, A_i, and standard deviations, σ_i. These calculated functions were subsequently compared to the experimentally derived functions. The agreement between simulated and experimental functions was quantified using a nonlinear least squares target function, χ^2, which was minimized by performing an iterative search of the parameter space using a multiparameter optimization algorithm.

In Fig. 7.19 is shown a summary of the results of the four-state kinetic network analysis applied to the +1, −1, and −2 $(iCy3)_2$ dimer-labeled ss–dsDNA constructs. In all cases, the agreement between simulated and experimental functions is very good. The top row of Fig. 7.19 shows the visibility PDFs overlaid with the four Gaussian functions that represent the conformational macrostates, S_1–S_4. The normalized areas of the Gaussians are equal to the equilibrium populations of the macrostates. In the second and third rows of Fig. 7.19, respectively, are shown the two- and three-point TCFs, which characterize the conformational dynamics of the $(iCy3)_2$ dimer-labeled ss–dsDNA constructs as a function of probe-labeling position. The two- and three-point TCFs are shown overlaid with their corresponding optimized functions from the network model analysis. It is important to emphasize that determination of the full set of optimization parameters requires the simultaneous fitting of all three statistical-dynamical functions.

As discussed in Sect. 7.3.4, the +1 construct exhibits the slowest overall decay. As the position of the $(iCy3)_2$ dimer probe is varied across the ss–dsDNA junction from +1 to −1, and again from −1 to −2, the relaxation dynamics become faster. This behavior is consistent with the view that the activation barriers that mediate the conformational transitions depend on the relative stabilities of the DNA bases and sugar-phosphate backbones immediately adjacent to the probe. The W-C base pairs within the duplex side of the ss–dsDNA junction are largely stacked, while the stacking interactions between the bases on the ssDNA side of the junction are much weaker due to the lack of complementary base pairing.

In Fig. 7.20 are shown the optimized kinetic and equilibrium parameters determined for the +1, −1, and −2 $(iCy3)_2$ dimer-labeled ss–dsDNA constructs and the corresponding free energy surfaces. In the left column of Fig. 7.20 are shown the kinetic network schemes for each of the three constructs. Macrostates S_1–S_4 are shown interconnected by the forward and backward time constants, t_{ij} (= k_{ij}^{-1}), along with their equilibrium populations, p_i^{eq}. From these parameters are obtained the Boltzmann-weighted relative stabilities, $G_{ij} = -k_B T \ln\left(p_i^{\text{eq}} / p_j^{\text{eq}}\right)$, and the relative Arrhenius activation energies, $G_{ij}^{\ddagger} = -k_B T \ln\left(t_{\min} / t_{ij}\right)$, where k_B is Boltzmann's constant.

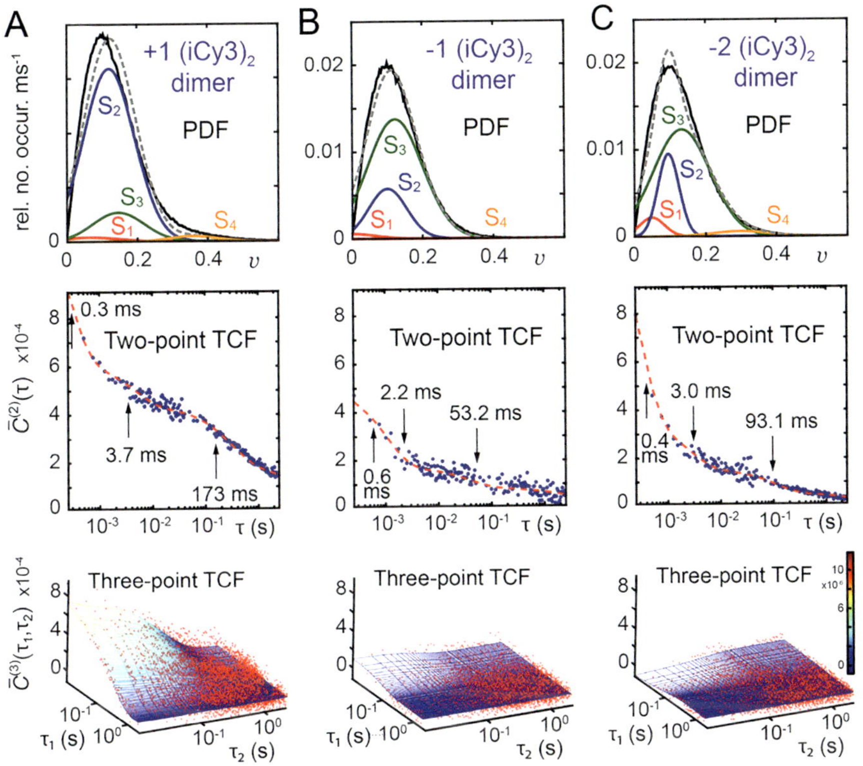

Fig. 7.19 Theoretical modeling results of PS-SMF measurements of +1 (**a**), −1 (**b**), and −2 (**c**) $(iCy3)_2$ dimer-labeled ss–dsDNA fork constructs in 100 mM NaCl, 6 mM $MgCl_2$, and 10 mM Tris at pH 8.0. The optimized parameters from a four-state kinetic network model are obtained by simultaneously fitting to the experimental PDF [$P(v)$, black curves, top row], the two-point TCF [$\bar{C}^{(2)}(\tau)$, blue dots, middle row], and the three-point TCF [$\bar{C}^{(3)}(\tau_1,\tau_2)$, red dots, bottom row]. The model PDFs are represented as sums of four Gaussian macrostates (red, blue, green, and orange curves, labeled S_1–S_4, respectively) with areas equal to the equilibrium probabilities, p_i^{eq}, with $i \in \{1, 2, 3, 4\}$. The model two-point TCFs are shown as red curves, and the model three-point TCFs are shown as solid surfaces. Adapted from Maurer et al. (2024)

The free energy surfaces shown in the right column of Fig. 7.20 are useful to visualize the relative stabilities and free energy barriers that separate the four macrostates, which change as the $(iCy3)_2$ dimer position is varied across the ss–dsDNA junctions. In plotting these free energy surfaces, a negative value has been assigned to the visibility of the S_3 macrostate corresponding to its left-handed twist, while the remaining macrostates are assigned positive values of the visibility corresponding to right-handed twists. The results indicate that for the +1 construct the S_2 macrostate, which is assigned to the W-C conformation, is thermodynamically more stable than the other macrostates, as expected. Furthermore, the kinetic barriers that mediate transitions between S_2 and the other macrostates are relatively high, indicating that the W-C conformation at the +1 position is the most mechanically stable. In

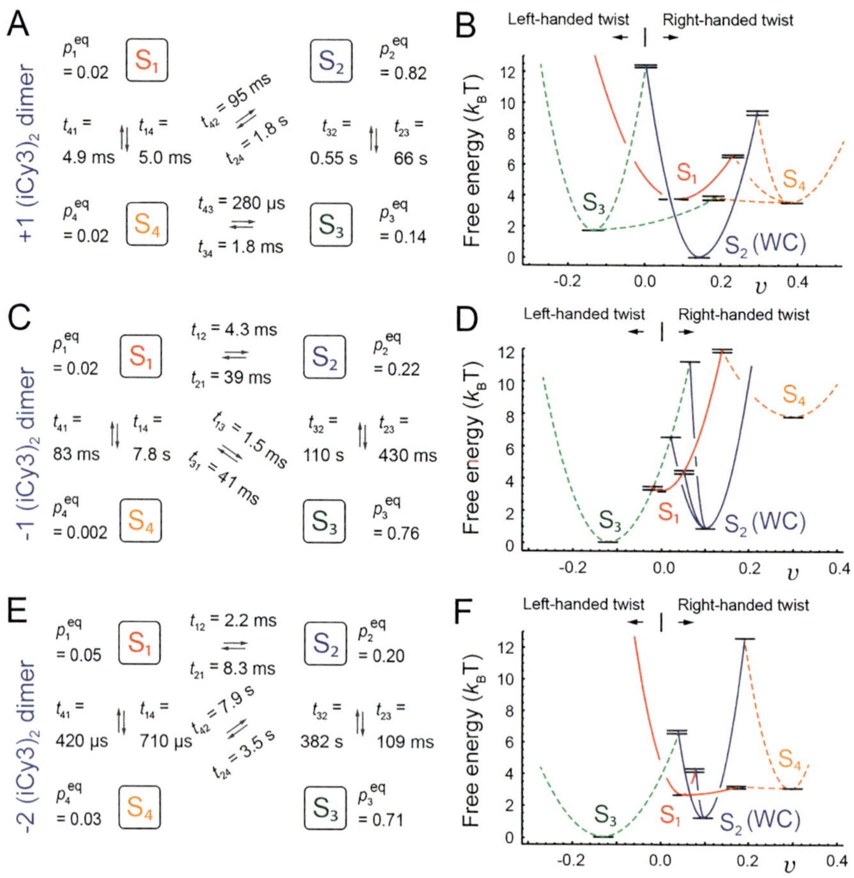

Fig. 7.20 Results of kinetic network model analysis of (**a**, **b**) +1, (**c**, **d**) -1, and (**e**, **f**) -2 (iCy3)$_2$ dimer-labeled ss–dsDNA fork constructs. Panels **a**, **c**, **e** show the kinetic network schemes and equilibrium population, p_i^{eq}, and time constant, t_{ij}, parameters, which are determined from the analysis of PS-SMF microscopy data. Panels **b**, **d**, **f** show the corresponding free energy surfaces derived from the optimized equilibrium and kinetic parameters obtained from the analysis. Adapted from Maurer et al. (2024)

contrast, for the −1 and −2 constructs, the left-handed S_3 macrostate is thermodynamically favored and the kinetic barriers that mediate transitions between S_3 and the other macrostates at these positions are relatively low. Thus, the left-handed S_3 macrostate is thermodynamically favored but mechanically unstable at the −1 and −2 positions.

The above studies show that the bases and sugar-phosphate backbones sensed by the (iCy3)$_2$ dimer probes can adopt four quasi-stable local conformations, whose relative stabilities and transition state barriers depend sensitively on the probe-labeling position relative to the ss–dsDNA fork junction. The +1 (iCy3)$_2$ dimer-labeled ss–dsDNA construct exhibits a relatively broad distribution of local base and backbone conformations within the duplex region of the ss–dsDNA junction. The conformational macrostates at this position are dominated by the S_2 (W-C) macrostate, which undergoes the slowest dynamics of the three constructs that we studied. This indicates that the S_2 macrostate, which is the most thermodynamically stable state within the duplex side of the fork junction, is also the most mechanically stable. In comparison, the −1 and −2 $(iCy3)_2$ dimer-labeled ss–dsDNA constructs exhibit progressively faster dynamics and a distinctly different distribution of conformational macrostates. At these latter positions, the left-handed S_3 macrostate is most stable, indicating that the thermodynamically favored conformations of bases and backbones sensed by the dimer probe within the ssDNA region of the fork junction are mechanically unstable and distinct from those of the duplex.

While the focus of the current chapter is on the behavior of DNA substrates in the absence of proteins, the structural and kinetic properties of the ss–dsDNA fork junction revealed by such experiments can, in principle, provide mechanistic insights into how replication and repair proteins that assemble at these sites carry out their biological functions. The significance of these studies is that they provide *position-specific* information about the relative stabilities and transition barriers that mediate conformational fluctuations at and near ss–dsDNA fork junctions where such protein–DNA complexes assemble and function.

7.4 Spectroscopic Studies of Local Base Stacking Conformation using Exciton-Coupled Fluorescent Nucleic Acid Base Dimer Probes

In previous sections, it was shown that spectroscopic studies performed on exciton-coupled $(iCy3)_2$ dimer-labeled ss–dsDNA constructs can provide detailed, albeit indirect, information about the local conformations and conformational "breathing" of DNA bases and sugar-phosphate backbones immediately adjacent to the probes. An alternative approach is to use fluorescent base analogs, which can be site-specifically substituted for native bases within a DNA construct. In principle, spectroscopic studies of fluorescent mononucleotide and dinucleotide analog-substituted DNA constructs can provide direct information about the local conformations of the base analog dimer probes themselves, and about the local conformations of DNA bases immediately adjacent to the base analog probe(s) (Camel et al. 2021; Ji et al. 2019; Jose et al. 2015; Widom et al. 2013b).

Base stacking is fundamentally important to the stability of double-stranded DNA. DNA exists in its most stable form as the iconic W-C double-helix in which flanking bases are fully stacked and the sugar-phosphate backbones adopt a right-handed twist (Watson and Crick 1953). While the W-C structure is the most stable conformation, on local length scales DNA may undergo "breathing" fluctuations to populate "thermally excited" conformations in which flanking bases are "twisted" relative to their neighbors and thus somewhat more or less stacked than in the

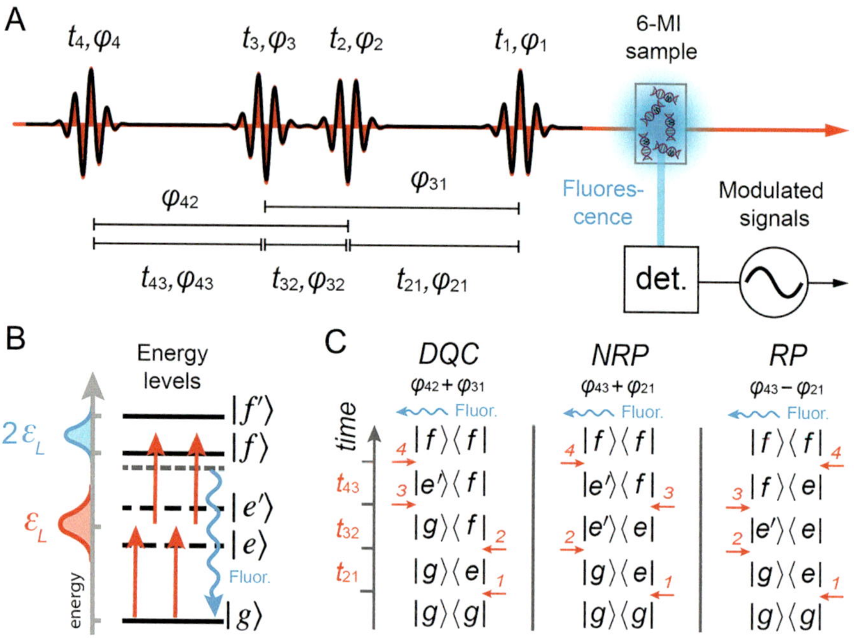

Fig. 7.21 (**a**) The 2PE-2DFS method employs a repeating sequence of four collinear ultrafast laser pulses, which excites the 6-MI sample. Fluorescence from the sample is detected using phase-sensitive methods. The relevant times and phases are indicated. (**b**) Energy level diagram indicating the laser pulse energy (centered at ε_L) and molecular state energies (centered at $2\varepsilon_L$). Fluorescence is spectrally separated from scattered laser excitation light. (**c**) Double-sided Feynman diagrams for 2PE-2DFS: DQC (double-quantum coherence, $\varphi_{42} + \varphi_{31}$), NRP (non-rephasing, $\varphi_{43} + \varphi_{21}$), and RP (rephasing, $\varphi_{43} - \varphi_{21}$). Adapted from Albrecht et al. (2024)

canonical W-C duplex. In such perturbed conformations, base pairs are separated, and thus inter-base hydrogen bonds are broken, and/or flanking bases are fully unstacked and un-hydrogen-bonded and thus locally exposed to the solvent environment. Such metastable, non-canonical conformations may aid in the recognition, assembly, and function of protein–DNA complexes involved in the core biological processes of DNA replication, transcription, and repair (von Hippel et al. 2013; von Hippel and McGhee 1972).

Nucleic acids labeled with fluorescent base analogs such as 6-methylisoxanthopterin (6-MI, an analog of guanine) and 2-aminopurine (2-AP, an analog of adenine) exhibit absorption and CD signals at lower energies than those of the canonical bases (see, e.g., Fig. 7.2f) and can thus provide information about the local conformations of the probe residues at specific positions within a base sequence (Datta et al. 2012; Camel et al. 2021; Ji et al. 2019; Widom et al. 2013b). For example, the CD spectrum of a fluorescent dinucleotide-substituted DNA construct, such as the $(6\text{-MI})_2$ dinucleotide-substituted DNA construct illustrated in Fig. 7.2e, can exhibit the spectral signatures of strong exciton coupling between flanking 6-MI residues (Datta et al. 2012). Furthermore, the fluorescence intensity of mononucleotide-substituted DNA constructs can provide additional information about local base stacking interactions and exposure to the aqueous solvent environment (Datta et al. 2012; Camel et al. 2021).

7.4.1 *Two-Photon Excitation Two-Dimensional Fluorescence Spectroscopy (2PE-2DFS) of the Fluorescent Guanine Analog 6-MI*

As discussed in Sect. 7.3.2, two-dimensional fluorescence spectroscopy (2DFS) is a fluorescence-detected Fourier transform (FT) optical method that is well suited to measure the electronic-vibrational structure of fluorescent molecular chromophores, such as 6-MI (Heussman et al. 2022; Kringle et al. 2018; Widom et al. 2013b). Like transmission-based multidimensional FT optical spectroscopies in the visible and IR regimes (Jonas 2003; Reppert and Tokmakoff 2016; Ghosh et al. 2017), 2DFS can reveal information about the optical transition pathways that are operative within a multi-level quantum molecular system. By virtue of its ability to sensitively monitor fluorescence against a dark background, 2DFS can detect relatively weak excited state population signals. Recent advances permit such phase-sensitive fluorescence-detected FT experiments to be carried out under weak signal conditions using single photon counting techniques (Tamimi et al. 2020).

Recently, a new experimental approach based on the principles of two-photon excitation (2PE) was introduced to perform 2DFS on the UV-absorbing fluorescent base analog 6-MI (Albrecht et al. 2024). Like the 2DFS experiments carried out on $(iCy3)_2$ dimer-labeled DNA constructs discussed in Sect. 7.3.2, similar experiments carried out on 6-MI substituted DNA constructs hold promise to provide direct information about local base conformation and conformational distributions. In prior work by Widom et al., 2DFS was used to measure the exciton coupling of $(2\text{-AP})_2$ dinucleotide in solution (Widom et al. 2013b). Unfortunately, such experiments on UV-absorbing chromophores are limited due to the difficulties of maintaining well-behaved broadband laser pulses in the UV spectral regime and detecting weak fluorescent signals above the background of scattered laser light. These difficulties are circumvented by using a broadband pulsed laser source in the visible regime with wavelength centered at ~675 nm, which is well separated from the wavelength of the fluorescent signal.

Two-photon excitation (2PE) occurs when the photon energy of the source laser is one half of that of the lowest energy transition between ground state, $|g\rangle$, and excited (final) state, $|f\rangle$ (see Fig. 7.21). While one-photon excitation (1PE) of the $|g\rangle \longrightarrow |f\rangle$ transition is not dipole-allowed, 2PE can be achieved through a perturbative sequence of density matrix elements that include coherences between $|g\rangle$, intermediate "virtual" states, $|e\rangle$ and $|e'\rangle$, and final states $|f\rangle$ and $|f'\rangle$ (Raymer et al. 2021). Here, the virtual states are assumed to have energies equal to one half of those of the excited states. Although the optical transitions involving virtual states do not absorb energy from the field, and thus do not lead to fluorescence, the virtual states do participate in coherences, e.g., $|g\rangle\langle e|$, $|e\rangle\langle f|$ or $|e\rangle\langle e'|$. Such 2PE processes have been used to study the electronically excited states of fluorescent DNA base analogs (Mikhaylov et al. 2018).

2PE-2DFS employs similar principles to those of 1PE-2DFS (Heussman et al. 2022; Tekavec et al. 2007; Perdomo et al. 2012; Lott et al. 2011; Karki et al. 2014; Widom et al. 2013b) in which the sample is illuminated by a repeating sequence of four collinear ultrafast laser pulses, as illustrated in Fig. 7.21a. In these experiments, the pulse repetition rate is 250 kHz, the center wavelength is ~675 nm, and the pulse full-width-at-half-maximum (FWHM) bandwidth is ~30 nm. The photon energy is set to approximately one-half of the lowest energy transition of the 6-MI molecule (with peak absorption wavelength ~340 nm). The pulses are characterized by their center times, t_i, phases, φ_i, inter-pulse delays, $t_{ij} = t_i - t_j$, and relative phases, φ_{ij} [$i, j \in \{1, 2, 3, 4\}$]. For a given measurement, the time delays t_{21}, t_{32} and t_{43} are held fixed while the relative phases are swept continuously at tens-of-kilohertz frequencies.

The four-pulse sequence induces one-photon transitions between ground state $|g\rangle$, intermediate ("virtual") states, $|e\rangle$ and $|e'\rangle$, and final excited states $|f\rangle$ and $|f\rangle$, as illustrated in Fig. 7.21b (Raymer et al. 2021). The ensuing 2PE fluorescence signals are proportional to the populations of the excited states that vary at distinct kilohertz modulation frequencies, which are isolated using phase-sensitive detection methods (Tamimi et al. 2020). In Fig. 7.21c are shown double-sided Feynman diagrams, which indicate the three possible pathways of perturbative fourth-order density matrix elements that generate excited state populations. Each of the three Feynman pathways depends on a specific sequence of field–matter interactions that give rise to a unique phase-modulation signature. The "double-quantum coherence (DQC)" pathway includes the intermediate coherence term $|g\rangle\langle f|$ between ground and excited states during the period t_{32}, and has phase signature $\varphi_{42} + \varphi_{31}$. In contrast, the "non-rephasing" (NRP) and "rephasing" (RP) pathways, which have phase signatures $\varphi_{43} + \varphi_{21}$ and $\varphi_{43} - \varphi_{21}$, respectively, include the intermediate coherence term $|e'\rangle\langle e|$ between distinct virtual states during the period t_{32}. The recorded signals are complex-valued response functions of the inter-pulse delays: $S_{DQC}(t_{21}, t_{32}, t_{43})$, $S_{NRP}(t_{21}, t_{32}, t_{43})$ and $S_{RP}(t_{21}, t_{32}, t_{43})$. Fourier transformation of the response functions with respect to the delay variables yields the frequency- and phase-dependent 2D spectra: $\hat{S}_{DQC}(\omega_{21}, \omega_{32}, \omega_{43})$, $\hat{S}_{NRP}(\omega_{21}, \omega_{32}, \omega_{43})$ and $\hat{S}_{RP}(\omega_{21}, \omega_{32}, \omega_{43})$ (Tekavec et al. 2007; Perdomo et al. 2012; Raymer et al. 2021).

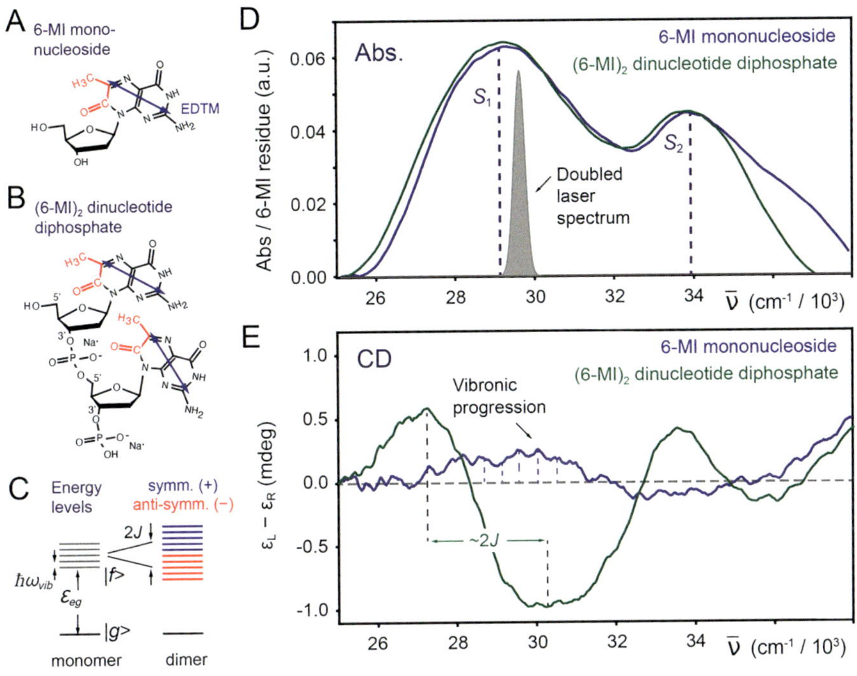

Fig. 7.22 (**a**) Molecular structure of the 6-MI nucleoside. Atoms that are additional to those of the natural base guanine are indicated in red. The blue double-headed arrow indicates the direction of the lowest energy electric dipole transition moment (EDTM). (**b**) Structure of the $(6\text{-MI})_2$ dinucleotide diphosphate. (**c**) Energy level diagram of the 6-MI nucleoside and the $(6\text{-MI})_2$ dinucleotide in the presence of electrostatic coupling, J, between EDTMs. The monomer electronic transition with energy, ε_{eg}, is coupled to a vibrational mode with energy spacing, $\hbar\omega_{vib}$. The coupling between monomers introduces symmetric (+, blue) and anti-symmetric (−, red) electronic-vibrational (vibronic) coherent states. (**d**) Absorption spectra of the 6-MI nucleoside (blue) and the $(6\text{-MI})_2$ dinucleotide (green) in aqueous buffer salt solution (10 mM TRIS, 100 mM NaCl, and 6 mM $MgCl_2$). Vertical dashed lines indicate the lowest energy electronic transition (S_1) at ~29,300 cm^{-1} and the second lowest transition (S_2) at 34,000 cm^{-1} of the 6-MI mononucleoside. The doubled laser spectrum used in the 2PE-2DFS experiments is shown in gray. (***E***) Circular dichroism (CD) spectra of the 6-MI nucleoside (blue) and $(6\text{-MI})_2$ dinucleotide diphosphate (green). Adapted from Albrecht et al. (2024)

7.4.2 Absorption and CD Experiments on 6-MI Mononucleoside and (6-MI)₂ Dinucleotide

In Fig. 7.22a, b are shown the molecular structures of 6-MI mononucleoside and $(6\text{-MI})_2$ dinucleotide diphosphate, respectively. The EDTM of the lowest energy $(\pi \longrightarrow \pi^*)$ S_1 transition is indicated as a double-headed blue arrow, which lies within the plane of the 6-MI nucleobase (Widom et al. 2012). The S_1 transition is coupled to at least one effective vibrational mode with energy, $\hbar\omega_{vib}$ = ~400 cm^{-1}. If the two base residues of the $(6\text{-MI})_2$ dimer are closely spaced—i.e., if they are "stacked"—the monomer EDTMs can couple electrostatically with Davydov splitting $2J$ to produce delocalized symmetric (+) and anti-symmetric (−) vibronic excitons, as illustrated in Fig. 7.22c. The precise geometric arrangement of the coupled EDTMs determines the dipole strengths and transition energies of the $(6\text{-MI})_2$ dimer.

In Fig. 7.22d, e are shown the experimental absorption and CD spectra, respectively, of the 6-MI mononucleoside (blue) and the $(6\text{-MI})_2$ dinucleotide (green). The peak absorption energy of the 6-MI nucleoside S_1 transition at ~29,300 cm^{-1} (~341 nm) is significantly lower than that of the second S_2 transition at ~34,000 cm^{-1} (~294 nm). The S_1 transition is relatively broad, and its underlying vibronic features are only barely perceptible. Stanley and co-workers resolved vibronic transitions of 6-MI using Stark spectroscopy (Kodali et al. 2012). The absorption spectrum of the $(6\text{-MI})_2$ dinucleotide exhibits additional broadening of the S_1 line shape in comparison to that of the 6-MI mononucleoside, which is due to the spectral shifts associated with the electrostatic coupling, J. Also shown in Fig. 7.22d is the frequency-doubled laser spectrum used in the 2PE-2DFS experiments described below.

The vibronic fine structure is more easily discerned in the CD spectra of the 6-MI mononucleoside and the $(6\text{-MI})_2$ dinucleotide, which are shown in Fig. 7.22e. In general, the CD spectrum is sensitive to the chiral environment immediately surrounding the EDTM (Nordén et al. 2010; Michl and Thulstrup 1995; Baiardi et al. 2013). If the direction of the EDTM depends on the nuclear coordinates of the 6-MI molecule (i.e., a violation of the Condon approximation and the presence of Herzberg–Teller coupling), the vibronic transitions will give rise to optical activity apparent in the CD spectrum (Baiardi et al. 2013; Kundu et al. 2022; Nooijen 2005). The CD of the 6-MI mononucleoside shown in Fig. 7.22e exhibits a weakly positive band of vibronic features over the range ~26,500 cm^{-1} to ~32,000 cm^{-1}, corresponding to the S_1 transition, and a weakly negative band of vibronic features over the range ~32,000 cm^{-1} to ~33,000 cm^{-1}, corresponding to the S_2 transition. In contrast, the CD of the $(6\text{-MI})_2$ DNTDP exhibits a strongly positive band (peaked at ~27,000 cm^{-1}) and a strongly negative band (peaked at ~30,000 cm^{-1}), which symmetrically splits the S_1 transition. Such behavior is characteristic of a right-handed Cotton effect that is due to the presence of electrostatic coupling between degenerate vibronic states of the component 6-MI monomers (Nordén et al. 2010). The peak-to-peak splitting is ~2800 cm^{-1}, suggesting an electrostatic coupling $J \lesssim 1400$ cm^{-1}. Like the effect seen for the S_1 transition, a right-handed Cotton effect appears to be operative for the S_2 transition.

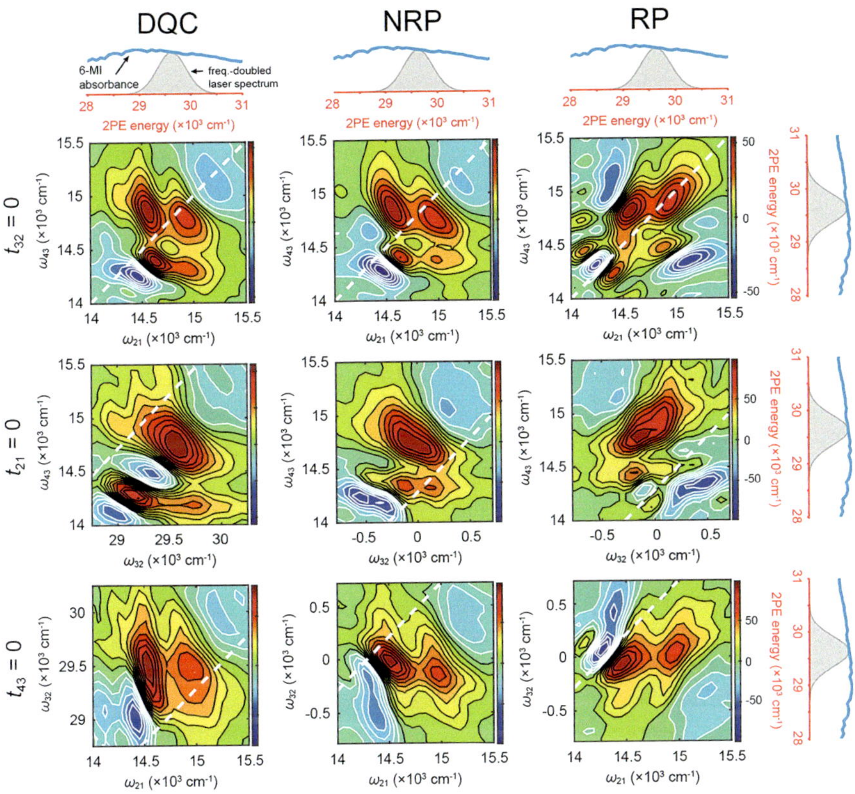

Fig. 7.23 2PE-2DFS experimental results for the 6-MI nucleoside in aqueous buffer salt solution (10 mM TRIS, 100 mM NaCl, and 6 mM $MgCl_2$). The real parts of the experimental spectra are plotted as 2D contour diagrams. Columns indicate a specific signal phase condition (from left to right: DQC, NRP, and RP) and rows indicate one of the three inter-pulse delays set to zero (from top to bottom: $t_{32} = 0$, $t_{21} = 0$, and $t_{43} = 0$). In the margins are shown the linear absorbance spectrum of the 6-MI MNS (blue) and the spectrum of the laser at twice its energy (gray). Adapted from Albrecht et al. (2024)

7.4.3 *2PE-2DFS Studies of 6-MI Nucleoside and (6-MI)$_2$ Dinucleotide*

In Fig. 7.23 are shown the results of 2PE-2DFS experiments carried out on the 6-MI nucleoside. In each panel is plotted the real part of the 2PE-2DFS spectrum for a specific phase condition (from left to right: DQC, NRP, and RP) and pulse delay scanning condition, where one of the inter-pulse delays is set to zero (from top to bottom: $t_{32} = 0$, $t_{21} = 0$, and $t_{43} = 0$) while the remaining delay variables are scanned. The 2D spectra are constructed by calculating the Fourier transforms of the measured response functions with respect to the scanned delay variables. The resulting 2D spectra are presented as contour diagrams, in which the horizontal and vertical axes are the frequencies of the scanned delays. In the horizontal and vertical margins of the 2D spectra are shown the linear absorption spectrum of the 6-MI mononucleoside and the frequency-doubled laser spectrum.

The 2D spectra for all nine of the experimental conditions shown in Fig. 7.23 exhibit peaks and cross-peaks associated with the double-sided Feynman pathways illustrated in Fig. 7.21c. The energy spacings of these features are consistent with the presence of vibronic transitions. For example, all the 2PE-2DFS spectra for $t_{32} = 0$ (top row) exhibit four prominent features with spacing ~400 cm^{-1}, which are shifted vertically below the diagonal (along the ω_{43} axis) by ~200 cm^{-1}. These features occur at approximately one half of the transition energies expected of vibronic features associated with the S_1 transition, which can be seen in the CD spectrum shown in Fig. 7.22e.

The energies and symmetries of the 2PE-2DFS line shapes [$\hat{S}_{DQC}\left(\omega_{21}, t_{32} = 0, \omega_{43}\right)$, $\hat{S}_{NRP}\left(\omega_{21}, t_{32} = 0, \omega_{43}\right)$ and $\hat{S}_{RP}\left(\omega_{21}, t_{32} = 0, \omega_{43}\right)$] are analogous to those of 1PE-2DFS described in Sect. 7.3.2, except that spectral features that occur at the one-photon transition energies (at the frequencies ω_{21} and ω_{43}) are associated with coherences between ground state and virtual states (e.g., $|g\rangle\langle e|$ and $|g\rangle\langle e'|$) and between virtual states and final states (e.g., $|e\rangle\langle f|$, $|e'\rangle\langle f|$). For the 2PE-2DFS spectra with $t_{21} = t_{43} = 0$ (second and third rows of Fig. 7.23, respectively), prominent spectral features for the DQC phase condition occur at two-photon transition energies (at the frequency ω_{32}) associated with coherences between ground state and final states (e.g., $|g\rangle\langle f|$, $|g\rangle\langle f|$). This contrasts with the NRP and RP phase conditions in which spectral features occur at relatively low energies at the frequency ω_{32} associated with coherences between virtual vibronic states (e.g., $|e\rangle\langle e'|$) with vibrational energy spacings. These results suggest that 2PE-2DFS experiments on the 6-MI MNS can be analyzed using a model Hamiltonian to extract quantitative information about the electronic-vibrational states of the chromophore in its monomeric form.

In Fig. 7.24 are shown the results of 2PE-2DFS experiments carried out on the $(6\text{-MI})_2$ dinucleotide. In general, the 2D spectral features of the $(6\text{-MI})_2$ dinucleotide are narrower, and in some cases split into additional peaks, than the 6-MI nucleoside (compare to Fig. 7.23). In particular, the 2D spectral features resolved along the ω_{32} axes for $t_{21} = t_{43} = 0$ (second and third rows) reflect the greatest sensitivity to the presence of exciton coupling. This sensitivity may be related to the absence of obscuring population contributions to the 2PE-2DFS signal during the t_{32} interval (see Fig. 7.21c). The absence of population contributions to the NRP and RP signals in 2PE-2DFS is therefore a uniquely advantageous situation. In comparison, in 1PE-2DFS the intermediate states are one-photon resonant, and population contributions to the fluorescence-detected signal tend to overwhelm any minor contributions associated with coherences. These results for the $(6\text{-MI})_2$ dinucleotide indicate presence of conformation-dependent electrostatic coupling, J, between the 6-MI nucleobases within the $(6\text{-MI})_2$ dinucleotide.

The 2PE-2DFS experiments presented in the current section demonstrate that this method can resolve vibronic fine structure of UV-absorbing fluorescent nucleic acid base analogs such as 6-MI, which are otherwise difficult to study using standard spectroscopic methods. The sensitivity of the 2PE-2DFS vibronic features to

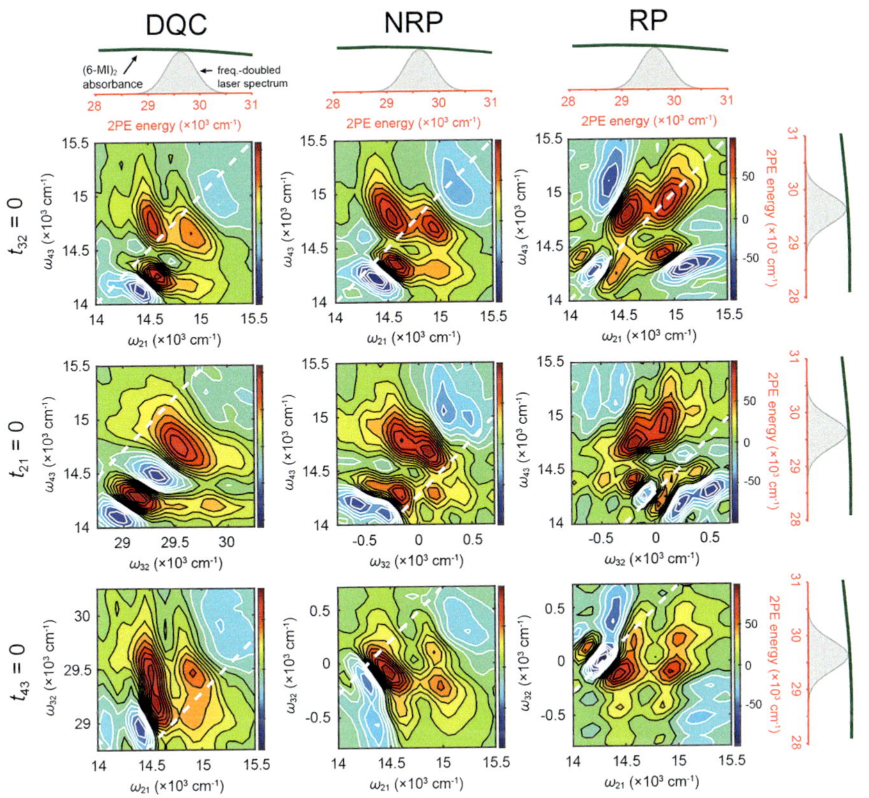

Fig. 7.24 2PE-2DFS experimental results for the $(6\text{-MI})_2$ dinucleotide in aqueous buffer salt solution (10 mM TRIS, 100 mM NaCl, and 6 mM $MgCl_2$). See Fig. 7.23 caption for further explanation. Adapted from Albrecht et al. (2024)

the presence of electrostatic coupling is evident from the differences between the 2D spectra of the 6-MI nucleoside and the $(6\text{-MI})_2$ dinucleotide (compare Fig. 7.23 to Fig. 7.24). This sensitivity is likely related to the lack of population contributions to the NRP and RP signals during the t_{32} interval (as illustrated in Fig. 7.21c), which is a unique advantage of the 2PE-2DFS approach. In principle, studies that compare experimental 2PE-2DFS data to theoretical models can determine detailed structural information about local base conformation and conformational disorder in these and related systems. These experiments serve as a proof-of-concept for future investigations of local base conformations of $(6\text{-MI})_2$ dinucleotide-substituted DNA constructs. Such studies hold promise to reveal important information about local base conformations at specific positions in DNA, and their influence on the protein–DNA interactions central to the core biochemical processes of DNA replication, recombination, and repair.

7.5 Outlook

This chapter reviews novel ensemble and single-molecule spectroscopic methods to study the local conformations, conformational disorder, and "breathing fluctuations" of ss–dsDNA fork junctions viewed through the "lens" of exciton-coupled dimer probes. These methods are highly sensitive to dimer-probe conformation due to the coherent coupling between the optical transition densities of the component monomers. Exciton-coupled $(iCy3)_2$ dimer probes, which are site-specifically positioned within the framework of a ss–dsDNA fork construct, provide indirect information about the local DNA bases and sugar-phosphate backbones immediately adjacent to the probes (Heussman et al. 2019, 2022, 2024; Kringle et al. 2018; Maurer et al. 2023, 2024). More direct information about the local conformations of the DNA bases can be obtained from DNA constructs in which one or more bases has been site-specifically substituted with a fluorescent base analog (Datta et al. 2012; Camel et al. 2021; Ji et al. 2019; Widom et al. 2013b; Albrecht et al. 2024).

These systems can be studied using single-molecule fluorescence measurements to directly probe the local conformational fluctuations of the bases and sugar-phosphate backbones of ss–dsDNA constructs over a broad range of time scales, ranging from microseconds to hundreds-of-milliseconds (Israels et al. 2021; Maurer et al. 2023, 2024; Phelps et al. 2016, 2017). In addition to providing detailed molecular level information about DNA "breathing," and how thermally excited local conformations mediate protein–DNA interactions, such methods can provide insights about the conformational distributions of biologically significant macrostates that exist for a functional protein–DNA complex. In principle, such single-molecule kinetic experiments can provide detailed mechanistic information about the rates of interconversion between stable macrostates, the identities and roles of metastable reactive intermediates, and the connected pathways of protein–DNA complex assembly processes.

Until recently, the relatively unfavorable spectroscopic properties of fluorescent base analog-substituted DNA constructs (e.g., low fluorescence quantum yields, low absorption cross-sections, inefficient photo-detection at near-UV wavelengths, etc.) have precluded single-molecule optical studies such as those discussed in the current chapter for cyanine probes. Nevertheless, the recent developments in phase-sensitive photon counting detection such as those discussed in the preceding sections (Tamimi et al. 2020) suggest that such single-molecule experiments on UV-absorbing molecular dimer probes may soon be feasible.

Acknowledgements The authors are grateful to their current and former laboratory colleagues for many helpful discussions. This work was supported by grants from the National Institutes of Health General Medical Sciences (Grant No. GM-15792 to A.H.M. and P.H.v.H.) and the National Science Foundation RAISE-TAQS Program (Grant No. PHY-1839216 to A.H.M.). S.M. and M.I.S. are supported by the National Science Foundation (Grant No. CHE-2303111). P.H.v.H. is an American Cancer Society Research Professor of Chemistry.

References

Acton TB, Xiao R, Anderson S, Aramini J, Buchwald WA, Ciccosanti C, Conover K, Everett J, Hamilton K, Huang YJ, Janjua H, Kornhaber G, Lau J, Lee DY, Liu G, Maglaqui M, Ma L, Mao L, Patel D, Rossi P, Sahdev S, Shastry R, Swapna GVT, Tang Y, Tong S, Wang D, Zhao L, Montelione GT (2011) Preparation of protein samples for NMR structure, function, and small-molecule screening studies. Meth Enzymol 493:21–60

Alberts BM (1987) Prokaryotic DNA replication mechanisms. Philos Trans R Soc Lond B Biol Sci 317:395–420

Albrecht CS, Scatena LF, von Hippel PH, Marcus AH (2024) Two-photon excitation two-dimensional fluorescence spectroscopy (2PE-2DFS) of the fluorescent nucleobase 6-MI. Proc SPIE 12863:1286305

Algar WR, Hildebrandt N, Vogel SS, Medintz IL (2019) FRET as a biomolecular research tool – understanding its potential while avoiding pitfalls. Nat Methods 16:815–829

Alia JD, Flack JA (2020) Unspecified verticality of Franck-Condon transition, absorption and emission spectra of cyanine dyes, and a classically inspired approximation. RSC Adv 10:43153–43167

Aydin M, Dede Ö, Akins DL (2011) Density functional theory and Raman spectroscopy applied to structure and vibrational mode analysis of 1,1′,3,3′-tetrachloro-benzimidazolocarbocyanine iodide and its aggregate. J Chem Phys 134:064325

Baiardi A, Bloino J, Barone V (2013) General time dependent approach to vibronic spectroscopy including Franck-Condon, Hertzberg-Teller, and Duschinsky effects. J Chem Theory Comput 9:4097–4115

Barone V, Bloino J, Biczysko M (2009) Vibrationally-resolved electronic spectra in GAUS-SIAN 09. Revision a 2:1–20

Barry J, Wong ML, Alberts B (2018) In vitro reconsitution of DNA replication initiated by genetic recombination: a T4 bacteriophage model for a type of DNA synthesis important for all cells. Molec Biol Cell 30:146–159

Benkovic SJ, Spiering MM (2017) Understanding DNA replication by the bacteriophage T4 replisome. J Biol Chem 292:18434–18442

Brown SP, Wimperis S (1995) Inhomogeneously broadening of two-dimensional NMR lineshapes. Chem Phys Lett 237:509–515

Camel BR, Jose D, Meze K, Dang A, von Hippel PH (2021) Mapping DNA conformations and interactions within the binding cleft of bacteriophage T4 single-stranded DNA binding protein (gp32) at single nucleotide resolution. Nucl Acids Res 49:916–927

Chandradoss SD, Haagsma AC, Lee YK, Hwang J-H, Nam J-M, Joo C (2014) Surface passivation for single-molecule protein studies. J Vis Exper 86:e50549

Czikklely V, Forsterling HD, Kuhn H (1970) Extended dipole model for aggregates of dye molecules. Chem Phys Lett 6(3):207–210

Datta K, Johnson NP, von Hippel PH (2010) DNA conformational changes at the primer-template junction regulate the fidelity of replication by DNA polymerase. Proc Natl Acad Sci USA 107(42):17980–17985

Datta K, Johnson NP, Villani G, Marcus AH, von Hippel PH (2012) Characterization of the 6-methyl isoxanthopterin (6-MI) base analog dimer, a spectroscopic probe for monitoring guanine base conformations at specific sites in nucleic acids. Nucl Acids Res 40(3):1191–1202

Delius H, Mantell NJ, Alberts B (1972) Characterization by electron microscopy of the complex formed between T4 bacteriophage gene 32-protein and DNA. J Mol Biol 67:341–350

Dhar M, Dickinson JA, Berg MA (2023) Efficient, nonparametric removal of noise and recovery of probability distributions from time series using nonlinear correlation functions: additive noise. J Chem Phys 159:054110

Didraga C, Pugžlys A, Hania PR, von Berlepsch H, Duppen K, Knoester J (2004) Structure, spectroscopy, and microscopic model of tubular carbocyanine dye aggregates. J Phys Chem B 108:14976–14985

Feng X, Spiering MM, Santos RLA, Benkovic SJ, Li H (2023) Structural basis of the T4 bacteriophage primosome assembly and primer synthesis. Nat Commun 14:4396

Ghosh A, Ostrander JS, Zanni MT (2017) Watching proteins wiggle: mapping structures with two-dimensional infrared spectroscopy. Chem Revs 117:10726–10759

Gozem S, Krylov A (2022) The ezSpectra suite: an easy-to-use toolkit for spectrscopy modeling. Wiley Interdiscip Rev Comput Mol Sci 12:e1546

Harada N, Berova N (2012) Spectroscopic analysis: exciton circular dichroism for chiral analysis. In: Carreira EM, Yamamoto H (eds) Comprehensive chirality. Elsevier, Amsterdam, pp 449–477

Hawkins ME (2001) Fluorescent pteridine nucleoside analogs: a window on DNA interactions. Cell Biochem Biophys 34(2):257–281

Heussman D, Kittell J, Kringle L, Tamimi A, von Hippel PH, Marcus AH (2019) Measuring local conformations and conformational disorder of $(Cy3)_2$ dimers labeled DNA fork junctions using absorbance, circular dichroism and two-dimensional fluorescence spectroscopy. Faraday Disc 216:211–235

Heussman D, Kittell J, von Hippel PH, Marcus AH (2022) Temperature-dependent local conformations and conformational distributions of cyanine dimer labeled single-stranded–double-stranded DNA junctions by 2D fluorescence spectroscopy. J Chem Phys 156(4):045101

Heussman D, Enkhbaatar L, Sorour MI, Kistler KA, von Hippel PH, Matsika S, Marcus AH (2024) Using transition density models to interpret experimental optical spectra of exciton-coupled cyanine $(iCy3)_2$ dimer probes of local DNA conformations at or near functional protein binding sites. Nucl Acids Res 52:1272–1289

Hochstrasser RM (2007) Two-dimensional spectroscopy at infrared and optical frequencies. Proc Nat Acad Sci 104:14190–14196

Hoeher JE, Sande NE, Widom JR (2023) Probing and perturbing riboswitch folding using a fluorescent base analogue. Photochem Photobiol 100:419

Israels B, Albrecht C, Dang A, Barney M, von Hippel PH, Marcus AH (2021) Sub-millisecond conformational transitions of single-stranded DNA lattices by photon correlation single-molecule FRET. J Phys Chem B 125:9426–9440

Ji H, Johnson NP, von Hippel PH, Marcus AH (2019) Local DNA base conformation and ligand intercalation in DNA constructs containing optical probes. Biophys J 117:1101–1115

Jia K, Wan Y, Xia A, Li S, Gong F, Yang G (2007) Characterization of photoinduced isomerization and intersystem crossing of the cyanine dye Cy3. J Phys Chem A 111:1593–1597

Jonas DM (2003) Optical analogs of 2D NMR. Science 300:1515–1517

Jose D, Datta K, Johnson NP, von Hippel PH (2009) Spectroscopic studies of position-specific DNA "breathing" fluctuations at replication forks and primer-template junctions. Proc Natl Acad Sci USA 106(11):4231–4236

Jose D, Weitzel SE, Baase WA, von Hippel PH (2015) Mapping the interactions of the single-stranded DNA binding protein of bacteriophage T4 (gp32) with DNA lattices at single nucleotide resolution: gp32 monomer binding. Nucl Acids Res 43:9276–9290

Joyce CM, Potapova O, Delucia AM, Huang X, Basu VP, Grindley ND (2008) Fingers-closing and other rapid conformational changes in DNA polymerase I (Klenow fragment) and their role in nucleotide selectivity. Biochemistry 47(23):6103–6116

Kao-Huang Y, Butler AP, O'Connor P, Noble DW, von Hippel PH (1977) Nonspecific DNA binding of genome-regulating proteins as a biological control mechanism: measurement of DNA-bound Escherichia coli lac repressor in vivo. Proc Nat Acad Sci 74:4228–4232

Karki KJ, Widom JR, Seibt J, Moody I, Lonergan MC, Pullerits T, Marcus AH (2014) Coherent two-dimensional photocurrent spectroscopy in a PbS quantum dot photocell. Nat Comm 5:5869-1-7

Kelch BA, Makino DL, O'Donnell M, Kuriyan J (2011) How a DNA polymerase clamp loader opens a sliding clamp. Science 334:1675–1680

Kistler KA, Pochas CM, Yamagata H, Matsika S, Spano FC (2011) Absorption, circular dichroism, and photoluminescence in perylene diimide bichromophores: polarization-dependent H- and J-aggregate behavior. J Phys Chem B 116:77–86
Kistler KA, Spano FC, Matsika S (2013) A benchmark of excitonic couplings derived from atomic transition charges. J Phys Chem B 117:2032–2044
Kodali G, Narayanan M, Stanley RJ (2012) Excited-state electronic properties of 6-methylisoxanthopterin (6-MI): an experimental and theoretical study. J Phys Chem B 116(9):2981–2989
Kringle L, Sawaya N, Widom JR, Adams C, Raymer MG, Aspuru-Guzik A, Marcus AH (2018) Temperature-dependent conformations of exciton-coupled Cy3 dimers in double-stranded DNA. J Chem Phys 148
Kundu S, Roy PP, Fleming GR, Makri N (2022) Franck-Condon and Hertzberg-Teller signatures in molecular absorption and emission spectra. J Phys Chem B 126:2899–2911
Laage D, Elsaesser T, Hynes JT (2017) Water dynamics in the hydration shells of biomolecules. Chem Revs 117:10694–10725
Lee W, von Hippel PH, Marcus AH (2014) Internally labeled Cy3 / Cy5 DNA constructs show greatly enhanced photostability in single-molecule FRET experiments. Nucl Acids Res 42:5967–5977
Lee W, Gillies JP, Jose D, von Hippel PH, Marcus AH (2016) Single-molecule FRET studies of cooperative and non-cooperative binding kinetics of the bacteriophage T4 single-stranded DNA binding protein (gp32) to ssDNA lattices at replication fork junctions. Nucl Acids Res 44:10691–10710
Levitus M, Ranjit S (2011) Cyanine dyes in biophysical research: the photophysics of polymethine fluorescent dyes in biomolecular environments. Q Rev Biophys 44(1):123–151
Lott GA, Perdomo-Ortiz A, Utterback JK, Widom JR, Aspuru-Guzik A, Marcus AH (2011) Conformation of self-assembled porphyrin dimers in liposome vesicles by phase-modulation 2D fluorescence spectroscopy. Proc Natl Acad Sci USA 108(40):16521–16526
Marcus AH, Heussman D, Maurer J, Albrecht CS, Herbert P, von Hippel PH (2023) Studies of local DNA conformational disorder using site-specific exciton-coupled dimer probe spectroscopy. Ann Revs Phys Chem 74:245–265
Maurer J, Albrecht CS, Herbert P, Heussman D, Chang A, von Hippel PH, Marcus AH (2023) Studies of DNA ‘breathing’ by polarization-sweep single-molecule fluorescence microscopy of exciton-coupled $(iCy3)_2$ dimer-labeled DNA fork constructs. J Phys Chem B 127:10730–10748
Maurer J, Albrecht CS, Herbert P, Heussman D, Chang A, von Hippel PH, Marcus AH (2024) Studies of DNA ‘breathing’ in exciton-coupled $(iCy3)_2$ dimer-labeled DNA constructs by polarization-sweep single-molecule fluorescence (PS-SMF) microscopy. Proc SPIE 12863:1286305
McConnell B, von Hippel PH (1970a) Hydrogen exchange as a probe of the dynamic structure of DNA. I. General acid-base catalysis. J Mol Biol 50(2):297–316
McConnell B, von Hippel PH (1970b) Hydrogen exchange as a probe of the dynamic structure of DNA. II. Effects of base composition and destabilizing salts. J Mol Biol 50(2):317–332
McGhee JD, von Hippel PH (1974) Theoretical aspects of DNA-protein interactions: cooperative and non-cooperative binding of large ligands to a one-dimensional homogeneous lattice. J Mol Biol 86:469–489
McGhee JD, von Hippel PH (1975a) Formaldehyde as a probe of DNA structure. I. Reaction with exocyclic amino groups of DNA bases. Biochemistry 14(6):1281–1296
McGhee JD, von Hippel PH (1975b) Formaldehyde as a probe of DNA structure. II. Reaction with endocyclic imino groups of DNA bases. Biochemistry 14(6):1297–1303
Michl J, Thulstrup EW (1995) Spectroscopy with polarized light: solute alignment by photoselection, in liquid crystals, polymers and membranes. VCH Publishers, Inc
Mikhaylov A, de Reguarati S, Pahapill J, Callis PR, Kohler B, Rebane A (2018) Two-photon absorption spectra of fluorescent isomorphic DNA base analogs. Biomed Optics Express 9:447–452

Mikulecky PJ, Feig AL (2006) Heat capacity changes associated with nucleic acid folding. Biopolymers 82:38–58
Murphy MC, Rasnik I, Cheng W, Lohman TM, Ha T (2004) Probing single-stranded DNA conformational flexibility using fluorescence spectroscopy. Biophys J 86:2530–2537
Nooijen M (2005) Investigation of Herzberg-Teller Franck-Condon approaches and classical simulations to include effects due to vibronic coupling in circular dichroism spectra: the case of dimethyloxirane continued. Int J Quant Chem 106:2489–2510
Nordén B, Rodger A, Dafforn T (2010) Linear dicrhoism and circular dichroism: a textbook on polarized-light spectroscopy. UK RSC Publishing, Cambridge
Perdomo A, Widom JR, Lott GA, Aspuru-Guzik A, Marcus AH (2012) Conformation and electronic population transfer in membrane supported self-assembled porphyrin dimers by two-dimensional fluorescence spectroscopy. J Phys Chem B 116:10757–10770
Phelps C, Lee W, Jose D, von Hippel PH, Marcus AH (2013) Single-molecule FRET and linear dichroism studies of DNA 'breathing' and helicase binding at replication fork junctions. Proc Natl Acad Sci USA 110:17320–17325
Phelps C, Israels B, Jose D, Marsh MC, von Hippel PH, Marcus AH (2016) Using multi-order time correlation functions (TCFs) to elucidate biomolecular reaction pathways from microsecond single-molecule fluorescence experiments. J Phys Chem B 120:13003–13016
Phelps C, Israels B, Jose D, Marsh MC, von Hippel PH, Marcus AH (2017) Using microsecond single-molecule FRET to determine the assembly pathways of T4 ssDNA binding protein onto model DNA replication forks. Proc Nat Acad Sci Plus USA 114:E3612–E3621
Plumridge A, Meisburger SP, Pollack L (2017a) Visualizing single-stranded nucleic acids in solution. Nucl Acids Res 45:1–13
Plumridge A, Meisburger SP, Andresen K, Pollack L (2017b) The impact of base stacking on the conformations and electrostatics of single-stranded DNA. Nucl Acids Res 45:3932–3943
Printz MP, von Hippel PH (1964) On the kinetics of hydrogen exchange in deoxyribonucleic acid. pH and salt effects. Biochemist 7:3194–3206
Printz MP, von Hippel PH (1965) Hydrogen exchange studies of DNA structure. Proc Nat Acad Sci 53:363–370
Privalov PL, Griko YV, Venyaminov SY (1986) Cold denaturation of myoglobin. J Mol Biol 190:487–498
Raymer MG, Landes T, Marcus AH (2021) Entangled two-photon absorption by atoms and molecules: a quantum optics tutorial. J Chem Phys 155:081501-1-25
Reichl LE (1998) A modern course in statistical physics. Wiley, New York
Reppert M, Tokmakoff A (2016) Computational amide I 2D IR spectroscopy as a probe of protein structure and dynamics. Ann Rev Phys Chem 67:359–386
Roy R, Hohng S, Ha T (2008) A practical guide to single-molecule FRET. Nat Meths 5:507–516
Sanstead PJ, Tokmakoff A (2018) Direct observation of activated kinetics and dowhill dynamics in DNA dehybridization. J Phys Chem B 122:3088–3100
Sanstead PJ, Stevenson P, Tokmakoff A (2016) Sequence-dependent mechanism of DNA oligonucleotide dehybridization resolved through infrared spectroscopy. J Am Chem Soc 138:11792–11801
Screenivasan R, Shkel IA, Chhabra M, Drennan A, Heitkamp S, Wang H-C, Sridevi MA, Plaskon D, McNerney C, Callies K, Cimperman CK, Record MT Jr (2020) Fluorescence-detected conformational changes in duplex DNA in open complex formation by Escherichia coli RNA polymerase: upstream wrapping and downstream bending precede clamp opening and insertion of the downstream duplex. Biochemistry 59:1565–1581
Slichter CP (1996) Principles of magnetic resonance. Springer series in solid-state sciences 1. Springer, Heidelberg
Sorour MI, Kistler KA, Marcus AH, Matsika S (2021) Accurate modeling of excitonc coupling in cyanine dye Cy3. J Phys Chem A 125:7852–7866
Sorour M, Marcus AH, Matsika S (2022) Modeling the electronic absorption spectra of the indocarbocyanine Cy3. Molecules 27:4062

Sorour M, Marcus AH, Matsika S (2023) Unravelling the origin of the vibronic spectral signatures in an excitonically coupled indocarbocyanine Cy3 dimer. J Chem Phys A 127:9530–9540
Spano FC (2010) The spectral signatures of Frenkel polarons in H- and J-aggregates. Acc Chem Res 43:429–439
Tamimi A, Landes T, Lavoie J, Raymer MG, Marcus AH (2020) Fluorescence-detected Fourier transform electronic spectroscopy by phase-tagged photon counting. Opt Express 28:25194–25214
Tekavec PF, Lott GA, Marcus AH (2007) Fluorescence-detected two-dimensional electronic coherence spectroscopy by acousto-optic phase modulation. J Chem Phys 127:214307
Vaitiekunas P, Crane-Robinson C, Privalov PL (2015) The energetic basis of the DNA double helix: a combined microcalorimetric approach. Nucl Acids Res 43:8577–8589
von Hippel PH (2007) From "simple" DNA-protein interactions to the macromolecular machines of gene expression. Annu Rev Biophys Biomol Struct 36:79–105
von Hippel PH, McGhee JD (1972) DNA-protein interactions. Annu Rev Biochem 41:231–300
von Hippel PH, Johnson NP, Marcus AH (2013) 50 years of DNA 'breathing': reflections on old and new approaches. Biopolymers 99:923–954
Watson JD, Crick FHC (1953) Molecular structure of nucleic acids. Nature 171:737–738
Widom JR, Rappoport D, Perdomo-Ortiz A, Thomsen H, Johnson NP, von Hippel PH, Aspuru-Guzik A, Marcus AH (2012) Electronic transition moments of 6-methyl isoxanthopterin (6-MI) – a fluorescent analog of guanine. Nucl Acids Res 41:995–1004
Widom JR, Perdomo-Ortiz A, Lee W, Rappoport D, Molinski TF, Aspuru-Guzik A, Marcus AH (2013a) Temperature-dependent conformations of a membrane supported 'zinc porphyrin tweezer' by 2D fluorescence spectroscopy. J Phys Chem A 117:6171–6184
Widom JR, Johnson NP, von Hippel PH, Marcus AH (2013b) Solution conformation of 2-aminopurine (2-AP) dinucleotide by ultraviolet 2D fluorescence spectroscopy (UV-2D FS). New J Phys 15:025028–025043
Zheng J, Kwak K, Fayer MD (2007) Ultrafast 2D IR vibrational echo spectroscopy. Acc Chem Res 40:75–83

Printed by Printforce, the Nether